Diffusion Processes and Partial Differential Equations

Diffusion Processes and Partial Differential Equations

Kazuaki Taira
Institute of Mathematics
University of Tsukuba
Ibaraki, Japan

ACADEMIC PRESS, INC.
Harcourt Brace Jovanovich, Publishers

Boston San Diego New York
Berkeley London Sydney
Tokyo Toronto

ACADEMIC PRESS, INC.
1250 Sixth Avenue, San Diego, CA 92101

United Kingdom Edition published by
ACADEMIC PRESS INC. (LONDON) LTD.
24-28 Oval Road, London NW1 7DX

Library of Congress Cataloging-in-Publication Data
Taira, Kazuaki.
Diffusion processes and partial differential equations/Kazuaki Taira
p. cm.
Bibliography: p.
Includes index.
ISBN 0-12-682220-4
1. Elliptic operators. 2. Markov processes. 3. Boundary value problems. I. Title.
QA329.42.T35 1988 87-27356
519.2′33—dc 19 CIP

88 89 90 91 9 8 7 6 5 4 3 2 1
Printed in the United States of America

To my parents,
Yasunori and Yasue Taira

Contents

Preface

This book is devoted to the functional analytic approach to the problem of construction of Markov processes in probability theory. It is well known that, by virtue of the Hille–Yosida theory of semigroups, the problem of construction of Markov processes can be reduced to the study of boundary value problems for degenerate elliptic (integro-)differential operators of second order. Several recent developments in the theory of partial differential equations have made possible further progresss in the study of boundary value problems and hence of the problem of construction of Markov processes. The presentation of these new results is the main purpose of this book. Unlike many other books on Markov processes, this book focuses on the relationship between Markov processes and elliptic boundary value problems, with emphasis on the study of maximum principles. Our approach is distinguished by the extensive use of the theory of partial differential equations.

This book grew out of lecture notes for graduate courses given by the author at Sophia University, Hokkaido University, Tôhoku University, Tokyo Metropolitan University and University of Tsukuba. It is addressed to graduate students and mathematicians with an interest in probability, functional analysis and partial differential equations. The contents of the book are divided into five principal parts.

The first part (Chapters 1–4) provides the elements of the Lebesgue theory of measure and integration, manifold theory, functional analysis and distribution theory which are used throughout the book. The material in these preparatory chapters is given for completeness, to minimize the necessity of consulting too many outside references. This makes the book fairly self-contained.

In the second part (Chapters 5–6), the basic definitions and results about Sobolev spaces are summarized, and the calculus of pseudo-differential operators—a modern theory of potentials—is developed. The theory of pseudo-differential operators forms a most convenient tool in the study of elliptic boundary value problems in Chapter 8.

Our subject proper starts with the third part (Chapter 7), where various maximum principles for degenerate elliptic differential operators of second order are studied. In particular, the underlying analytical mechanism of propagation of maximums is revealed here. This plays an important role in the interpretation and study of Markov processes in terms of partial differential equations in Chapter 10.

The fourth part (Chapter 8) is devoted to general boundary value problems for second-order elliptic differential operators. The basic questions of existence, uniqueness and regularity of solutions of general boundary value problems with spectral parameter are studied in the framework of Sobolev spaces, using the calculus of pseudo-differential operators. Our approach is not far removed from the classical potenial approach. A fundamental existence and uniqueness theorem is proved here. The importance of such a theorem is visible in constructing Markov processes in Chapter 10.

The fifth and final part (Chapters 9–10) is devoted to the functional analytic approach to the problem of construction of Markov processes. General existence theorems for Markov processes in terms of boundary value problems are proved in Chapter 9, and then the construction of Markov processes is carried out in Chapter 10, by solving general boundary value problems with spectral parameter.

Bibliographical references are discussed primarily in Notes at the end of the chapters. These notes are intended to supplement the text and place it in better perspective.

To make the material in Chapters 7–10 accessible to a broad spectrum of readers, I have added *Introduction and Summary*. In this introductory chapter, I have included ten elementary (but important) examples of diffusion processes, and further I have attempted to state our problems and results in such a fashion that a broad spectrum of readers could understand, and also to describe how these problems can be solved, using the mathematics I present in Chapters 1–6.

I hope that this book will lead to a better insight into the study of three interrelated subjects of analysis: Markov processes, semigroups and elliptic boundary value problems, and further that the reader will appreciate intimate connections between partial differential equations and Markov processes.

Kazuaki Taira
Tsukuba, Japan

Acknowledgments

In preparing this book, I am indebted to many people. In particular, I would like to express my hearty thanks to Jacques Camus, Daisuke Fujiwara, Bernard Helffer, Yuji Kasahara, Charles Rockland, Masao Tanikawa and Kôichi Uchiyama. Kasahara and Tanikawa helped me to learn the material that is presented in this book. I am deeply indebted to Professors Kiyosi Itô, Minoru Motoo, Tadashi Ueno and Shinzo Watanabe for their constant interest in my work.

This work was begun at Ecole Normale Supérieure d'Ulm and Université de Paris-Sud (1976/78) with the financial support of the French Government while I was on leave from Tokyo Institute of Technology, and a major part of the work was done at University of Tsukuba (1978/79) with the aid of a grant-in-aid for scientific research and at the Institute for Advanced Study (1980/81) with the financial support of the National Science Foundation while I was on leave from University of Tsukuba. I take this opportunity to express my gratitude to all these institutions.

I would like to extend my warmest thanks to Professor Richard Bellman who originally suggested that my work be published in book form. Thanks are also due to the editorial and production staff of Academic Press Boston and Tokyo for their unfailing helpfulness and cooperation during the production of the book.

Last, but not least, I owe a great debt of gratitude to my wife, Naomi, who not only typed a part of the manuscript but also gave me moral support during the preparation of this book.

Notation and Conventions

The following notation is used for sets of numbers:

$\mathbf{N}$	the positive integers,
$\mathbf{Z}$	the integers,
$\mathbf{R}$	the real numbers,
$\mathbf{C}$	the complex numbers,
$[a, b]$	the closed interval $\{x \in \mathbf{R};\ a \leq x \leq b\}$,
$[a, b)$	the semiclosed interval $\{x \in \mathbf{R};\ a \leq x < b\}$,
(a, b)	the open interval $\{x \in \mathbf{R};\ a < x < b\}$.

The notation for set-theoretic concepts is standard, and is described on pages 37 and 38. Other symbols introduced in the text are listed on pages 431 through 442.

We shall use without explanation the following:

$\equiv$	defined as (or "be identically equal to" as usual)
$\Rightarrow$	implication sign
$\Leftrightarrow$	two-sided implication sign (if and only if)

- ■ end of a proof
- ▼ proof of a lemma is done, but the proof of the theorem or proposition goes on.

Numbers in square brackets, e.g. [1], refer to the bibliography.

Definitions, results, remarks and examples are numbered within sections. For example, in Section $n.m$ (where n denotes the chapter), the theorems are indexed by Theorem $n.m.k$. Formulas and conditions are also numbered within sections, and those in Section $n.m$ are indexed by (k). When in another section we refer to such a formula (k) or condition (k), we designate it by $(n.m.k)$. The really important ones are called theorems, and the slightly less important ones propositions.

Introduction and Summary

I. Markov Processes and Semigroups

I.1 Brownian Motion

In 1828, the English botanist R. Brown observed that pollen grains suspended in water move chaotically, incessantly changing their direction of motion. The physical explanation of this phenomenon is that a single grain suffers innumerable collisions with the randomly moving molecules of the surrounding water.

A mathematical theory for Brownian motion was put forward by A. Einstein in 1905. Let $p(t, x, y)$ be the probability density function that a one-dimensional Brownian particle starting at position x will be found at position y at time t. Einstein derived the following formula from statistical mechanical considerations:

$$p(t, x, y) = \frac{1}{\sqrt{2\pi Dt}} \exp\left[-\frac{(y-x)^2}{2Dt} \right].$$

Here D is a positive constant determined by the radius of the particle, the interaction of the particle with surrounding molecules, temperature and the Boltzmann constant. This gives an accurate method of measuring Avogadro's number by observing particles undergoing Brownian motion. Einstein's theory was experimentally tested by J. Perrin between 1906 and 1909.

Brownian motion was put on a firm mathematical foundation for the first time by N. Wiener in 1923. Let Ω be the space of continuous functions $\omega: [0, \infty) \to \mathbf{R}$ with coordinates $x_t(\omega) = \omega(t)$ and let $\mathscr{F}$ be the smallest σ-algebra in Ω which contains all sets of the form $\{\omega \in \Omega; a \leq x_t(\omega) < b\}$, $t \geq 0$, $a < b$. Wiener constructed probability measures P_x, $x \in \mathbf{R}$, on $\mathscr{F}$ for which the following formula holds:

$$\begin{aligned} P_x\{\omega \in \Omega; a_1 \leq x_{t_1}(\omega) < b_1, a_2 \leq x_{t_2}(\omega) < b_2, \ldots, a_n \leq x_{t_n}(\omega) < b_n\} \\ = \int_{a_1}^{b_1} \int_{a_2}^{b_2} \cdots \int_{a_n}^{b_n} p(t_1, x, y_1) p(t_2 - t_1, y_1, y_2) \cdots \\ \times p(t_n - t_{n-1}, y_{n-1}, y_n)\, dy_1\, dy_2 \cdots dy_n, \\ 0 < t_1 < t_2 < \cdots < t_n < \infty. \end{aligned} \tag{1}$$

This formula expresses the "starting afresh" property of Brownian motion that if a Brownian particle reaches a position, then it behaves subsequently as though that position had been its initial position. The measure P_x is called the *Wiener measure* starting at x.

P. Lévy found another construction of Brownian motion, and gave a profound description of qualitative properties of the individual Brownian path in his book *Processus stochastiques et mouvement brownien* (1948).

I.2 Markov Processes

Markov processes are an abstraction of the idea of Brownian motion. Let K be a locally compact, separable metric space and $\mathscr{B}$ the σ-algebra of all Borel sets in K, that is, the smallest σ-algebra containing all open sets in K. (The reader may content himself with thinking of $\mathbf{R}$ while reading about K.) Let $(\Omega, \mathscr{F}, P)$ be a probability space. A function X defined on Ω taking values in K is called a *random variable* if it satisfies

$$\{X \in E\} = X^{-1}(E) \in \mathscr{F} \qquad \text{for all} \quad E \in \mathscr{B}.$$

We express this by saying that X is $\mathscr{F}/\mathscr{B}$-measurable. A family $\{x_t\}_{t \geq 0}$ of random variables is called a *stochastic process*, and may be thought of as the motion in time of a physical particle. The space K is called the *state space* and Ω the *sample space*. For a fixed $\omega \in \Omega$, the function $x_t(\omega)$, $t \geq 0$, defines in the state space K a *trajectory* or *path* of the process corresponding to the sample point ω.

In this generality the notion of a stochastic process is of course not so interesting. The most important class of stochastic processes is the class of Markov processes which is characterized by the Markov property. Intuitively, the (temporally homogeneous) *Markov property* is that the prediction of subsequent motion of a particle, knowing its position at time t, depends neither on the value of t nor on what has been observed during the time interval $[0, t]$; that is, a particle starts afresh.

Now we introduce a class of Markov processes which we will deal with in this book (Definition 9.1.3).

Suppose that we are given the following:

1) A locally compact, separable metric space K and the σ-algebra $\mathscr{B}$ of all Borel sets in K. A point ∂ is adjoined to K as the point at infinity if K is not compact, and as an isolated point if K is compact. We let

$$\begin{aligned} K_\partial &= K \cup \{\partial\}, \\ \mathscr{B}_\partial &= \text{the } \sigma\text{-algebra in } K_\partial \text{ generated by } \mathscr{B}. \end{aligned}$$

2) The space Ω of all mappings $\omega: [0, \infty] \to K_\partial$ such that $\omega(\infty) = \partial$ and that if $\omega(t) = \partial$ then $\omega(s) = \partial$ for all $s \geq t$. We let ω_∂ be the constant map $\omega_\partial(t) = \partial$ for all $t \in [0, \infty]$.

3) For each $t \in [0, \infty]$, the coordinate map x_t defined by $x_t(\omega) = \omega(t)$, $\omega \in \Omega$.

4) For each $t \in [0, \infty]$, a mapping $\varphi_t: \Omega \to \Omega$ defined by $\varphi_t\omega(s) = \omega(t + s)$, $\omega \in \Omega$. Note that $\varphi_\infty \omega = \omega_\partial$ and $x_t \circ \varphi_s = x_{t+s}$ for all $t, s \in [0, \infty]$.

5) A σ-algebra $\mathscr{F}$ in Ω and an increasing family $\{\mathscr{F}_t\}_{0 \leq t \leq \infty}$ of sub-σ-algebras of $\mathscr{F}$.

6) For each $x \in K_\partial$, a probability measure P_x on $(\Omega, \mathscr{F})$.

We say that these elements define a (temporally homogeneous) *Markov process* $\mathscr{X} = (x_t, \mathscr{F}, \mathscr{F}_t, P_x)$ if the following four conditions are satisfied:

(i) For each $0 \leq t < \infty$, the function x_t is $\mathscr{F}_t/\mathscr{B}_\partial$-measurable, that is,

$$\{x_t \in E\} \in \mathscr{F}_t \qquad \text{for all} \quad E \in \mathscr{B}_\partial.$$

(ii) For each $0 \leq t < \infty$ and $E \in \mathscr{B}$, the function

$$p_t(x, E) = P_x\{x_t \in E\} \tag{2}$$

is a Borel measurable function of $x \in K$.

(iii) $P_x\,\{\omega \in \Omega; x_0(\omega) = x\} = 1$ for each $x \in K_\partial$.

(iv) For all $t, h \in [0, \infty]$, $x \in K_\partial$ and $E \in \mathscr{B}_\partial$, we have

$$P_x\{x_{t+h} \in E \mid \mathscr{F}_t\} = p_h(x_t, E), \tag{3}$$

or equivalently

$$P_x(A \cap \{x_{t+h} \in E\}) = \int_A p_h(x_t(\omega), E)\, dP_x(\omega), \qquad A \in \mathscr{F}_t. \tag{3'}$$

Here is an intuitive way of thinking about the above definition of a Markov process. The sub-σ-algebra $\mathscr{F}_t$ may be interpreted as the collection of events that are observed during the time interval $[0, t]$. The value $P_x(A)$, $A \in \mathscr{F}$, may be interpreted as the probability of the event A under the condition that a particle starts at position x; hence the value $p_t(x, E)$ expresses the transition probability that a particle starting at position x will be found in the set E at time t. The function p_t is called the *transition function* of the process $\mathscr{X}$. The transition function p_t specifies the probability structure of the process. The intuitive meaning of the crucial condition (iv) is that the future behavior of a particle, knowing its history up to time t, is the same as the behavior of a particle starting at $x_t(\omega)$, that is, a particle starts afresh. A particle moves in the space K until it "dies" at which time it reaches the point ∂; hence the point ∂ is called the *terminal point*.

With this interpretation in mind, we let

$$\zeta(\omega) = \inf\{t \in [0, \infty];\ x_t(\omega) = \partial\}.$$

The random variable ζ is called the *lifetime* of the process $\mathscr{X}$.

Using the Markov property (3′) repeatedly, we easily obtain the following formula, analogous to formula (1):

$$P_x\{\omega \in \Omega;\ x_{t_1}(\omega) \in A_1, x_{t_2}(\omega) \in A_2, \ldots, x_{t_n}(\omega) \in A_n\}$$
$$= \int_{A_1}\int_{A_2}\cdots\int_{A_n} p_{t_1}(x, dy_1)p_{t_2-t_1}(y_1, dy_2)\cdots p_{t_n-t_{n-1}}(y_{n-1}, dy_n),$$
$$0 < t_1 < t_2 < \cdots < t_n < \infty, \qquad A_1, A_2, \ldots, A_n \in \mathscr{B}.$$

I.3 Transition Functions

From the viewpoint of analysis, the transition function is something more convenient than the Markov process itself. In fact, it can be shown that the transition functions of Markov processes generate solutions of certain parabolic partial differential equations such as the classical diffusion equation and, conversely, these differential equations can be used to construct and study the transition functions and the Markov processes themselves.

First we give the precise definition of a transition function which is adapted to analysis (Definition 9.1.4):

Let K be a locally compact, separable metric space and $\mathscr{B}$ the σ-algebra of all Borel sets in K. A function $p_t(x, E)$, defined for all $t \geq 0$, $x \in K$ and $E \in \mathscr{B}$, is called a (temporally homogeneous) *Markov transition function* on K if it satisfies the following four conditions:

(a) $p_t(x, \cdot)$ is a measure on $\mathscr{B}$ and $p_t(x, K) \leq 1$ for each $t \geq 0$ and $x \in K$.
(b) $p_t(\cdot, E)$ is a Borel measurable function for each $t \geq 0$ and $E \in \mathscr{B}$.
(c) $p_0(x, \{x\}) = 1$ for each $x \in K$.
(d) For any $t, s \geq 0$ and $E \in \mathscr{B}$, we have

$$p_{t+s}(x, E) = \int_K p_t(x, dy) p_s(y, E). \tag{4}$$

Equation (4), called the *Chapman–Kolmogorov equation*, expresses the idea that a transition from the position x to the set E in time $t + s$ is composed of a transition from x to some position y in time t, followed by a transition from y to the set E in the remaining time s; the latter transition has probability $p_s(y, E)$ which depends only on y (see Figure 0-1). Thus it is just condition (d) which reflects the Markov property that a particle starts afresh.

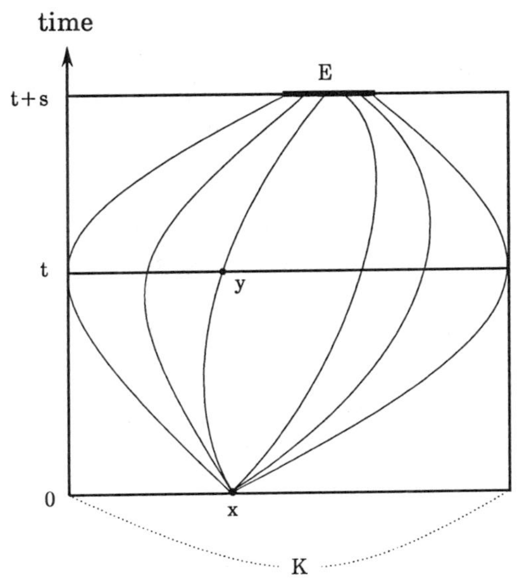

Figure 0-1

The Chapman–Kolmogorov equation (4) tells us that the function $p_t(x, K)$ is monotonically increasing as $t \downarrow 0$, so that the limit $p_{+0}(x, K) = \lim_{t \downarrow 0} P_t(x, K)$ exists.

A Markov transition function p_t is said to be *normal* if it satisfies

$$p_{+0}(x, K) = 1 \qquad \text{for all} \quad x \in K.$$

The next result justifies our definition of a transition function, and hence it will be fundamental for our further study of Markov processes (Theorem 9.1.6).

For every Markov process, the function p_t, defined by formula (2), is a Markov transition function. Conversely, every normal Markov transition function corresponds to some Markov process.

Here are some important examples of normal transition functions on $\mathbf{R}$:

Example 1 (uniform motion). If $t \geq 0$, $x \in \mathbf{R}$ and $E \in \mathscr{B}$, we let

$$p_t(x, E) = \chi_E(x + vt), \tag{5}$$

where v is a constant, and $\chi_E(y) = 1$ if $y \in E$ and $= 0$ if $y \notin E$.

This process, starting at x, moves deterministically with constant velocity v.

Example 2 (Poisson process). If $t \geq 0$, $x \in \mathbf{R}$ and $E \in \mathscr{B}$, we let

$$p_t(x, E) = e^{-\lambda t} \sum_{n=0}^{\infty} \frac{(\lambda t)^n}{n!} \chi_E(x + n), \tag{6}$$

where λ is a positive constant.

This process, starting at x, advances one unit by jumps, and the probability of n jumps in time t is equal to $e^{-\lambda t}(\lambda t)^n/n!$.

Example 3 (Brownian motion). If $t > 0$, $x \in \mathbf{R}$ and $E \in \mathscr{B}$, we let

$$p_t(x, E) = \frac{1}{\sqrt{2\pi t}} \int_E \exp\left[-\frac{(y - x)^2}{2t}\right] dy, \tag{7}$$

and

$$p_0(x, E) = \chi_E(x).$$

This is a mathematical model of one-dimensional Brownian motion.

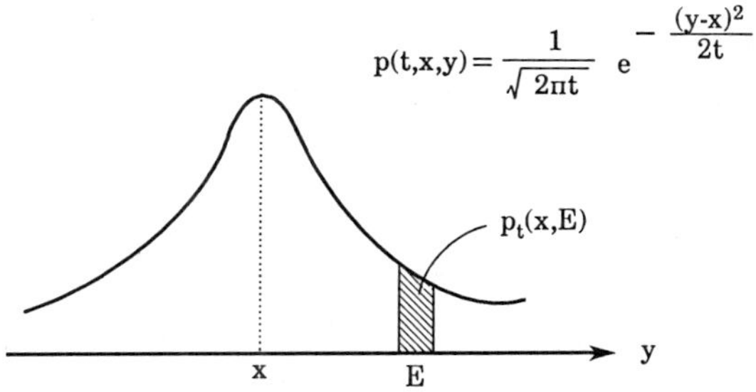

Figure 0-2

Example 4 (Brownian motion with constant drift). If $t > 0$, $x \in \mathbf{R}$ and $E \in \mathscr{B}$, we let

$$p_t(x, E) = \frac{1}{\sqrt{2\pi t}} \int_E \exp\left[-\frac{(y - mt - x)^2}{2t} \right] dy, \tag{8}$$

and

$$p_0(x, E) = \chi_E(x),$$

where m is a constant.

This represents Brownian motion with constant drift m; the process can be represented as $\{x_t + mt\}$, where $\{x_t\}$ is Brownian motion.

Example 5 (Cauchy process). If $t > 0$, $x \in \mathbf{R}$ and $E \in \mathscr{B}$, we let

$$p_t(x, E) = \frac{1}{\pi} \int_E \frac{t}{t^2 + (y - x)^2} dy, \tag{9}$$

and

$$p_0(x, E) = \chi_E(x).$$

This process can be thought of as the "trace" on the real line of trajectories of two-dimensional Brownian motion, and it moves by jumps.

I.4 Kolmogorov's Equations

In the first works devoted to Markov process, the most fundamental was A. N. Kolmogorov's work (1931) where the general concept of a Markov transition function was introduced for the first time and an analytic method of describing Markov transition functions was proposed.

We now take a close look at Kolmogorov's work. Let p_t be a Markov transition function on $\mathbf{R}$, and suppose that the following conditions are satisfied:

(i) For each $\varepsilon > 0$, we have

$$\lim_{t\downarrow 0} \frac{1}{t} \sup_{x\in\mathbf{R}} p_t(x, \mathbf{R}\setminus(x-\varepsilon, x+\varepsilon)) = 0.$$

(ii) The limits

$$\lim_{t\downarrow 0} \frac{1}{t} \int_{x-\varepsilon}^{x+\varepsilon} p_t(x, dy)(y-x)^2 = a(x),$$

$$\lim_{t\downarrow 0} \frac{1}{t} \int_{x-\varepsilon}^{x+\varepsilon} p_t(x, dy)(y-x) = b(x)$$

and

$$\lim_{t\downarrow 0} \frac{1}{t} (p_t(x, \mathbf{R}) - 1) = c(x)$$

exist for each $x \in \mathbf{R}$.

Physically, the limit $a(x)$ may be interpreted as a variance instantaneous (with respect to t) velocity at position x, and the limit $b(x)$ has a similar interpretation as a mean. The transition functions (5), (7) and (8) satisfy conditions (i) and (ii) with $a(x) = 0$, $b(x) = v$, $c(x) = 0$; $a(x) = 1$, $b(x) = c(x) = 0$; $a(x) = 1$, $b(x) = m$, $c(x) = 0$ respectively, whereas the transition functions (6) and (9) do not satisfy condition (i).

Further suppose that the transition function p_t has a density $p(t, x, y)$ with respect to the Lebesgue measure dy. Intuitively, the density $p(t, x, y)$ represents the state of the process at position y at time t, starting from the initial state that a unit mass is at position x. Under certain regularity conditions, Kolmogorov showed that the density $p(t, x, y)$ is, for fixed y, the fundamental solution of the Cauchy problem:

$$\begin{cases} \dfrac{\partial p}{\partial t} = \dfrac{a(x)}{2}\dfrac{\partial^2 p}{\partial x^2} + b(x)\dfrac{\partial p}{\partial x} + c(x)p, & t > 0, \\ \lim\limits_{t\downarrow 0} p(t, x, y) = \delta(x-y), \end{cases} \tag{10}$$

and is, for fixed x, the fundamental solution of the Cauchy problem:

$$\begin{cases} \dfrac{\partial p}{\partial t} = \dfrac{\partial^2}{\partial y^2}\left(\dfrac{a(y)}{2} p\right) - \dfrac{\partial}{\partial y}(b(y)p) + c(y)p, & t > 0, \\ \lim\limits_{t\downarrow 0} p(t, x, y) = \delta(y-x). \end{cases} \tag{11}$$

Here δ is the Dirac measure, and $\delta(x - y)$ (resp. $\delta(y - x)$) represents a unit mass at position y (resp. x). Equation (10) is called *Kolmogorov's backward equation*, since we consider the terminal state (the variable y) to be fixed and vary the initial state (the variable x). In this context, equation (11) is called *Kolmogorov's forward equation*. These equations are also called the *Fokker–Planck partial differential equations*. In the case of Brownian motion (Example 3), equations (10) and (11) become the classical diffusion (or heat) equation:

$$\frac{\partial p}{\partial t} = \frac{1}{2}\frac{\partial^2 p}{\partial x^2}, \qquad t > 0.$$

$$\frac{\partial p}{\partial t} = \frac{1}{2}\frac{\partial^2 p}{\partial y^2}, \qquad t > 0.$$

Conversely, Kolmogorov raised the problem of construction of Markov transition functions by solving the given Fokker–Planck partial differential equations (10) and (11).

It is worth pointing out here that the forward equation (11) is given in a more intuitive form than the backward equation (10), but regularity conditions on the functions a and b are more stringent than those needed in the backward case. This suggests that the backward approach is more convenient than the forward approach from the viewpoint of analysis.

In 1936, W. Feller treated this problem by classical analytic methods, and proved that equation (10) (or (11)) has a unique solution $p(t, x, y)$ under certain regularity conditions on the functions a, b and c, and that this solution $p(t, x, y)$ determines a Markov process. In 1943, R. Fortet proved that these solutions correspond to Markov processes with continuous paths.

On the other hand, S. N. Bernstein (1938) and P. Lévy (1948) made probabilistic approaches to this problem, by using stochastic differential equations.

I.5 Feller Semigroups

In the 1950s, the theory of Markov processes entered a new period of intensive development. The Hille–Yosida theory of semigroups in functional analysis made possible further progress in the study of Markov processes.

Kolmogorov's backward and forward equations (10) and (11) can be formulated in terms of semigroup theory, which we now do.

Let K be a locally compact, separable metric space and $B(K)$ the space of real-valued, bounded Borel measurable functions on K; $B(K)$ is a Banach space with the supremum norm

$$\|f\| = \sup_{x \in K} |f(x)|.$$

We can associate with each Markov transition function p_t on K a family $\{T_t\}_{t \geq 0}$ of linear operators acting on $B(K)$ in the following way:

$$T_t f(x) = \int_K p_t(x, dy) f(y), \qquad f \in B(K). \tag{12}$$

Then the operators T_t are non-negative and contractive on $B(K)$:

$$f \in B(K),\ 0 \leq f \leq 1 \text{ on } K \Rightarrow 0 \leq T_t f \leq 1 \text{ on } K.$$

Further the Chapman–Kolmogorov equation (4) implies that the family $\{T_t\}$ forms a *semigroup*:

$$T_{t+s} = T_t \cdot T_s, \qquad t, s \geq 0.$$

We also have

$$T_0 = I = \text{the identity operator.}$$

The Hille–Yosida theory of semigroups requires the strong continuity of $\{T_t\}_{t \geq 0}$:

$$\lim_{t \downarrow 0} \|T_t f - f\| = 0, \qquad f \in B(K). \tag{13}$$

That is,

$$\lim_{t \downarrow 0} \sup_{x \in K} \left| \int_K p_t(x, dy) f(y) - f(x) \right| = 0, \qquad f \in B(K). \tag{13'}$$

We define the infinitesimal generator $\mathfrak{A}$ of the semigroup $\{T_t\}_{t \geq 0}$ by the formula

$$\mathfrak{A} f = \lim_{t \downarrow 0} \frac{T_t f - f}{t},$$

provided that the limit exists in $B(K)$. Then the Hille–Yosida theory tells us that the semigroup $\{T_t\}$ can be written as

$$T_t = e^{t\mathfrak{A}},$$

with a suitable interpretation of the exponential, and that the infinitesimal generator $\mathfrak{A}$ determines completely the semigroup $\{T_t\}$. The exponential differential equation associated with $\{T_t\}$,

$$\frac{d}{dt}(T_t f) = \mathfrak{A}(T_t f),$$

is a generalization of Kolmogorov's backward equation (10).

On the other hand, let $M(K)$ be the space of real Borel measures on K; $M(K)$ is a Banach space with the total variation norm. If $\mu \in M(K)$, we let

$$U_t\mu(E) = \int_K \mu(dx)p_t(x, E), \qquad E \in \mathscr{B}.$$

Then the operators U_t also form a contraction semigroup on $M(K)$. The semigroup $\{U_t\}$ has the probabilistic interpretation that if μ is the initial probability distribution, then $U_t\mu$ may be interpreted as the probability distribution at time t. The differential equation

$$\frac{d}{dt}\left(\int_K U_t\mu(dx)f(x)\right) = \int_K U_t\mu(dx)\mathfrak{A}f(x)$$

is a generalization of Kolmogorov's forward equation (11).

Although the semigroup $\{T_t\}$ appears less natural than the semigroup $\{U_t\}$, as the further development of the theory has shown, it is the more convenient one from the viewpoint of functional analysis. For technical reasons, we will concentrate on the semigroup $\{T_t\}$.

If p_t is the transition function of a Markov process $\mathscr{X}$, then the infinitesimal generator $\mathfrak{A}$ of the associated semigroup $\{T_t\}_{t \geq 0}$ is called the infinitesimal generator of the process $\mathscr{X}$.

Now, taking $f = \chi_{\{x\}} \in B(K)$ in formula (13′), we obtain that

$$\lim_{t \downarrow 0} p_t(x, \{x\}) = 1, \qquad x \in K. \tag{14}$$

But, the Brownian motion transition function (7), the most important and interesting example, does not satisfy condition (14). Thus we shift our attention to continuous functions, instead of measurable functions.

Let $C(K)$ be the space of real-valued, bounded continuous functions on K; $C(K)$ is a Banach space with the supremum (maximum) norm

$$\|f\| = \sup_{x \in K} |f(x)|.$$

We say that a function $f \in C(K)$ converges to $a \in \mathbf{R}$ as $x \to \partial$ if, for each $\varepsilon > 0$, there exists a compact subset E of K such that

$$|f(x) - a| < \varepsilon \qquad \text{for all} \quad x \in K \setminus E,$$

and write $\lim_{x \to \partial} f(x) = a$. Let $C_0(K)$ be the subspace of $C(K)$ which consists of all functions satisfying $\lim_{x \to \partial} f(x) = 0$; the space $C_0(K)$ is a closed subspace of $C(K)$. We remark that $C_0(K)$ may be identified with $C(K)$ if K is compact.

Now we introduce a useful convention:

> Any real-valued function f on K is extended to $K_\partial = K \cup \{\partial\}$ by setting $f(\partial) = 0$.

From this point of view, the space $C_0(K)$ is identified with the subspace of $C(K_\partial)$ which consists of all functions f satisfying $f(\partial) = 0$. Further we extend a transition function P_t on K to a transition function p_t' on K_∂ as follows:

$$\begin{cases} p_t'(x, E) = p_t(x, E), & x \in K,\ E \in \mathscr{B}, \\ p_t'(x, \{\partial\}) = 1 - p_t(x, K), & x \in K, \\ p_t'(\partial, K) = 0, & p_t'(\partial, \{\partial\}) = 1. \end{cases}$$

Note that our convention is consistent, since $T_t f(\partial) = f(\partial) = 0$.

A Markov transition function p_t is called a *Feller function* if the function

$$T_t f(x) = \int_K p_t(x, dy) f(y)$$

is a continuous function of $x \in K$ whenever f is bounded and continuous on K. That is, the Feller property is equivalent to saying that the space $C(K)$ is an invariant subspace of $B(K)$ for the operators T_t. We say that p_t is a C_0-*function* if the space $C_0(K)$ is an invariant subspace of $C(K)$ for the operators T_t. For example, the transitions functions in Examples 1–5 are all Feller and C_0-functions.

The next result states the most important relationship between Feller transition functions and semigroups on $C(K)$ (Theorems 9.2.1 and 9.2.2):

If p_t is a Feller transition function on K, then its associated operators $\{T_t\}_{t \geq 0}$, defined by formula (12), *form a non-negative and contraction semigroup on $C(K)$. Conversely, if $\{T_t\}_{t \geq 0}$ is a non-negative and contraction semigroup on $C_0(K)$, then there exists a unique C_0-transition function p_t on K such that formula* (12) *holds.* (15)

The Feller property deals with continuity of a transition function $p_t(x, E)$ in x, and does not, by itself, have any concern with continuity in t. Now we give a necessary and sufficient condition on $p_t(x, E)$ in order that its associated semigroup $\{T_t\}_{t \geq 0}$ be strongly continuous in t on the space $C_0(K)$:

$$\lim_{s \to 0} \| T_{t+s} f - T_t f \| = 0, \qquad f \in C_0(K).$$

A Markov transition function p_t on K is said to be *uniformly stochastically continuous* on K if the following condition is satisfied:

For each $\varepsilon > 0$ and each compact $E \subset K$, we have

$$\lim_{t \downarrow 0} \sup_{x \in E} [1 - p_t(x, U_\varepsilon(x))] = 0,$$

where $U_\varepsilon(x)$ is an ε-neighborhood of x.

For example, the transition functions in Examples 1–5 are all uniformly stochastically continuous.

Then we have the following result (Theorem 9.2.3):

Let p_t be a C_0-transition function on K. The associated semigroup $\{T_t\}_{t \geq 0}$ is strongly continuous in t on $C_0(K)$ if and only if p_t is uniformly stochastically continuous on K and satisfies: (16)

(L) *For each $s > 0$ and compact $E \subset K$, we have*

$$\lim_{x \to \partial} \sup_{0 \leq t \leq s} p_t(x, E) = 0.$$

A strongly continuous, non-negative and contraction semigroup $\{T_t\}_{t \geq 0}$ on $C_0(K)$ is called a *Feller semigroup*. Therefore, the above results (15) and (16) can be summarized as follows (Theorem 9.2.6):

If p_t is a uniformly stochastically continuous C_0-transition function on K and satisfies condition (L), then its associated operators $\{T_t\}_{t \geq 0}$, defined by formula (12), form a Feller semigroup on K. Conversely, if $\{T_t\}_{t \geq 0}$ is a Feller semigroup on K, then there exists a uniformly stochastically continuous C_0-transition function p_t on K, satisfying condition (L), such that formula (12) holds. (17)

The most important applications of result (17) are of course in the second statement.

I.6 Path Functions of Markov Processes

It is naturally interesting and important to ask the following question:

Given a Markov transition function p_t, under which conditions on p_t does there exist a Markov process with transition function p_t whose paths are almost surely *continuous*?

A Markov process $\mathscr{X} = (x_t, \mathscr{F}, \mathscr{F}_t, P_x)$ is said to be *right-continuous* provided that for each $x \in K$ we have

$$P_x\{\omega \in \Omega; \text{ the mapping } t \to x_t(\omega) \text{ is a right-continuous function from } [0, \infty) \text{ into } K_\partial\} = 1.$$

Further we say that $\mathscr{X}$ is *continuous* provided that for each $x \in K$ we have

$$P_x\{\omega \in \Omega; \text{ the maping } t \to x_t(\omega) \text{ is a continuous function from } [0, \zeta) \text{ into } K\} = 1.$$

Here ζ is the lifetime of the process $\mathscr{X}$.

Now we give some useful criteria for path-continuity in terms of transition functions (Theorem 9.1.9):

Let K be a locally compact, separable metric space and p_t a normal Markov transition function on K.

(i) *Suppose that the following two conditions are satisfied:*

(L) *For each $s > 0$ and each compact $E \subset K$, we have*

$$\lim_{x \to \partial} \sup_{0 \le t \le s} p_t(x, E) = 0.$$

(M) *For each $\varepsilon > 0$ and each compact $E \subset K$, we have*

$$\lim_{t \to 0} \sup_{x \in E} p_t(x, K \setminus U_\varepsilon(x)) = 0.$$

Then there exists a right-continuous *Markov process $\mathscr{X}$ with transition function p_t.*

(ii) *Suppose that condition* (L) *and the following condition (replacing condition (M)) are satisfied:*

(N) *For each $\varepsilon > 0$ and each compact $E \subset K$, we have*

$$\lim_{t \downarrow 0} \frac{1}{t} \sup_{x \in E} p_t(x, K \setminus U_\varepsilon(x)) = 0.$$

Then there exists a continuous *Markov process $\mathscr{X}$ with transition function p_t.*

For example, the Poisson process (Example 2) and the Cauchy process (Example 5) are right-continuous Markov processes; uniform motion (Example 1), Brownian motion (Example 3) and Brownian motion with constant drift (Example 4) are continuous Markov processes.

We remark that condition (L) is trivially satisfied, if the state space K is compact.

I.7 Strong Markov Processes

A Markov process is called a *strong Markov process* if the "starting afresh" property holds not only for every fixed moment but also for suitable random times. (For the precise definition of this "strong" Markov property, see Definition 9.1.11.)

We state a useful criterion for the strong Markov property (Theorem 9.1.12):

Every right-continuous Markov process whose transition function has the C_0-property is a strong Markov process.

Combining this result with the criterion for path-continuity in Subsection I.6, we have the following simple criterion in terms of transition functions (Theorem 9.1.14):

Every uniformly stochastically continuous C_0-transition function which satisfies condition (L) is the transition function of some strong Markov process. (18)

For example, the transition functions in Examples 1–5 correspond to strong Markov processes.

A continuous strong Markov process is called a *diffusion process.*

The next result gives a sufficient condition for the existence of a diffusion process with a prescribed Markov transition function (Theorem 9.1.15):

Every uniformly stochastically continuous C_0-transition function which satisfies conditions (L) and (N) is the transition function of some diffusion process.

For example, the transition functions in Example 1 (uniform motion) and Examples 3 and 4 (Brownian motion) correspond to diffusion processes.

Here are two more examples of diffusion processes on the half line $[0, \infty)$ in which we must take account of the effect of the boundary point 0:

Example 6 (reflecting barrier Brownian motion). If $t > 0$, $x \in [0, \infty)$ and $E \in \mathscr{B}$, we let

$$p_t(x, E) = \frac{1}{\sqrt{2\pi t}}\left(\int_E \exp\left[-\frac{(y-x)^2}{2t}\right] dy + \int_E \exp\left[-\frac{(y+x)^2}{2t}\right] dy\right), \tag{19}$$

and

$$p_0(x, E) = \chi_E(x).$$

This represents Brownian motion with reflecting barrier at $x = 0$; the process may be represented as $\{|x_t|\}$, where $\{x_t\}$ is Brownian motion on $\mathbf{R}$.

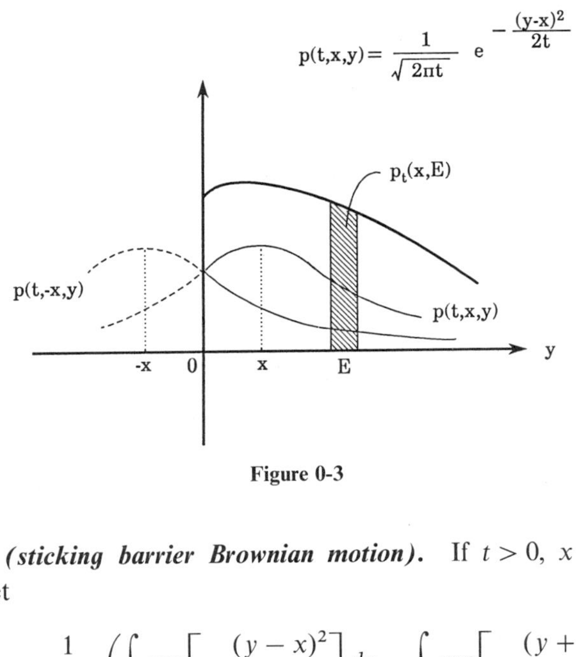

Figure 0-3

Example 7 (sticking barrier Brownian motion). If $t > 0$, $x \in [0, \infty)$ and $E \in \mathscr{B}$, we let

$$p_t(x, E) = \frac{1}{\sqrt{2\pi t}}\left(\int_E \exp\left[-\frac{(y-x)^2}{2t}\right] dy - \int_E \exp\left[-\frac{(y+x)^2}{2t}\right] dy\right)$$
$$+ \left[1 - \frac{1}{\sqrt{2\pi t}} \int_{-x}^{x} \exp\left(-\frac{z^2}{2t}\right) dz\right] \chi_E(0), \tag{20}$$

and

$$p_0(x, E) = \chi_E(x).$$

This represents Brownian motion with sticking barrier at $x = 0$; when a Brownian particle reaches $x = 0$ for the first time, it sticks there forever.

It is easy to verify that the transition functions (19) and (20) are uniformly stochastically continuous C_0-transition functions which satisfy conditions (L) and (N).

I.8 Infinitesimal Generators of Feller Semigroups

Now we return to the consideration of a Feller semigroup.

Let K be a locally compact, separable metric space and $C_0(K)$ the space of continuous functions on K vanishing at the point at infinity ∂. If p_t is a uniformly stochastically continuous C_0-transition function on K, then its associated operators defined by

$$T_t f(x) = \int_K p_t(x, dy) f(y), \qquad f \in C_0(K),$$

form a Feller semigroup on K. Recall that the infinitesimal generator $\mathfrak{A}$ of the semigroup $\{T_t\}_{t \geq 0}$ is defined by the formula

$$\mathfrak{A}f = \lim_{t \downarrow 0} \frac{T_t f - f}{t}, \tag{21}$$

provided that the limit exists in $C_0(K)$. The domain $D(\mathfrak{A})$ of $\mathfrak{A}$ consists of all functions $f \in C_0(K)$ for which the limit (21) exists.

First we write down explicitly the infinitesimal generators of Feller semigroups associated with the transition functions in Examples 1–7.

Example 1 (uniform motion). $K = \mathbf{R}$.

$$\begin{cases} D(\mathfrak{A}) = \{f \in C_0(K);\ f' \in C_0(K)\}, \\ \mathfrak{A}f = vf', \qquad f \in D(\mathfrak{A}). \end{cases}$$

Example 2 (Poisson process). $K = \mathbf{R}$.

$$\begin{cases} D(\mathfrak{A}) = C_0(K), \\ \mathfrak{A}f(x) = \lambda(f(x+1) - f(x)), \qquad f \in D(\mathfrak{A}). \end{cases}$$

We remark that the operator $\mathfrak{A}$ is not "local"; the value $\mathfrak{A}f(x)$ depends on the values $f(x)$ and $f(x+1)$. This reflects the fact that the Poisson process changes state by jumps.

Example 3 (Brownian motion). $K = \mathbf{R}$.

$$\begin{cases} D(\mathfrak{A}) = \{f \in C_0(K);\ f' \in C_0(K),\ f'' \in C_0(K)\}, \\ \mathfrak{A}f = \dfrac{1}{2} f'', \qquad f \in D(\mathfrak{A}). \end{cases}$$

The operator $\mathfrak{A}$ is "local", that is, the value $\mathfrak{A}f(x)$ is determined by the values of f in an arbitrary small neighborhood of x. This reflects the fact that Brownian motion changes state by continuous motion.

Example 4 (Brownian motion with constant drift). $K = \mathbf{R}$.

$$\begin{cases} D(\mathfrak{A}) = \{f \in C_0(K);\ f' \in C_0(K),\ f'' \in C_0(K)\}, \\ \mathfrak{A}f = \dfrac{1}{2} f'' + mf', \qquad f \in D(\mathfrak{A}). \end{cases}$$

Example 5 (Cauchy process). $K = \mathbf{R}$. The domain $D(\mathfrak{A})$ contains C^2 functions on K with compact support, and the infinitesimal generator $\mathfrak{A}$ is of the form

$$\mathfrak{A}f(x) = \frac{1}{\pi} \int_0^\infty [f(x+y) + f(x-y) - 2f(x)] \frac{dy}{y^2}.$$

Example 6 (reflecting barrier Brownian motion). $K = [0, \infty)$.

$$\begin{cases} D(\mathfrak{A}) = \{f \in C_0(K);\ f' \in C_0(K),\ f'' \in C_0(K),\ f'(0) = 0\}, \\ \mathfrak{A}f = \dfrac{1}{2} f'', \qquad f \in D(\mathfrak{A}). \end{cases}$$

Example 7 (sticking barrier Brownian motion). $K = [0, \infty)$.

$$\begin{cases} D(\mathfrak{A}) = \{f \in C_0(K);\ f' \in C_0(K),\ f'' \in C_0(K),\ f''(0) = 0\}, \\ \mathfrak{A}f = \dfrac{1}{2} f'', \qquad f \in D(\mathfrak{A}). \end{cases}$$

Here are two more examples where it is difficult to begin with a transition function and the infinitesimal generator is the basic tool of describing the process:

Example 8 (sticky barrier Brownian motion). $K = [0, \infty)$.

$$\begin{cases} D(\mathfrak{A}) = \{f \in C_0(K);\ f' \in C_0(K),\ f'' \in C_0(K),\ f'(0) - \alpha f''(0) = 0\}, \\ \mathfrak{A}f = \dfrac{1}{2} f'', \qquad f \in D(\mathfrak{A}). \end{cases}$$

Here α is a positive constant.

This process may be thought of as a "combination" of the reflecting and sticking Brownian motions. The reflecting and sticking cases are obtained by letting $\alpha \to 0$ and $\alpha \to \infty$, respectively.

Example 9 (absorbing barrier Brownian motion). $K = [0, \infty)$ where the boundary point 0 is identified with the point at infinity ∂.

$$\begin{cases} D(\mathfrak{A}) = \{f \in C_0(K);\ f' \in C_0(K),\ f'' \in C_0(K),\ f(0) = 0\}, \\ \mathfrak{A}f = \dfrac{1}{2} f'', \qquad f \in D(\mathfrak{A}). \end{cases}$$

This represents Brownian motion with absorbing barrier at $x = 0$; a Brownian particle "dies" at the first moment when it hits the boundary $x = 0$.

It is worth pointing out here that a strong Markov process cannot stay at a single position for a positive length of time and then leave that position by continuous motion; it must either jump away or leave instantaneously.

We give a simple example of a strong Markov process which changes state not by continuous motion but by jumps when the motion reaches the boundary:

Example 10. $K = [0, \infty)$.

$$\begin{cases} D(\mathfrak{A}) = \Big\{f \in C_0(K);\ f' \in C_0(K),\ f'' \in C_0(K),\ f''(0) \\ \qquad = 2c \displaystyle\int_0^\infty (f(y) - f(x))\, dF(y)\Big\}, \\ \mathfrak{A}f = \dfrac{1}{2} f'', \qquad f \in D(\mathfrak{A}). \end{cases}$$

Here c is a positive constant and F is a distribution function on $(0, \infty)$.

This process may be interpreted as follows: when a Brownian particle reaches the boundary $x = 0$, it stays there for a positive length of time and then jumps back to a random point, chosen with the function F, in the interior $(0, \infty)$. The constant c is the parameter in the "waiting time" distribution at the boundary $x = 0$. We remark that the boundary condition

$$f''(0) = 2c \int_0^\infty (f(y) - f(x))\, dF(y)$$

depends on the values of f far away from the boundary $x = 0$, unlike the boundary conditions in Examples 6–9.

I.9 One-dimensional Diffusion Processes

A Markov process is said to be one-dimensional or multidimensional according as the state space is a subset of $\mathbf{R}$ or $\mathbf{R}^n$ $(n \geq 2)$.

In the early 1950s, W. Feller characterized completely the analytic structure of one-dimensional diffusion processes; he gave an intrinsic representation of the infinitesimal generator $\mathfrak{A}$ of a one-dimensional diffusion process and determined all possible boundary conditions which describe the domain $D(\mathfrak{A})$. The probabilistic meaning of Feller's work was clarified by E. B. Dynkin, K. Itô, H. P. McKean, Jr., D. B. Ray and others. One-dimensional diffusion processes are completely studied both from analytic and probabilistic viewpoints.

Now we take a close look at Feller's work. Let $\mathscr{X} = (x_t, \mathscr{F}, \mathscr{F}_t, P_x)$ be a one-dimensional Markov process with state space K. A point x of K is called a *right* (resp. *left*) *singular point* if $x_t(\omega) \geq x$ (resp. $x_t(\omega) \leq x$) for all $t \in [0, \zeta(\omega))$ with P_x-measure one. A right and left singular point is called a *trap*. For example, the point at infinity ∂ is a trap. A point which is neither right nor left singular is called a *regular point*.

For simplicity, we suppose that the state space K is the half line,

$$K = [0, \infty),$$

and all its interior points are regular. Feller proved that there exist a strictly increasing, continuous function s on $(0, \infty)$ and Borel measures m and k on $(0, \infty)$ such that the infinitesimal generator $\mathfrak{A}$ of the process $\mathscr{X}$ can be expressed as

$$\mathfrak{A}f(x) = \lim_{y \downarrow x} \frac{f^+(y) - f^+(x) - \int_{(x, y]} f(z)\, dk(z)}{m(x, y]}. \tag{22}$$

Here:

1. $f^+(x) = \lim_{\varepsilon \downarrow 0} (f(x+\varepsilon) - f(x))/(s(x+\varepsilon) - s(x))$, the right-derivative of f with respect to s.
2. The measure m is positive for non-empty open subsets, and is finite for compact subsets.
3. The measure k is finite for compact subsets.

The function s is called a *canonical scale*, and the measures m and k are called a *canonical measure* (or *speed measure*) and a *killing measure* for the process $\mathscr{X}$, respectively. They determine the behavior of a Markovian particle in the interior of the state space K.

We remark that the right-hand side of (22) is a generalization of the second-order differential operator

$$af'' + bf' + cf,$$

where $a > 0$ and $c \leq 0$ on K. For example, the formula

$$\mathfrak{A}f = af'' + bf'$$

can be written in the form (22), if we take

$$\begin{aligned} s(x) &= \int_0^x \exp\left[-\int_0^y \frac{b(z)}{a(z)}\,dz\right] dy,\\ dm(x) &= \frac{1}{a(x)} \exp\left[\int_0^x \frac{b(y)}{a(y)}\,dy\right] dx,\\ dk(x) &= 0. \end{aligned}$$

The boundary point 0 is called a *regular boundary* if it satisfies: For an arbitrary point $r \in (0, \infty)$, we have

$$\int_{(0,r)} [s(r) - s(x)][dm(x) + dk(x)] < \infty,$$

$$\int_{(0,r)} [m((x, r)) + k((x, r))]\, ds(x) < \infty.$$

It can be shown that this notion is independent of the point r used. Intuitively, the regularity of the boundary point means that a Markovian particle approaches the boundary in finite time with positive probability, and also enters the interior from the boundary.

The behavior of a Markovian particle at the boundary point is characterized by boundary conditions. In the case of regular boundary points, Feller determined all possible boundary conditions which are satisfied by the functions f in the domain $D(\mathfrak{A})$ of $\mathfrak{A}$. A general boundary condition is of the form

$$\gamma f(0) - \delta \mathfrak{A} f(0) + \mu f^{+}(0) = 0, \tag{23}$$

where γ, δ and μ are constants such that

$$\gamma \leq 0, \qquad \delta \geq 0, \qquad \mu \geq 0, \qquad \mu + \delta > 0.$$

If we admit jumps from the boundary into the interior, then a general boundary condition takes the form

$$\gamma f(0) - \delta \mathfrak{A} f(0) + \mu f^{+}(0) + \int_{(0,\infty)} [f(x) - f(0)]\, d\nu(x) = 0, \tag{24}$$

where ν is a Borel measure with respect to which the function $\min(1, s(x) - s(+0))$ is integrable. We remark that boundary condition (24) is a "combination" of the boundary conditions in Examples 6–10 if we take $s(x) = x$, $dm(x) = 2\,dx$, $dk(x) = 0$.

I.10 Multidimensional Diffusion Processes

The main purpose of this book is to generalize Feller's work to the multidimensional case.

In 1959, A. D. Ventcel' studied the problem of determining all possible boundary conditions for multidimensional diffusion processes, which we now state.

Let D be a *bounded* domain in $\mathbf{R}^N$ with smooth boundary ∂D and $C(\bar{D})$ the space of real-valued continuous functions on $\bar{D} = D \cup \partial D$. A Feller semigroup on $\bar{D}$ is a strongly continuous, non-negative and contraction semigroup $\{T_t\}_{t \geq 0}$ on $C(\bar{D})$. Results (17) and (18) tell us that there corresponds to a Feller semigroup $\{T_t\}_{t \geq 0}$ on $\bar{D}$ a strong Markov process $\mathscr{X}$ on $\bar{D}$ whose transition function $p_t(x, dy)$ satisfies

$$T_t f(x) = \int_{\bar{D}} p_t(x, dy) f(y), \qquad f \in C(\bar{D}).$$

Under certain continuity hypotheses concerning $p_t(x, dy)$, such as condition (N) in Subsection I.6, Ventcel' showed that the infinitesimal generator $\mathfrak{A}$ of $\{T_t\}$ is described analytically as follows (Theorems 9.4.1 and 9.5.1):

(i) Let x be a point of the interior D of the state space $\bar{D}$. For all $u \in D(\mathfrak{A}) \cap C^2(\bar{D})$, we have

$$\begin{aligned}\mathfrak{A}u(x) &= Au(x)\\ &\equiv \sum_{i,j=1}^{N} a^{ij}(x)\frac{\partial^2 u}{\partial x_i\,\partial x_j}(x) + \sum_{i=1}^{N} b^i(x)\frac{\partial u}{\partial x_i}(x) + c(x)u(x),\end{aligned}$$

where the matrix $(a^{ij}(x))$ is positive semi-definite and $c(x) \leq 0$.

(ii) Let x' be a point of the boundary ∂D of the state space $\bar{D}$, and choose a local coordinate system $x = (x_1, x_2, \ldots, x_{N-1}, x_N)$ in a neighborhood of x' such that $x \in D$ if $x_N > 0$ and $x \in \partial D$ if $x_N = 0$. Then every function $u \in D(\mathfrak{A}) \cap C^2(\bar{D})$ satisfies the following boundary condition, analogous to boundary condition (23):

$$\begin{aligned}Lu(x') = &\sum_{i,j=1}^{N-1} \alpha^{ij}(x')\frac{\partial^2 u}{\partial x_i\,\partial x_j}(x') + \sum_{i=1}^{N-1} \beta^i(x')\frac{\partial u}{\partial x_i}(x')\\ &+ \gamma(x')u(x') + \mu(x')\frac{\partial u}{\partial x_N}(x') - \delta(x')Au(x')\\ = &\ 0.\end{aligned}$$

Here the matrix $(\alpha^{ij}(x'))$ is positive semi-definite and $\gamma(x') \leq 0$, $\mu(x') \geq 0$, $\delta(x') \geq 0$. The boundary condition L is called a *Ventcel' boundary condition.*

Probabilistically, the above result may be interpreted as follows: a Markovian particle of the diffusion process $\mathscr{X}$ on $\bar{D}$ is governed by a degenerate elliptic differential operator A of second order in the interior D of the state space $\bar{D}$, and it obeys a Ventcel' boundary condition L on the boundary ∂D of $\bar{D}$. The terms

$$\sum_{i,j=1}^{N-1} \alpha^{ij}(x')\frac{\partial^2 u}{\partial x_i\,\partial x_j}(x') + \sum_{i=1}^{N-1} \beta^i(x')\frac{\partial u}{\partial x_i}(x'), \qquad \gamma(x')u(x'),$$

$$\mu(x')\frac{\partial u}{\partial x_N}(x') \quad \text{and} \quad \delta(x')Au(x')$$

of L are supposed to correspond to the diffusion along the boundary, absorption, reflection and viscosity phenomena, respectively (cf. Examples 6–9 in Subsection I.8 and Figure 0-4).

Analytically, via the Hille–Yosida theory of semigroups, it may be interpreted as follows: a Feller semigroup $\{T_t\}_{t \geq 0}$ on $\bar{D}$ is described by a degenerate elliptic differential operator A of second order and a Ventcel' boundary condition L, if the paths of its corresponding strong Markov

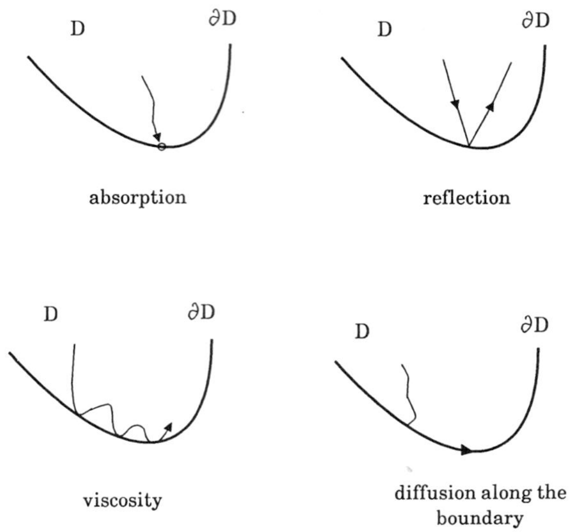

Figure 0-4

process $\mathscr{X}$ are continuous. We are thus reduced to the study of (non-)elliptic boundary value problems for (A, L) in the theory of partial differential equations.

In this book we shall consider, conversely, the following:

Problem. Construct a Feller semigroup $\{T_t\}_{t \geq 0}$ on $\bar{D}$ with prescribed analytic data (A, L).

II. Propagation of Maximums

Now we pause from our main development in order to study an intimate connection between Markov processes and partial differential equations. This will play an important role in the study of Markov processes in terms of partial differential equations.

We begin with the following elementary result:

Let I be an open interval of $\mathbf{R}$. If $u \in C^2(I)$, $d^2u/dx^2 \geq 0$ in I and u takes its maximum at a point of I, then u is a constant.

This result can be extended to the N-dimensional case, with the operator d^2/dx^2 replaced by the Laplacian $\Delta = \sum_{i=1}^{N} \partial^2/\partial x_i^2$:

Let D be a connected open subset of $\mathbf{R}^N$. If $u \in C^2(D)$, $\Delta u \geq 0$ in D and u takes its maximum at a point of D, then u is a constant. (1)

Result (1) is well known by the name of the *strong maximum principle* for the Laplacian.

Now we study the underlying *analytical* mechanism of propagation of maximums for degenerate elliptic differential operators of second order, which will reveal an intimate connection between partial differential equations and Markov processes.

Let A be a second-order, *degenerate elliptic* differential operator with real coefficients such that

$$A = \sum_{i,j=1}^{N} a^{ij}(x) \frac{\partial^2}{\partial x_i \, \partial x_j} + \sum_{i=1}^{N} b^i(x) \frac{\partial}{\partial x_i},$$

where the coefficients a^{ij}, b^i satisfy:

1. The a^{ij} are C^2 functions on $\mathbf{R}^N$ all of whose derivatives of order ≤ 2 are bounded in $\mathbf{R}^N$, $a^{ij} = a^{ji}$ and

$$\sum_{i,j=1}^{N} a^{ij}(x)\xi_i \xi_j \geq 0, \qquad x \in \mathbf{R}^N, \ \xi \in \mathbf{R}^N.$$

2. The b^i are C^1 functions on $\mathbf{R}^N$ with bounded derivatives in $\mathbf{R}^N$.

We consider the following:

Problem. Let D be a connected open subset of $\mathbf{R}^N$ and x a point of D. Then determine the largest connected, relatively closed subset $D(x)$ of D, containing x, such that:

$$\begin{cases} \text{If } u \in C^2(D),\ Au \geq 0 \text{ in } D,\ \sup_D u = M < +\infty \\ \text{and } u(x) = M, \text{ then } u \equiv M \text{ throughout } D(x). \end{cases} \tag{$*$}$$

The set $D(x)$ is called the *propagation* set of x in D.

We now give a coordinate-free description of the set $D(x)$ in terms of subunit vectors whose notion is introduced by Fefferman–Phong.

A tangent vector $X = \sum_{j=1}^{N} \gamma^j(\partial/\partial x_j)$ at $x \in D$ is said to be *subunit* for the operator $A^0 = \sum_{i,j=1}^{N} a^{ij}(\partial^2/\partial x_i\, \partial x_j)$ if it satisfies

$$\left(\sum_{j=1}^{N} \gamma^j \eta_j\right)^2 \leq \sum_{i,j=1}^{N} a^{ij}(x)\eta_i\eta_j, \qquad \eta = \sum_{j=1}^{N} \eta_j\, dx_j \in T_x^*(D),$$

where $T_x^*(D)$ is the cotangent space of D at x. Note that this notion is coordinate-free. So we rotate the coordinate axes so that the matrix (a^{ij}) is diagonalized at x:

$$(a^{ij}(x)) = (\lambda_i \delta_{ij}), \qquad \lambda_1 > 0, \ldots, \lambda_r > 0,\ \lambda_{r+1} = \cdots = \lambda_N = 0.$$

Here $r = \operatorname{rank}(a^{ij}(x))$. Then it is easy to see that the vector X is subunit for A^0 if and only if it is contained in the following ellipsoid of dimension r (see Figure 0-5):

$$\frac{(\gamma^1)^2}{\lambda_1} + \cdots + \frac{(\gamma^r)^2}{\lambda_r} \leq 1, \qquad \gamma^{r+1} = \cdots = \gamma^N = 0. \tag{2}$$

A *subunit trajectory* is a Lipschitz path $\gamma\colon [t_1, t_2] \to D$ such that the tangent vector $\dot{\gamma}(t) = (d/dt)(\gamma(t))$ is subunit for A^0 at $\gamma(t)$ for almost every t. We remark that if $\dot{\gamma}(t)$ is subunit for A^0, so is $-\dot{\gamma}(t)$; hence subunit trajectories are not oriented.

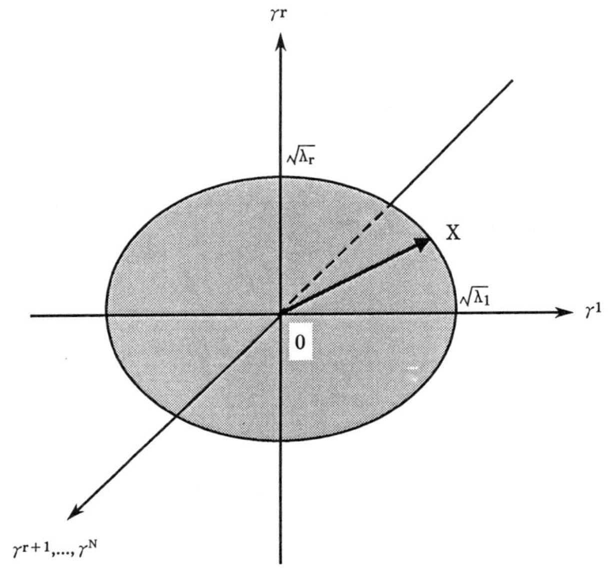

Figure 0-5

We let

$$X_0 = \sum_{i=1}^{N} \left(b^i - \sum_{j=1}^{N} \frac{\partial a^{ij}}{\partial x_j} \right) \frac{\partial}{\partial x_i}.$$

The vector field X_0 is called the drift vector field in probability theory, while it is the so-called subprincipal part of the operator A in terms of the theory of partial differential equations.

A *drift trajectory* is a curve $\theta: [t_1, t_2] \to D$ such that

$$\dot{\theta}(t) = X_0(\theta(t)) \text{ on } [t_1, t_2],$$

and this curve is oriented in the direction of increasing t.

Our main result is the following (Theorem 7.2.1):

The propagation set $D(x)$ of x in D contains the closure $D'(x)$ in D of all points $y \in D$ which can be joined to x by a finite number of subunit and drift trajectories. (3)

This result tells us that if the matrix (a^{ij}) is non-degenerate at x, that is, if $r = \operatorname{rank}(a^{ij}(x)) = N$, then the maximum propagates in an open neighborhood of x; but if the matrix (a^{ij}) is degenerate at x, then the maximum propagates only in a "thin" ellipsoid of dimension r (cf. formula (2)), and in the direction of X_0. Now we see the reason why the strong maximum principle (1) holds for the Laplacian.

We consider a few simple examples in the case when D is the square $(-1, 1) \times (-1, 1)$, $(N = 2)$.

Example 1. $A_1 = \partial^2/\partial x^2 + x^2(\partial^2/\partial y^2)$. The subunit vector fields for A_1 are generated by

$$\left(\frac{\partial}{\partial x}, x \frac{\partial}{\partial y} \right).$$

Hence we have:

$$\text{The set } D'((x, y)) \text{ is equal to } D \text{ for every } (x, y) \in D. \tag{4}$$

That is, the strong maximum principle (1) remains valid for the operator A_1.

Example 2. $A_2 = x^2(\partial^2/\partial x^2) + \partial^2/\partial y^2$. The subunit vector fields for A_2 are generated by

$$\left(x \frac{\partial}{\partial x}, \frac{\partial}{\partial y} \right). \tag{5}$$

Thus we have:

$$D'((x, y)) = \begin{cases} [0, 1) \times (-1, 1) & \text{if } x > 0, \\ \{0\} \times (-1, 1) & \text{if } x = 0, \\ (-1\ 0] \times (-1, 1) & \text{if } x < 0. \end{cases}$$

It can be shown that the strong maximum principle (1) does not hold for the operator A_2 (cf. Remark 7.2.3).

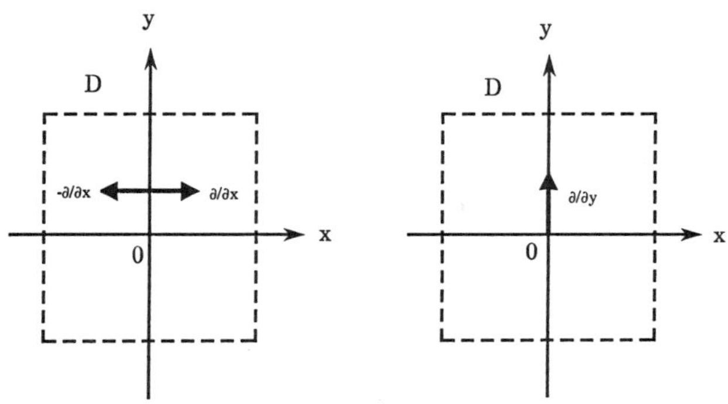

Example 1 Example 2

Figure 0-6

Example 3. $A_3 = x^2(\partial^2/\partial x^2) + \partial^2/\partial y^2 + y(\partial/\partial x)$. The subunit vector fields for $A_3^0 = A_2$ are generated by (5), and the drift vector field is

$$(y - 2x)\frac{\partial}{\partial x}.$$

Thus, by virtue of the drift vector field, we have assertion (4), and so the strong maximum principle (1) remains valid for the operator A_3.

Example 4. $A_4 = x^2(\partial^2/\partial x^2) + \partial^2/\partial y^2 + \partial/\partial x$. The subunit vector fields for $A_4^0 = A_2$ are generated by (5), and the drift vector field is

$$(1 - x)\frac{\partial}{\partial x}.$$

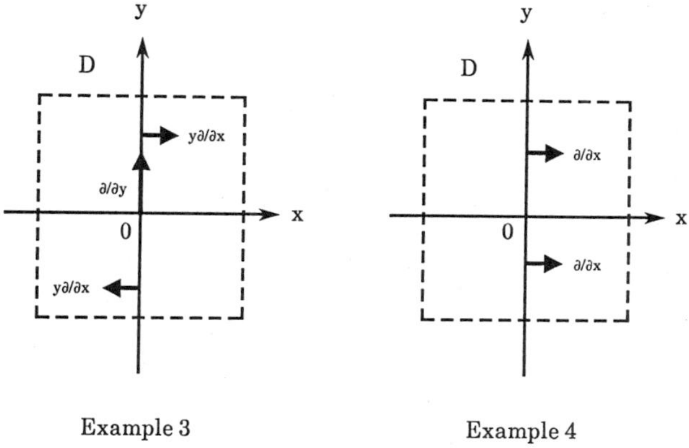

Figure 0-7

Hence we have:

$$D'((x, y)) = \begin{cases} D & \text{if } x < 0, \\ [0, 1) \times (-1, 1) & \text{if } x \geq 0. \end{cases}$$

It can also be shown that the strong maximum principle (1) does not hold for the operator A_4.

It is worth pointing out here that the propagation set $D'(x)$ coincides with the *support* of the Markov process corresponding to the operator A, which is the closure of the collection of all possible trajectories of a Markovian particle, starting at x, with generator A.

In the case when the operator A is written as the sum of squares of vector fields, we can give another (equivalent) description of the set $D'(x)$.

Suppose that the operator A is of the form

$$A = \sum_{k=1}^{r} Y_k^2 + Y_0,$$

where the Y_k are real C^2 vector fields on $\mathbf{R}^N$ and Y_0 is a real C^1 vector field on $\mathbf{R}^N$. *Hill's diffusion trajectory* is a curve $\beta\colon [t_1, t_2] \to D$ such that

$$\dot{\beta}(t) = Y_k(\beta(t)), \qquad \dot{\beta}(t) \neq 0 \text{ on } [t_1, t_2].$$

Hill's diffusion trajectories are not oriented; they may be traversed in either direction. *Hill's drift trajectories* are defined similarly, with Y_k replaced by Y_0, but they are oriented in the direction of increasing t.

In this case, our main result (3) can be restated as follows (Theorem 7.2.4):

The propagation set $D'(x)$ coincides with the closure in D of all points $y \in D$ which can be joined to x by a finite number of Hill's diffusion and drift trajectories. (3′)

Further, result (3) may be reformulated in various ways. For example, we have the following result (Theorem 7.2.1′):

Let c be a continuous function on D such that $c \leq 0$ in D. If $u \in C^2(D)$, $(A + c)u \geq 0$ in D and u attains its positive *maximum M at a point x of D, then $u \equiv M$ throughout $D'(x)$.*

III. Construction of Feller Semigroups

Now we return to the problem of construction of Feller semigroups. First we give a general existence theorem for Feller semigroups in terms of boundary value problems.

Let D be a bounded domain in $\mathbf{R}^N$ with C^∞ boundary ∂D, and let A be a second-order *elliptic* differential operator with real coefficients such that

$$Au(x) = \sum_{i,j=1}^{N} a^{ij}(x) \frac{\partial^2 u}{\partial x_i \, \partial x_j}(x) + \sum_{i=1}^{N} b^i(x) \frac{\partial u}{\partial x_i}(x) + c(x)u(x)$$

where:

1. $a^{ij} \in C^\infty(\mathbf{R}^N)$, $a^{ij} = a^{ji}$ and there exists a constant $a_0 > 0$ such that

$$\sum_{i,j=1}^{N} a^{ij}(x)\xi_i \xi_j \geq a_0 |\xi|^2, \qquad x \in \mathbf{R}^N, \ \xi \in \mathbf{R}^N. \tag{1}$$

2. $b^i \in C^\infty(\mathbf{R}^N)$.
3. $c \in C^\infty(\mathbf{R}^N)$ and $c \leq 0$ on $\bar{D}$.

Let L be a Ventcel' boundary condition such that

$$Lu(x') = \sum_{i,j=1}^{N-1} \alpha^{ij}(x') \frac{\partial^2 u}{\partial x_i \, \partial x_j}(x') + \sum_{i=1}^{N-1} \beta^i(x') \frac{\partial u}{\partial x_i}(x')$$
$$+ \gamma(x')u(x') + \mu(x') \frac{\partial u}{\partial \mathbf{n}}(x') - \delta(x')Au(x')$$

where:

1. The α^{ij} are the components of a C^∞ symmetric contravariant tensor of type $\binom{2}{0}$ on ∂D and

$$\sum_{i,j=1}^{N-1} \alpha^{ij}(x')\eta_i\eta_j \geq 0, \qquad x' \in \partial D,\ \eta = \sum_{j=1}^{N-1} \eta_j\, dx_j \in T^*_{x'}(\partial D),$$

   where $T^*_{x'}(\partial D)$ is the cotangent space of ∂D at x'.
2. $\beta^i \in C^\infty(\partial D)$. (2)
3. $\gamma \in C^\infty(\partial D)$ and $\gamma \leq 0$ on ∂D.
4. $\mu \in C^\infty(\partial D)$ and $\mu \geq 0$ on ∂D.
5. $\delta \in C^\infty(\partial D)$ and $\delta \geq 0$ on ∂D.
6. $\mathbf{n}$ is the unit interior normal to ∂D at x'.

A Ventcel' boundary condition L is said to be *transversal* on ∂D if it satisfies

$$\mu(x) + \delta(x) > 0 \qquad \text{on } \partial D.$$

Intuitively, the transversality condition implies that either reflection or viscosity phenomenon occurs at each point of ∂D. Probabilistically, this means that every Markov process on ∂D is the "trace" on ∂D of trajectories of some Markov process on $\bar{D}$.

The next result states sufficient conditions for the existence of a Feller semigroup in terms of boundary value problems (Theorem 9.6.22):

Let the differential operator A satisfy condition (1) *and let the boundary condition L satisfy condition* (2), *and be transversal on ∂D. Suppose that the following two conditions are satisfied*:

[I] (Existence) *For some constants $\alpha \geq 0$ and $\lambda \geq 0$, the boundary value problem*

$$\begin{cases} (\alpha - A)u = 0 \text{ in } D, \\ (\lambda - L)u = \varphi \text{ on } \partial D \end{cases} \tag{$*$}$$

has a solution u in $C(\bar{D})$ for any φ in some dense subset of $C(\partial D)$.

[II] (Uniqueness) *For some constant $\alpha > 0$, we have*:

$$u \in C(\bar{D}),\ (\alpha - A)u = 0 \text{ in } D,\ Lu = 0 \text{ on } \partial D \qquad \Rightarrow \qquad u = 0 \text{ in } D.$$

Then there exists a Feller semigroup $\{T_t\}_{t\geq 0}$ *on* $\bar{D}$ *whose infinitesimal generator* $\mathfrak{A}$ *is characterized as follows*:

1. *The domain* $D(\mathfrak{A})$ *of* $\mathfrak{A}$ *is the space*

$$D(\mathfrak{A}) = \{u \in C(\bar{D});\ Au \in C(\bar{D}),\ Lu = 0\}.$$

2. $\mathfrak{A}u = Au, \qquad u \in D(\mathfrak{A}).$

The probabilistic meaning of the unique solvability of problem (∗) is that there exists a Markov process $\mathscr{Y}$ (with discontinuous paths) on the boundary ∂D. But, the transversality condition for L implies that every Markov process on ∂D is the "trace" on ∂D of trajectories of some Markov process on $\bar{D}$. Hence we can "piece together" the process $\mathscr{Y}$ with A-diffusion in D to construct a Markov process $\mathscr{X}$ on $\bar{D}$ and hence a Feller semigroup $\{T_t\}_{t\geq 0}$ on $\bar{D}$.

Thus we are reduced to the study of the boundary value problem (∗) with spectral parameter α. In Chapter 8, using a method essentially due to Agmon–Nirenberg, we study the basic questions of existence and uniqueness of solutions of general boundary value problems for second-order elliptic differential operators with spectral parameter. The Agmon–Nirenberg method is a technique of treating a spectral parameter as a second-order elliptic differential operator of an extra variable and relating the old problem to a new one with additional variable. (For details, see Section 8.4.)

Now we state our existence theorems for Feller semigroups.

As in Section II, we say that a tangent vector

$$v = \sum_{j=1}^{N-1} v^j \frac{\partial}{\partial x_j} \in T_{x'}(\partial D)$$

is subunit for the operator,

$$L^0 = \sum_{i,j=1}^{N-1} \alpha^{ij} \frac{\partial^2}{\partial x_i\, \partial x_j}$$

if it satisfies

$$\left(\sum_{j=1}^{N-1} v^j \eta_j\right)^2 \leq \sum_{i,j=1}^{N-1} \alpha^{ij}(x')\eta_i\eta_j, \qquad \eta = \sum_{j=1}^{N-1} \eta_j\, dx_j \in T^*_{x'}(\partial D).$$

If $\rho > 0$, we define a "non-Euclidean" ball $B_{L^0}(x', \rho)$ of radius ρ about x' as follows:

$B_{L^0}(x', \rho) =$ the set of all points $y \in \partial D$ which can be joined to x' by a Lipschitz path $\gamma: [0, \rho] \to \partial D$ for which the tangent vector $\dot{\gamma}(t)$ of ∂D at $\gamma(t)$ is subunit for L^0 for almost every t.

Also we let

$B_E(x', \rho) =$ the ordinary Euclidean ball of radius ρ about x'.

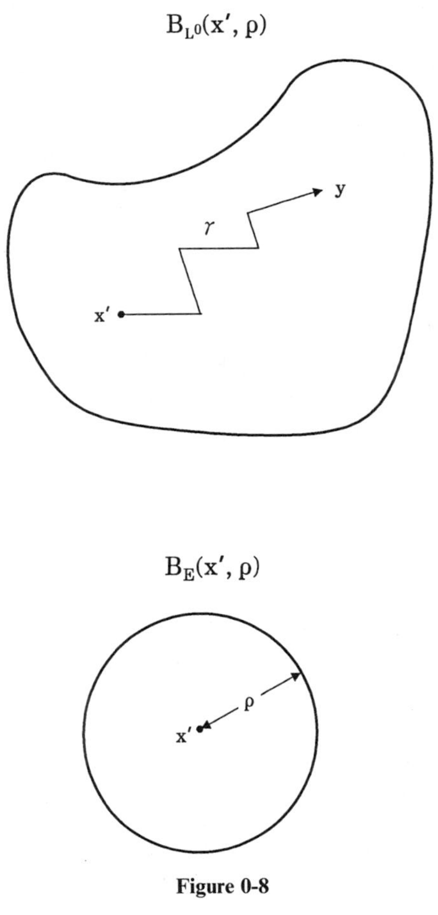

Figure 0-8

Our main result is the following (Theorem 10.1.1):

Let the differential operator A satisfy condition (1) *and let the boundary condition L satisfy condition* (2) *and be transversal on ∂D. Suppose that:*

(A.1) *There exist constants $0 < \varepsilon \leq 1$ and $C > 0$ such that for all sufficiently small $\rho > 0$ we have*

$$B_E(x', \rho) \subset B_{L^0}(x', C\rho^\varepsilon), \qquad x' \in M = \{x' \in \partial D;\ \mu(x') = 0\}.$$

Then there exists a Feller semigroup $\{T_t\}_{t \geq 0}$ on $\bar{D}$ whose infinitesimal generator $\mathfrak{A}$ is characterized as follows:

1. *The domeain $D(\mathfrak{A})$ of $\mathfrak{A}$ is the space*

$$D(\mathfrak{A}) = \{u \in C(\bar{D});\ Au \in C(\bar{D}),\ Lu = 0\}.$$

2. $\mathfrak{A}u = Au, \qquad u \in D(\mathfrak{A}).$

Furthermore, the generator $\mathfrak{A}$ coincides with the minimal closed extension in $C(\bar{D})$ of the restriction of A to the space $\{u \in C^2(\bar{D});\ Lu = 0\}$.

Result (3) in Section II tells us that the non-Euclidean ball $B_{L^0}(x', \rho)$ may be interpreted as the set of all points where a Markovian particle with generator L^0, starting at x', diffuses during the time interval $[0, \rho]$. Hence the intuitive meaning of hypothesis (A.1) is that a Markovian particle with generator L^0 goes through the set M where no reflection phenomenon occurs in *finite* time (cf. Figure 0-9). Therefore, the above result may be stated as follows: if a Markovian particle goes through the set where no reflection phenomenon occurs in finite time, then there exists a Feller semigroup corresponding to such a diffusion phenomenon.

Further we consider the case when $\alpha^{ij} \equiv 0$ on ∂D:

$$\begin{aligned} Lu(x') = {} & \sum_{i=1}^{N-1} \beta^i(x') \frac{\partial u}{\partial x_i}(x') + \gamma(x')u(x') \\ & + \mu(x') \frac{\partial u}{\partial \mathbf{n}}(x') - \delta(x')Au(x'). \end{aligned} \tag{3}$$

Here $\beta = \sum_{i=1}^{N-1} \beta^i(\partial/\partial x_i)$ is a C^∞ vector field on ∂D.

Then we have the following result (Theorem 10.1.3):

Let A and L be as above, L being of the form (3). *Suppose that:*

(A.2) *The vector field β is non-zero on the set $M = \{x' \in \partial D;\ \mu(x') = 0\}$ and any maximal integral curve of β is not* entirely *contained in M.*

Then we have the same conclusion as in the above result.

Note that the vector field β is the drift vector field. Hence result (3) in Section II tells us that hypothesis (A.2) has an intuitive meaning similar to hypothesis (A.1).

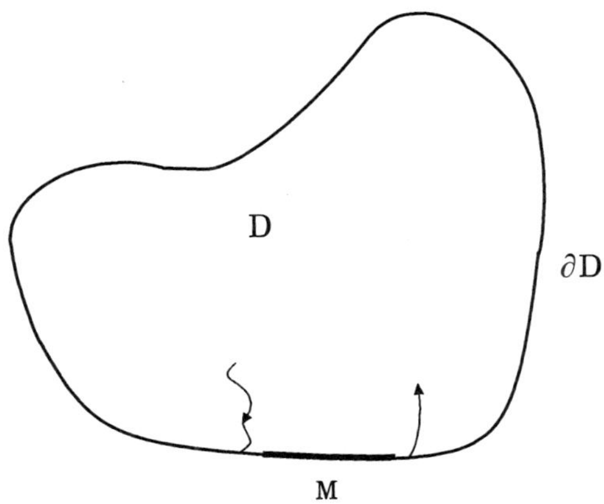

Figure 0-9

1 Preparatory Material

This chapter is a summary of the basic definitions and results about topological spaces, linear spaces and measure spaces which will be used throughout the book. Most of the material will be quite familiar to the reader and may be omitted. This chapter, included for the sake of completeness, should serve to settle questions of notation and such.

1.1. Sets

A *set* is a collection of elements, and is described either by listing their members or by expressions of the form $\{x; P\}$, which denote the set of those elements x satisfying property P. The *empty set* is the set with no element, and is denoted by $\varnothing$. The words *collection, family* and *class* will be used synonymously with set.

The notation $x \in A$ (or $A \ni x$) means that x is a member or element of the set A. We also say that x belongs to the set A. The elements of A are frequently called *points*, and in this case the set A is referred to as the *space*. If every element of a set A is also an element of a set B, then A is said to be a *subset* of B, and we write $A \subset B$ or $B \supset A$. Two sets A and B are said to be equal if $A \subset B$ and $B \subset A$, and we write $A = B$. The negations of $\in$, $\subset$ and $=$

are denoted by $\notin$, $\not\subset$ and $\neq$, respectively. If $A \subset B$ but $A \neq B$, then A is called a *proper subset* of B.

The *difference* between two sets A and B is the set of all those elements of A which do not belong to B, and is denoted by $A \setminus B$. If A is a subset of a fixed set X, then the difference $X \setminus A$ is called the *complement* of A, and is denoted by A^c,

We will often consider a collection $\{A_\lambda; \lambda \in \Lambda\}$ of sets A_λ indexed by the set Λ. The *union* of the sets A_λ is the set of all those elements which belong to at least one of the A_λ, and is denoted by $\bigcup_{\lambda \in \Lambda} A_\lambda$. The *intersection* of the sets A_λ is the set of all those elements which belong to every A_λ, and is denoted by $\bigcap_{\lambda \in \Lambda} A_\lambda$. A collection $\{A_\lambda\}$ of sets is said to be *disjoint* if every two distinct sets of the A_λ have no element in common. In this case, the union of the sets A_λ is called a *disjoint union*, and is denoted by $\sum_{\lambda \in \Lambda} A_\lambda$.

The *Cartesian product* $A_1 \times \cdots \times A_n$ of sets $A_1, \ldots, A_n$ is the set of all ordered n-tuples $(a_1, \ldots, a_n)$ with $a_i \in A_i$ for each i.

1.2. Mappings

Let X and Y be two sets. A correspondence f which assigns to each element x of X an element $f(x)$ of Y is called a *mapping* or *map* of X into Y, and we write $f: X \to Y$. When describing a mapping f by describing its effect on individual elements, we use the special arrow $\mapsto$, and write "$x \mapsto f(x)$". The terms mapping, *function* and *transformation* will be used synonymously.

If A is a subset of X, the set $f(A) = \{f(x); x \in A\}$ is called the *image* of A under f. If B is a subset of Y, the set $f^{-1}(B) = \{x \in X; f(x) \in B\}$ is called the *inverse image* of B under f. The *domain* $D(f)$ of f is the set X and the *range* $R(f)$ of f is the set $f(X)$.

If, for each element y of $f(X)$, there exists only one element x of X such that $f(x) = y$, then f is called a *one-to-one map* or *injection* of X into Y. We also say that f is *one-to-one* or *injective*. If f is injective, then the *inverse (mapping)* f^{-1}, defined by $x = f^{-1}(y) = f^{-1}(\{y\})$, is a mapping with domain $f(X)$ and range X. A mapping f is called an *onto map* or *surjection* if $f(X) = Y$. We also say that f is *onto* or *surjective*. If f is both an injection and a surjection, then it is called a *bijection*. We also say that f is *bijective*.

If $f: X \to Y$ and $g: Y \to Z$ are two mappings, the *composite mapping* $g \circ f: X \to Z$ is defined by the formula $(g \circ f)(x) = g(f(x))$, $x \in X$.

1.3. Topological Spaces

Let X be a non-empty set. A collection $\mathcal{O}$ of subsets of X is said to be a *topology* on X if it satisfies the following conditions:

($T1$) The empty set $\varnothing$ and the set X itself belong to $\mathcal{O}$.
($T2$) If O_1, O_2 are members of $\mathcal{O}$, then the intersection $O_1 \cap O_2$ belongs to $\mathcal{O}$.
($T3$) If $\{O_\lambda\}_{\lambda \in \Lambda}$ is an arbitrary collection of members of $\mathcal{O}$, then the union $\bigcup_{\lambda \in \Lambda} O_\lambda$ belongs to $\mathcal{O}$.

The pair $(X, \mathcal{O})$ is called a *topological space* and the members of $\mathcal{O}$ are called *open sets* in X; their complements are called *closed sets.*

Let $(X, \mathcal{O})$ be a topological space. A *neighborhood* of a point x of X is an open set which contains x, and the *neighborhood system* $\mathcal{U}(x)$ of x is the collection of all neighborhoods of x. A subcollection $\mathcal{U}^*(x)$ of $\mathcal{U}(x)$ is called a *fundamental neighborhood system* of x if it has the following property:

(FV) For any $U \in \mathcal{U}(x)$, there exists $V \in \mathcal{U}^*(x)$ such that $V \subset U$.

A topology on X can be formulated in terms of fundamental neighborhood systems as follows:

1) A family $\{\mathcal{U}^*(x)\}_{x \in X}$ of fundamental neighborhood systems of a topological space $(X, \mathcal{O})$ enjoys the following properties:

($V1$) If $V \in \mathcal{U}^*(x)$, then $x \in V$.
($V2$) For $V_1, V_2 \in \mathcal{U}^*(x)$, there exists $V_3 \in \mathcal{U}^*(x)$ such that $V_3 \subset V_1 \cap V_2$.
($V3$) If $V \in \mathcal{U}^*(x)$, then for each $y \in V$ there exists $W \in \mathcal{U}^*(y)$ such that $W \subset V$.

2) Conversely, suppose that we are given for each point x of a set X a collection $\mathcal{U}^*(x)$ of subsets of X, and that the family $\{\mathcal{U}^*(x)\}_{x \in X}$ satisfies conditions ($V1$), ($V2$) and ($V3$). We let

$$\mathcal{O} = \{O \subset X; \text{For every point } x \text{ of } O \text{ there exists } V \in \mathcal{U}^*(x) \text{ such that } V \subset O\}.$$

Then it is easy to verify that the collection $\mathcal{O}$ satisfies axioms ($T1$), ($T2$) and ($T3$) of a topology and that $\mathcal{U}^*(x)$ is a fundamental neighborhood system of x in the topological space $(X, \mathcal{O})$.

Let $\mathcal{O}_1$ and $\mathcal{O}_2$ be two topologies on the same set X. Then $\mathcal{O}_1$ is said to be *stronger* than $\mathcal{O}_2$ if every $\mathcal{O}_2$-open set is an $\mathcal{O}_1$-open set. We also say that $\mathcal{O}_2$ is *weaker* than $\mathcal{O}_1$.

Let X be a topological space (we often omit $\mathcal{O}$ and refer to X as a topological space). A point x of X is called an *accumulation point* of a subset A of X if every neighborhood of x contains at least one point of A different from x. A subset of X is closed if and only if it contains all its accumulation points. The *closure* $\bar{A}$ of a subset A of X is the smallest closed subset of X which contains A. The *interior* A^o of A is the largest open subset of X contained in A. The set $\bar{A} \setminus A^o$ is called the *boundary* of A.

A subset A of X is said to be *everywhere dense* or simply *dense* in X if $\bar{A} = X$. A topological space is said to be *separable* if it contains a countable, dense subset.

A topological space X is said to satisfy the *first axiom of countability* if, for each point x of X, there exists a fundamental neighborhood system of x which has countably many members.

A family of open sets in X is called an *open base* for X if every open set can be expressed as a union of members of this family. A topological space X is said to satisfy the *second axiom of countability* if there exists an open base for X which has countably many members. A topological space with a countable open base is separable.

A topological space X is called a *Hausdorff space* if, for two arbitrary distinct points x, y of X, there exist a neighborhood U of x and a neighborhood V of y such that $U \cap V = \varnothing$. We also say that X is Hausdorff.

Let Y be a subset of a topological space $(X, \mathcal{O})$. We let

$$\mathcal{O}_Y = \{O \cap Y;\ O \in \mathcal{O}\}.$$

Then the collection $\mathcal{O}_Y$ of subsets of Y satisfies axioms $(T1)$, $(T2)$ and $(T3)$ of a topology; hence $\mathcal{O}_Y$ is a topology on Y. This topology is called the *relative topology* of Y as a subset of $(X, \mathcal{O})$, and $(Y, \mathcal{O}_Y)$ is called a *topological subspace* of $(X, \mathcal{O})$.

If $X_1, \ldots, X_n$ are topological spaces, then a topology is defined on the Cartesian product $X_1 \times \cdots \times X_n$ by taking as a fundamental neighborhood system of a point $(x_1, \ldots, x_n)$ all sets of the form $U_1 \times \cdots \times U_n$, where U_i is a neighborhood of x_i for each i. This topology is called the *product topology* and $X_1 \times \cdots \times X_n$ is called the *product topological space*.

1.4. Compactness

A collection $\{U_\lambda\}_{\lambda \in \Lambda}$ of open sets of a topological space X is called an *open covering* of X if $X = \bigcup_{\lambda \in \Lambda} U_\lambda$. A topological space X is said to be *compact* if every open covering $\{U_\lambda\}$ of X contains some finite subcollection of $\{U_\lambda\}$

which still covers X. If a subset of X is compact considered as a topological subspace of X, then it is called a *compact subset* of X.

A subset of a topological space X is said to be *relatively compact* if its closure is a compact subset of X. A topological space X is said to be *locally compact* if every point of X has a relatively compact neighborhood.

A subset of a topological space X is called a *σ-compact subset* if it is a countable union of compact sets.

Compactness is such a useful property that, given a non-compact space $(X, \mathcal{O})$, it is worthwhile constructing a compact space $(X' \mathcal{O}')$ with X being a dense subset. Such a space is called a *compactification* of $(X, \mathcal{O})$. The simplest way in which this can be achieved is by adjoining one extra point ∞ to the space X; a topology $\mathcal{O}'$ can be defined on $X' = X \cup \{\infty\}$ in such a way that $(X', \mathcal{O}')$ is compact and that $\mathcal{O}$ is the relative topology induced on X by $\mathcal{O}'$. The topological space $(X', \mathcal{O}')$ is called the *one-point compactification* of $(X, \mathcal{O})$,

1.5. Connectedness

A topological space X is said to be *connected* if there do not exist two non-empty open subsets O_1, O_2 of X such that $O_1 \cap O_2 = \varnothing$ and $X = O_1 \cup O_2$. A subset Y of X is called a *connected subset* if it is connected considered as a topological subspace of X. For each point x of X, there exists a maximal connected subset C_x of x which contains x. The subset C_x is called the connected component of X which contains x. If y is a point of C_x, then $C_x = C_y$. A connected subset C of X is called a *connected component* of X if $C = C_x$ for each $x \in C$. If C and C' are connected components of X, then $C = C'$ or $C \neq C'$ according as $C \cap C' \neq \varnothing$ or $C \cap C' = \varnothing$.

1.6. Metric Spaces

A set X is called a *metric space* if there is defined a real-valued function ρ on the Cartesian product $X \times X$ such that:

(D1) $0 \leq \rho(x, y) < +\infty$.
(D2) $\rho(x, y) = 0$ if and only if $x = y$.
(D3) $\rho(x, y) = \rho(y, x)$.
(D4) $\rho(x, y) \leq \rho(x, z) + \rho(y, z)$ (triangle inequality).

The function ρ is called a *metric* or *distance function* on X.

If $x \in X$ and $\varepsilon > 0$, then $B(x; \varepsilon)$ will denote the open ball of radius ε about x, that is, $B(x; \varepsilon) = \{y \in X; \rho(x, y) < \varepsilon\}$. The countable family $\{B(x; 1/n); n \in \mathbf{N}\}$ of open balls forms a fundamental neighborhood system of x; hence a metric space satisfies the first axiom of countability.

A topological space X is said to be *metrizable* if one can introduce a metric ρ on X in such a way that the induced topology on X by ρ is just the original topology on X.

Two metrics ρ_1 and ρ_2 on the same set X are said to be *equivalent* if, for each $\varepsilon > 0$, there exists $\delta > 0$ such that

$$\begin{cases} \rho_1(x, y) < \delta \Rightarrow \rho_2(x, y) < \varepsilon, \\ \rho_2(x, y) < \delta \Rightarrow \rho_1(x, y) < \varepsilon. \end{cases}$$

Equivalent metrics induce the same topology.

If x is a point of X and A is a subset of X, then we define the *distance* $\operatorname{dist}(x, A)$ from x to A by the formula

$$\operatorname{dist}(x, A) = \inf_{a \in A} \rho(x, a).$$

1.7. Baire's Category Theorem

Let X be a topological space. A subset of X is said to be *nowhere dense* in X if its closure does not contain a non-empty open subset of X. Any countable union of nowhere dense sets is called a set of the *first category*; all other subsets of X are of the *second category*.

Let (X, ρ) be a metric space. A sequence $\{x_n\}$ in X is called a *Cauchy sequence* if it satisfies Cauchy's convergence condition

$$\lim_{n, m \to \infty} \rho(x_n, x_m) = 0.$$

A metric space X is said to be *complete* if every Cauchy sequence in X converges to a point in X.

The next theorem about complete metric spaces is one of the fundamental theorems in analysis.

1.7.1 Theorem (Baire–Hausdorff). *A non-empty complete metric space is of the second category.*

1.8. Continuous Mappings

Let X and Y be topological spaces. A mapping $f: X \to Y$ is said to be *continuous at a point* x_0 of X if, to every neighborhood V of $f(x_0)$, there corresponds a neighborhood U of x_0 such that $f(U) \subset V$. For metric spaces, this definition of continuity is equivalent to the usual epsilon-delta definition.

If $f: X \to Y$ is continuous at every point of X, we say that f is *continuous*. A necessary and sufficient condition for f to be continuous is that the inverse image $f^{-1}(V)$ of every open set V in Y is an open set in X.

If $f: X \to Y$ is a bijection and both f and f^{-1} are continuous, then f is called a *homeomorphic map* or *homeomorphism* of X onto Y. Two topological spaces are said to be *homeomorphic* if there is a homeomorphism between them.

Let X and Y be locally compact, Hausdorff topological spaces. A continuous mapping $f: X \to Y$ is said to be *proper* if the inverse image $f^{-1}(K)$ of every compact set K in Y is a compact set in X.

1.9. Linear Spaces

Let the symbol **K** denote the real number field **R** or the complex number field **C**. A set X is called a *linear space* or *vector space* over **K** if two operations, called *addition* and *scalar multiplication*, are defined in X with the following properties:

(i) To every pair of elements x, y of X, there is associated an element $x + y$ of X in such a way that:

(a) $x + y = y + x$,
(b) $(x + y) + z = x + (y + z)$,
(c) There exists a unique element 0 of X, called the *zero vector*, such that

$$x + 0 = x, \qquad x \in X.$$

(d) For each element x of X, there exists a unique element $-x$ of X, called the *inverse element* of x, such that

$$x + (-x) = 0.$$

(ii) To any element x of X and each $\alpha \in \mathbf{K}$, there is associated an element αx of X in such a way that:

(a) $(\alpha\beta)x = \alpha(\beta x)$,
(b) $1x = x$,

(c) $\alpha(x + y) = \alpha x + \alpha y$,
(d) $(\alpha + \beta)x = \alpha x + \beta x$.

The elements of X are called *vectors* and the elements of $\mathbf{K}$ are called *scalars*. We also say that $\mathbf{K}$ is the coefficient field of the linear space X. A linear space X is said to be *real* or *complex* according as $\mathbf{K} = \mathbf{R}$ or $\mathbf{K} = \mathbf{C}$.

Let X be a linear space over $\mathbf{K}$. If $x_1, \ldots, x_n$ are vectors of X, a vector of the form $\alpha_1 x_1 + \cdots + \alpha_n x_n$ with $\alpha_1, \ldots, \alpha_n \in \mathbf{K}$ is called a *linear combination* of $x_1, \ldots, x_n$. The vectors $x_1, \ldots, x_n$ are said to be *linearly independent* if $\alpha_1 x_1 + \cdots + \alpha_n x_n = 0$ implies that $\alpha_1 = \cdots = \alpha_n = 0$. We also say that the set $\{x_1, \ldots, x_n\}$ is linearly independent. The vectors $x_1, \ldots, x_n$ are said to be *linearly dependent* if $\alpha_1 x_1 + \cdots + \alpha_n x_n = 0$, with some $\alpha_i \neq 0$.

If a linear space X contains n linearly independent vectors, but any $n + 1$ or more vectors are linearly dependent, then X is said to be *n-dimensional* or to have *dimension n*; we then write $\dim X = n$. If the number of linearly independent vectors in X is not finite, then X is said to be *infinite dimensional.*

A set $\{x_1, \ldots, x_n\}$ of n linearly independent vectors in an n-dimensional linear space X is called a *basis* of X. Then an arbitrary vector x of X can be written uniquely as

$$x = \alpha_1 x_1 + \cdots + \alpha_n x_n.$$

The scalars $\alpha_1, \ldots, \alpha_n$ are called the *components* of x with respect to the basis $\{x_1, \ldots, x_n\}$.

A subset M of a linear space X is called a *linear subspace*, or simply a *subspace*, of X if it is a linear space with respect to the addition and scalar multiplication defined in X. A subset M of X is a subspace if and only if $x + y \in M$ and $\alpha x \in M$ whenever $x, y \in M$ and $\alpha \in \mathbf{K}$. For a subset A of X, there exists a smallest subspace $[A]$ of X which contains A. In fact, the space $[A]$ is the intersection of all linear subspaces of X which contain A or it is the totality of finite linear combinations of elements of A. The space $[A]$ is called the *subspace spanned by A*.

Let M, N be two subspaces of a linear space X. The linear subspace spanned by the union $M \cup N$ is called the *sum* of M and N, and is denoted by $M + N$. If $M \cap N = \{0\}$, then the sum $M + N$ is called the *direct sum* of M and N, and is denoted by $M \dotplus N$. An arbitrary element x of the direct sum $M \dotplus N$ can be expressed uniquely in the form

$$x = y + z, \qquad y \in M, \qquad z \in N.$$

A set A in a linear space X is said to be *convex* if all points of the form

$$\alpha x + (1 - \alpha)y, \qquad 0 < \alpha < 1,$$

are in A whenever $x, y \in A$. For example, all linear subspaces are convex.

1.10. Linear Topological Spaces

A *linear topological space* or a *topological vector space* is a linear space, and at the same time a Hausdorff topological space such that the linear space operations of addition and scalar multiplication are continuous. That is, if X is a linear topological space over the real or complex number field $\mathbf{K}$, then the two mappings

$$X \times X \ni \{x, y\} \mapsto x + y \in X$$

and

$$\mathbf{K} \times X \ni \{\alpha, x\} \mapsto \alpha x \in X$$

are both continuous. We remark that the topology on X is translation invariant; this means that a subset A of X is open if and only if each of its translates $x + A = \{x + a; a \in A\}$ is open. Hence the topology on X is completely determined by a fundamental neighborhood system of the origin (the zero vector).

A linear topological space X is called a *locally convex linear topological space* if there exists a fundamental neighborhood system of the origin consisting of convex sets.

1.11. Factor Spaces

Let X be a linear space and M a linear subspace of X. We say that two elements x_1 and x_2 of X are *equivalent* modulo M if $x_1 - x_2 \in M$; we then write $x_1 \sim x_2 \pmod{M}$. The relation $\sim$ enjoys the so-called equivalence laws:

(*E*1) $x \sim x$ (reflexivity).
(*E*2) If $x_1 \sim x_2$, then $x_2 \sim x_1$ (symmetry).
(*E*3) If $x_1 \sim x_2$ and $x_2 \sim x_3$, then $x_1 \sim x_3$ (transitivity).

For each $x \in X$, we let

$$\tilde{x} = \{x' \in X; x' \sim x\}.$$

Then we have

$$\tilde{x} = \{x + m; m \in M\},$$

and hence

$$\begin{cases} x_1 \sim x_2 \Leftrightarrow \tilde{x}_1 = \tilde{x}_2, \\ x_1 \nsim x_2 \Leftrightarrow \tilde{x}_1 \cap \tilde{x}_2 = \varnothing. \end{cases} \tag{1}$$

The set $\tilde{x}$ is called an *equivalence class* modulo M, and each element of $\tilde{x}$ is called a *representative* of the class $\tilde{x}$. Assertion (1) implies that the space X can be decomposed into equivalence classes modulo M.

We denote by X/M the totality of equivalence classes modulo M. In the set X/M we can define addition and scalar multiplication as follows:

$$\begin{cases} \tilde{x} + \tilde{y} = \widetilde{x + y}, \\ \alpha\tilde{x} = \widetilde{\alpha x}, \qquad \alpha \in \mathbf{K}. \end{cases}$$

In fact, it is easy to verify that the above definitions do not depend on the choice of representatives x, y of the equivalence classes $\tilde{x}$, $\tilde{y}$, respectively. Therefore, the set X/M is a linear space and is called the *factor space* of X modulo M.

If the factor space X/M has finite dimension, then we say that the subspace M has *finite codimension*, and $\dim X/M$ is called the *codimension* of M and is denoted by codim M. It is easy to see that the subspace M has finite codimension n if and only if there exists an n-dimensional linear subspace N of X such that $M \dotplus N = X$.

1.12. Algebras and Modules

A linear space $\mathfrak{A}$ over a field K is called an (*associative*) *algebra* if, to every pair of elements a, b of $\mathfrak{A}$, there is associated an element $a \circ b$ of $\mathfrak{A}$ in such a way that:

$$\begin{aligned} &\alpha(a \circ b) = (\alpha a) \circ b = a \circ (\alpha b), \\ &(a + b) \circ c = a \circ c + b \circ c, \\ &a \circ (b + c) = a \circ b + a \circ c, \qquad\qquad (\alpha \in K;\ a, b, c \in \mathfrak{A}). \\ &a \circ (b \circ c) = (a \circ b) \circ c \text{ (associative law)}, \end{aligned}$$

If $a \circ b = b \circ a$ for every pair a, $b \in \mathfrak{A}$, then $\mathfrak{A}$ is said to be *commutative*. A subset J of a commutative algebra $\mathfrak{A}$ is called an *ideal* of $\mathfrak{A}$ if it is a linear subspace of $\mathfrak{A}$ and satisfies

$$a \in \mathfrak{A}, \qquad b \in J \Rightarrow a \circ b \in J.$$

For example, $\mathfrak{A}$ itself and $\{0\}$ are ideals of $\mathfrak{A}$.

Let $\mathfrak{A}$ be an algebra. A linear space $\mathscr{M}$ over K is called an $\mathfrak{A}$-*module* if, to every pair of an element a of $\mathfrak{A}$ and an element x of $\mathscr{M}$, there is associated an element ax of $\mathscr{M}$ in such a way that:

$$\begin{aligned} a(\alpha x) &= (\alpha a)x = \alpha(ax), \\ a(x+y) &= ax + ay, \\ (a+b)x &= ax + bx, \qquad (\alpha \in K;\, a, b \in \mathfrak{A},\, x, y \in \mathscr{M}). \\ a(bx) &= (a \circ b)x, \end{aligned}$$

1.13. Linear Operators

Let X, Y be linear spaces over the same scalar field $\mathbf{K}$. A mapping T, defined on a linear subspace D of X and taking values in Y, is said to be *linear* if it preserves the operations of addition and scalar multiplication:

$$\begin{cases} T(x_1 + x_2) = Tx_1 + Tx_2, & x_1, x_2 \in D; \\ T(\alpha x) = \alpha T x, & x \in D, \quad \alpha \in \mathbf{K}. \end{cases} \tag{1}$$

We often write Tx, rather than $T(x)$, if T is linear. We let

$$D(T) = D,$$

$$R(T) = \{Tx;\, x \in D(T)\},$$

$$N(T) = \{x \in D(T);\, Tx = 0\},$$

and call them the *domain*, the *range* and the *null space* of T, respectively. The mapping T is called a *linear operator* from $D(T) \subset X$ into Y. We also say that T is a linear operator from X into Y with domain $D(T)$. In the particular case when $Y = \mathbf{K}$, the mapping T is called a *linear functional* on $D(T)$. In other words, a linear functional is a $\mathbf{K}$-valued function on $D(T)$ which satisfies condition (1).

If a linear operator T is a one-to-one map of $D(T)$ onto $R(T)$, then the inverse mapping T^{-1} is a linear operator on $R(T)$ onto $D(T)$. The mapping T^{-1} is called the *inverse operator* or simply the *inverse* of T. A linear operator T admits the inverse T^{-1} if and only if $Tx = 0$ implies that $x = 0$.

Let T_1 and T_2 be linear operators from a linear space X into a linear space Y with domains $D(T_1)$ and $D(T_2)$, respectively. Then $T_1 = T_2$ if and only if $D(T_1) = D(T_2)$ and $T_1 x = T_2 x$ for all $x \in D(T_1) = D(T_2)$. If $D(T_1) \subset D(T_2)$ and $T_1 x = T_2 x$ for all $x \in D(T_1)$, then we say that T_2 is an *extension* of T_1, and also that T_1 is a *restriction* of T_2, and we write $T_1 \subset T_2$.

1.14. Differentiable Mappings

Let $\mathbf{R}^n$ be the n-dimensional Euclidean space. If $p = (p^1, \ldots, p^n)$ is a point of $\mathbf{R}^n$, we let

$$x^i(p) = p^i, \qquad 1 \leq i \leq n.$$

Then each x^i is a function on $\mathbf{R}^n$ into $\mathbf{R}$. The n-tuple $(x^1, \ldots, x^n)$ of functions is called the *standard coordinate system* of $\mathbf{R}^n$.

The notation $(x^1, \ldots, x^n)$ denotes the standard coordinate system of $\mathbf{R}^n$, but it is convenient to use the same notation $x = (x^1, \ldots, x^n)$ to denote a point of $\mathbf{R}^n$. In this book, we shall use $(x^1, \ldots, x^n)$ in either sense, but there will be no confusion.

Let U be an open subset of $\mathbf{R}^n$ and k a positive integer. A real-valued function f defined on U is said to be *differentiable of class* C^k, or simply *of class* C^k, on U if all its partial derivatives up to order k exist and are continuous on U. We also say that f is a C^k *function* on U. If f is of class C^k on U for every positive integer k, then it is said to be *of class* C^∞ on U. We also say that f is a C^∞ *function* on U.

If f is a mapping of an open subset U of $\mathbf{R}^n$ into $\mathbf{R}^m$, we write it in component form:

$$f(x) = (f^1(x), \ldots, f^m(x)), \qquad x = (x^1, \ldots, x^n) \in U.$$

Then the mapping f is said to be *differentiable of class* C^r, or simply *of class* C^r, on U if each component function f^i is differentiable of class C^r on U, where r is a positive integer or $r = \infty$. We also say that f is a C^r *mapping* on U.

Now let f be a C^1 mapping on U. If $v \in \mathbf{R}^n$, then at each point x of U the limit

$$\lim_{t \to 0} \frac{f(x + tv) - f(x)}{t}$$

exists in $\mathbf{R}^m$, and is equal to the vector

$$\left(\frac{df^1}{dt}(x + tv)\bigg|_{t=0}, \ldots, \frac{df^m}{dt}(x + tv)\bigg|_{t=0} \right).$$

This is denoted by $f'(x)(v)$, and is called the *derivative of* f *in the direction* v *at* x. By virtue of the chain rule, it follows that the correspondence

$$f'(x)\colon v \mapsto f'(x)v$$

is a linear mapping of $\mathbf{R}^n$ into $\mathbf{R}^m$. In fact, if we write the elements of $\mathbf{R}^n$ and $\mathbf{R}^m$ as column vectors, then we have

$$f'(x)\begin{pmatrix} v^1 \\ \vdots \\ v^n \end{pmatrix} = \begin{pmatrix} u^1 \\ \vdots \\ u^m \end{pmatrix}, \tag{1}$$

with

$$u^i = \sum_{j=1}^{n} \frac{\partial f^i}{\partial x^j}(x)\, v^j. \tag{2}$$

The linear map $f'(x)$ is called the *derivative* of f at x. It follows from formulas (1) and (2) that the matrix of $f'(x)$ with respect to the standard bases in $\mathbf{R}^n$ and $\mathbf{R}^m$ is given by

$$Jf(x) = \begin{pmatrix} \dfrac{\partial f^1}{\partial x^1}(x) & \cdots & \dfrac{\partial f^1}{\partial x^n}(x) \\ \vdots & & \vdots \\ \dfrac{\partial f^m}{\partial x^1}(x) & \cdots & \dfrac{\partial f^m}{\partial x^n}(x) \end{pmatrix}.$$

The matrix $Jf(x)$ is called the *Jacobian matrix* of f at x. If $m = n$, then the determinant of $Jf(x)$ is called the *Jacobian determinant* of f at x.

We remark that the usual chain rule can be restated in terms of Jacobian matrices.

Let U, V be two open subsets of $\mathbf{R}^n$. If $f: U \to V$ is a homeomorphism such that both f and f^{-1} are of class C^r $(1 \le r \le \infty)$, then we say that f is a C^r *diffeomorphism* of U onto V.

1.15. Vector Fields and Integral Curves

Let U be an open subset of $\mathbf{R}^n$. A *vector field* on U is a mapping X of U into $\mathbf{R}^n$, which we interpret as assigning a vector to each point of U. Let x_0 be a point of U. An *integral curve of* X *at* x_0 is a C^1 map c from an open interval I of $\mathbf{R}$ containing 0 into U such that

$$\begin{cases} \dot{c}(t) = X(c(t)), \\ c(0) = x_0, \end{cases}$$

where $\dot{c} = dc/dt$.

Let D be a subset of $\mathbf{R}^n$. A mapping f of D into $\mathbf{R}^n$ is said to be *Lipschitz continuous* on D if there exists a constant $K > 0$ such that

$$|f(x) - f(y)| \leq K|x - y|, \qquad x, y \in D.$$

The constant K is called a *Lipschitz constant* for f. We say that f is *locally Lipschitz continuous* on D if it is Lipschitz continuous on compact subsets of D. By the mean value theorem, we see that a C^1 mapping is locally Lipschitz continuous.

The next theorem is one of the fundamental theorems in the theory of ordinary differential equations.

1.15.1 Theorem. *Let U be an open subset of $\mathbf{R}^n$ and $X: U \to \mathbf{R}^n$ a Lipschitz continuous vector field with a Lipschitz constant K. Let $x_0 \in U$, and suppose that the closed ball $\overline{B(x_0; 2a)}$ of radius $2a$ about x_0 is contained in U and that the vector field X is bounded by a constant $L > 0$ on the ball $\overline{B(x_0; 2a)}$. If $b = \min(1/K, 2a/L)$, then there exists a unique C^1 map $x: (-b, b) \to U$ such that*

$$\begin{cases} \dot{x}(t) = X(x(t)), \\ x(0) = x_0. \end{cases}$$

Furthermore, if we denote by α_x the solution of the problem

$$\begin{cases} \dot{x}(t) = X(x(t)), \\ x(0) = x, \end{cases}$$

then the mapping $x \mapsto \alpha_x$ of the open ball $B(x_0; a)$ of radius a about x_0 into U is Lipschitz continuous.

We restate this theorem in terms of integral curves:

1.15.2 Theorem. *Let U be an open subset of $\mathbf{R}^n$ and $X: U \to \mathbf{R}^n$ a Lipschitz continuous vector field. Then we have:*

(i) *For each $x_0 \in U$, there exists an integral curve of X at x_0.*

(ii) *If $c_1: I_1 \to U$ and $c_2: I_2 \to U$ are two integral curves of X at the same point of U, then $c_1 = c_2$ on $I_1 \cap I_2$.*

(iii) *There exist an open subset U_0 of U containing x_0, an open interval I_0 containing zero and a continuous mapping*

$$\alpha: U_0 \times I_0 \to U,$$

such that for each $x \in U_0$ the mapping $\alpha_x: I_0 \to U$, defined by $\alpha_x(t) = \alpha(x, t)$, is an integral curve of X at x. Furthermore, the mapping α is Lipschitz continuous in the variable x and is of class C^1 in the variable t.

For each $x \in U$, we let

$I(x) =$ the union of all open intervals containing zero on which integral curves of X at x are defined.

Parts (i) and (ii) of Theorem 1.15.2 allow us to define the integral curve uniquely on all of $I(x)$.

Further we let

$$\mathscr{D}_X = \text{the set of those } (x, t) \in U \times \mathbf{R} \text{ such that } t \in I(x),$$

and define a *global flow* of X to be the map

$$\alpha: \mathscr{D}_X \to U$$

such that, for each $x \in U$, the mapping $\alpha_x: I(x) \to U$, given by $\alpha_x(t) = \alpha(x, t)$, is an integral curve of X at x. The curve α_x is called the *maximal integral curve* of X at x.

The next theorem describes the set $\mathscr{D}_X$ and the mapping α.

1.15.3 Theorem. *Let U be an open subset of $\mathbf{R}^n$ and $X: U \to \mathbf{R}^n$ a C^r vector field with $1 \leq r \leq \infty$. Then we have:*

(i) *$\mathscr{D}_X \supset U \times \{0\}$ and $\mathscr{D}_X$ is open in $U \times \mathbf{R}$.*
(ii) *The mapping $\alpha: \mathscr{D}_X \to U$ is of class C^r.*
(iii) *For $(x, t) \in \mathscr{D}_X$, the pair $(\alpha(x, t), s)$ is in $\mathscr{D}_X$ if and only if the pair $(x, t + s)$ is in $\mathscr{D}_X$. In this case, we have $\alpha(x, t + s) = \alpha(\alpha(x, t), s)$.*

1.16. Measurable Spaces

Let X be a non-empty set. A collection $\mathscr{M}$ of subsets of X is said to be a *σ-algebra* in X if it has the following properties:

($S1$) The empty set $\varnothing$ belongs to $\mathscr{M}$.
($S2$) If $A \in \mathscr{M}$, then its complement $A^c = X \setminus A$ belongs to $\mathscr{M}$.
($S3$) If $\{A_n\}$ is an arbitrary countable collection of members of $\mathscr{M}$, then the union $\bigcup_{n=1}^{\infty} A_n$ belongs to $\mathscr{M}$.

The pair $(X, \mathscr{M})$ is called a *measurable space* and the members of $\mathscr{M}$ are called *measurable sets* in X.

For any collection $\mathscr{F}$ of subsets of X, there exists a smallest σ-algebra $\sigma(\mathscr{F})$ in X which contains $\mathscr{F}$. This $\sigma(\mathscr{F})$ is sometimes called the *σ-algebra generated by $\mathscr{F}$*.

A collection $\mathcal{F}$ of subsets of X is called:

(i) a *π-system* in X if it is closed under finite intersections;
(ii) a *d-system* in X if it has the following properties:

(a) The set X itself belongs to $\mathcal{F}$.
(b) If $A, B \in \mathcal{F}$ and $A \subset B$, then the difference $B \setminus A$ belongs to $\mathcal{F}$.
(c) If $\{A_n\}$ is an increasing sequence of members of $\mathcal{F}$, then the union $\bigcup_{n=1}^{\infty} A_n$ belongs to $\mathcal{F}$.

We remark that a collection $\mathcal{F}$ is a σ-algebra if and only if it is both a π-system and a d-system. For any collection $\mathcal{F}$ of subsets of X, there exists a smallest d-system $d(\mathcal{F})$ which contains $\mathcal{F}$.

The next theorem gives a useful criterion for the d-system $d(\mathcal{F})$ to be a σ-algebra.

1.16.1 Theorem (the monotone class theorem). *If a collection $\mathcal{F}$ of subsets of X is a π-system, then we have $d(\mathcal{F}) = \sigma(\mathcal{F})$.*

1.17. Measurable Functions

We let

$$\bar{\mathbf{R}} = \{-\infty\} \cup \mathbf{R} \cup \{+\infty\}$$

with the obvious ordering. The topology on $\bar{\mathbf{R}}$ is defined by declaring that the open sets in $\bar{\mathbf{R}}$ are those which are unions of segments of the types

$$(a, b), \qquad [-\infty, a), \qquad (a, +\infty].$$

The elements of $\bar{\mathbf{R}}$ are called *extended real numbers.*

Let $(X, \mathcal{M})$ be a measurable space. An extended real-valued function f, defined on a set $A \in \mathcal{M}$, is said to be *$\mathcal{M}$-measurable*, or simply *measurable*, if for every $a \in \mathbf{R}$ the set

$$\{x \in A;\ f(x) > a\}$$

is in $\mathcal{M}$.

If A is a subset of X, we let

$$\chi_A(x) = \begin{cases} 1 & \text{if } x \in A, \\ 0 & \text{if } x \notin A. \end{cases}$$

The function χ_A is called the *characteristic function* of A.

A real-valued function f on X is called a *simple function* if it takes on only a finite number of values. Thus, if $a_1, \ldots, a_m$ are the distinct values of f, then f can be written as

$$f = \sum_{j=1}^{m} a_j \chi_{A_j},$$

where $A_j = \{x \in X; f(x) = a_j\}$. We remark that the function f is measurable if and only if each A_j is measurable.

The next theorem characterizes measurable functions in terms of simple functions.

1.17.1 Theorem. *An extended real-valued function defined on a measurable set is measurable if and only if it is a pointwise limit of a sequence of measurable simple functions. Furthermore, every non-negative measurable function is a pointwise limit of an increasing sequence of non-negative measurable simple functions.*

The next theorem is a version of the monotone class theorem (Theorem 1.16.1), and will be useful for the study of measurability of functions.

1.17.2 Theorem. *Let $\mathscr{F}$ be a π-system in X, and let $\mathscr{H}$ be a linear space of real-valued functions on X. Suppose that:*

(i) $1 \in \mathscr{H}$ *and* $\chi_A \in \mathscr{H}$ *for all* $A \in \mathscr{F}$.
(ii) *If* $\{f_n\}$ *is an increasing sequence of non-negative functions in* $\mathscr{H}$ *such that* $f = \sup_n f_n$ *is bounded, then* $f \in \mathscr{H}$.

Then the linear space $\mathscr{H}$ contains all real-valued, bounded $\sigma(\mathscr{F})$-measurable functions on X.

1.18. Measures

Let $(X, \mathscr{M})$ be a measurable space. An extended real-valued function μ defined on $\mathscr{M}$ is called a *non-negative measure*, or simply a *measure*, if it has the following properties:

$(M1)$ $0 \le \mu(A) \le \infty, \qquad A \in \mathscr{M}.$
$(M2)$ $\mu(\varnothing) = 0.$

($M3$) The function μ is *countably additive*, that is,

$$\mu\left(\sum_{i=1}^{\infty} A_i\right) = \sum_{i=1}^{\infty} \mu(A_i)$$

for any disjoint countable collection $\{A_i\}$ of members of $\mathcal{M}$.

The triple $(X, \mathcal{M}, \mu)$ is called a *measure space*. In other words, a measure space is a measurable space which has a non-negative measure defined on the σ-algebra of its measurable sets. If $\mu(X) < +\infty$, then the measure μ is called a *finite measure* and the space $(X, \mathcal{M}, \mu)$ is called a finite measure space. If X is a countable union of sets of finite measure, then the measure μ is said to be *σ-finite* on X. We also say that the measure space $(X, \mathcal{M}, \mu)$ is σ-finite.

Lebesgue measures

The next theorem is one of the fundamental theorems in measure theory.

1.18.1 Theorem. *There exist a σ-algebra $\mathcal{M}$ in $\mathbf{R}^n$ and a non-negative measure μ on $\mathcal{M}$ having the following properties:*

(i) *Every open set in R^n is in $\mathcal{M}$.*
(ii) *If $A \subset B$, $B \in \mathcal{M}$ and $\mu(B) = 0$, then $A \in \mathcal{M}$ and $\mu(A) = 0$.*
(iii) *If $A = \{x \in \mathbf{R}^n;\ a_j \le x_j \le b_j\ (1 \le j \le n)\}$, then $A \in \mathcal{M}$ and $\mu(A) = \prod_{j=1}^{n} (b_j - a_j)$.*
(iv) *The measure μ is* translation invariant, *that is, if $x \in \mathbf{R}^n$ and $A \in \mathcal{M}$, then the set $x + A = \{x + y;\ y \in A\}$ is in $\mathcal{M}$ and $\mu(x + A) = \mu(A)$.*

The elements of $\mathcal{M}$ are called *Lebesgue measurable sets* in $\mathbf{R}^n$ and the measure μ is called the *Lebesgue measure* on $\mathbf{R}^n$.

Signed Measures

Let $(X, \mathcal{M})$ be a measurable space. A real-valued function μ defined on $\mathcal{M}$ is called a *signed measure* or *real measure* if it is countably additive, that is, if

$$\mu\left(\sum_{i=1}^{\infty} A_i\right) = \sum_{i=1}^{\infty} \mu(A_i)$$

for any disjoint countable collection $\{A_i\}$ of members of $\mathcal{M}$.

We remark that every rearrangement of the series $\sum_i \mu(A_i)$ also converges, since the disjoint union $\sum_i A_i$ is not changed if the subscripts are permuted. A signed measure takes its values in $(-\infty, +\infty)$, but a non-negative measure may take $+\infty$; hence the non-negative measures do not form a subclass of the signed measures.

If μ is a signed measure, we define a function $|\mu|$ on $\mathscr{M}$ as follows:

$$|\mu|(A) = \sup \sum_i |\mu(A_i)|, \qquad A \in \mathscr{M}.$$

Here the supremum is taken over all countable partitions $\{A_i\}$ of A into members of $\mathscr{M}$. Then the function $|\mu|$ is a finite non-negative measure on $\mathscr{M}$. The measure $|\mu|$ is called the *total variation measure* of μ, and the quantity $|\mu|(X)$ is called the *total variation* of μ. Note that

$$|\mu(A)| \leq |\mu|(A) \leq |\mu|(X), \qquad A \in \mathscr{M}. \tag{1}$$

Borel Measures

Let X be a locally compact Hausdorff space. There exists a smallest σ-algebra $\mathscr{B}$ in X which contains all open sets in X. The members of $\mathscr{B}$ are called *Borel sets* in X. A signed measure defined on $\mathscr{B}$ is called a *real Borel measure* on X. A non-negative Borel measure μ is said to be *regular* if, for every $B \in \mathscr{B}$, we have

$$\begin{aligned} \mu(B) &= \sup\{\mu(F); F \subset B, F \text{ compact}\} \\ &= \inf\{\mu(G); B \subset G, G \text{ open}\}. \end{aligned}$$

We give a useful criterion for regularity of μ:

1.18.2 Theorem. *Let X be a locally compact Hausdorff space in which every open set is σ-compact. If μ is a non-negative Borel measure on X such that $\mu(K) < +\infty$ for every compact set $K \subset X$, then it is regular.*

Product Measures

Let $(X, \mathscr{M})$ and $(Y, \mathscr{N})$ be measurable spaces. We let

$\mathscr{M} \times \mathscr{N}$ = the smallest σ-algebra in $X \times Y$ which contains all sets of the form $A \times B$ where $A \in \mathscr{M}$ and $B \in \mathscr{N}$.

Then $(X \times Y, \mathscr{M} \times \mathscr{N})$ is a measurable space.

For the product of measure spaces, we have:

1.18.3 Theorem. *Let $(X, \mathcal{M}, \mu)$ and $(Y, \mathcal{N}, \nu)$ be σ-finite measure spaces. Then there exists a unique σ-finite, non-negative measure λ on $\mathcal{M} \times \mathcal{N}$ such that*

$$\lambda(A \times B) = \mu(A)\nu(B), \qquad A \in \mathcal{M}, \qquad B \in \mathcal{N}.$$

The measure λ is called the *product measure* of μ and ν, and is denoted by $\mu \times \nu$.

Direct Image of Measures

Let $(X, \mathcal{M})$ and $(Y, \mathcal{N})$ be measurable spaces. A mapping f of X into Y is said to be *measurable* if the inverse image $f^{-1}(B)$ of every $B \in \mathcal{N}$ is in $\mathcal{M}$.

Let $(X, \mathcal{M}, \mu)$ be a measure space and $(Y, \mathcal{N})$ a measurable space. If $f: X \to Y$ is a measurable mapping, then we can define a measure ν on $(Y, \mathcal{N})$ by the formula

$$\nu(B) = \mu(f^{-1}(B)), \qquad B \in \mathcal{N}.$$

We then write $\nu = f_* \mu$. The measure $f_* \mu$ is called the *direct image* of μ under f.

1.19. Integrals

Let $(X, \mathcal{M}, \mu)$ be a measure space. If A is a measurable subset of X, and if f is a non-negative measurable simple function on A of the form

$$f = \sum_{j=1}^{m} a_j \chi_{A_j}, \qquad a_j \geq 0,$$

then we let

$$\int_A f(x)\, d\mu(x) = \sum_{j=1}^{m} a_j \mu(A_j). \tag{1}$$

The convention: $0 \cdot \infty = 0$ is used here; it may happen that $a_j = 0$ and $\mu(A_j) = \infty$. If f is a non-negative measurable function on A, we let

$$\int_A f(x)\, d\mu(x) = \sup \int_A s(x)\, d\mu(x), \tag{2}$$

where the supremum is taken over all measurable simple functions s on A such that $0 \leq s(x) \leq f(x)$, $x \in A$. We remark that if f is a non-negative simple function, then the two definitions (1) and (2) of $\int_A f(x)\, d\mu(x)$ coincide.

If f is a measurable function on A, we can write it in the form

$$f = f^+ - f^-,$$

where

$$\begin{cases} f^+ = \max\{f, 0\}, \\ f^- = \max\{-f, 0\}. \end{cases}$$

Both f^+ and f^- are non-negative measurable functions on A. Then we define the *integral* of f by the formula

$$\int_A f(x)\, d\mu(x) = \int_A f^+(x)\, d\mu(x) - \int_A f^-(x)\, d\mu(x),$$

provided at least one of the integrals on the right-hand side is finite. If both integrals are finite, we say that f is *μ-integrable* or simply *integrable* on A.

For simplicity, we abbreviate

$$\int_A f\, d\mu = \int_A f(x)\, d\mu(x).$$

If μ is the Lebesgue measure on $\mathbf{R}^n$, we customarily write

$$\int_A f(x)\, d\mu$$

instead of $\int_A f(x)\, d\mu(x)$.

A proposition concerning the points of a measurable set A is said to hold *μ-almost everywhere* (μ-a.e.), or simply *almost everywhere* (a.e.) on A, if there exists a measurable set N of measure zero such that the proposition holds for all $x \in A \setminus N$. For example, if f and g are measurable functions on A, and if

$$\mu(\{x \in A;\ f(x) \neq g(x)\}) = 0,$$

then we say that $f = g$ a.e. on A.

The next three theorems are concerned with the interchange of integration and limit process.

1.19.1 Theorem (the monotone convergence theorem). *If $\{f_n\}$ is an increasing sequence of non-negative measurable functions on a measurable set A, then we have*

$$\lim_{n\to\infty} \int_A f_n \, d\mu = \int_A \left(\lim_{n\to\infty} f_n \right) d\mu.$$

1.19.2 Theorem (Fatou's lemma). *If $\{f_n\}$ is a sequence of non-negative measurable functions on a measurable set A, then we have*

$$\int_A \left(\liminf_{n\to\infty} f_n \right) d\mu \le \liminf_{n\to\infty} \int_A f_n \, d\mu.$$

1.19.3 Theorem (the dominated convergence theorem). *Let $\{f_n\}$ be a sequence of measurable functions on a measurable set A which converges pointwise to a function f on A. If there exists a non-negative integrable function g on A such that $|f_n(x)| \le g(x)$, $x \in A$, $n = 1, 2, \ldots$, then the function f is integrable on A and we have*

$$\int_A f \, d\mu = \lim_{n\to\infty} \int_A f_n \, d\mu.$$

Let $(X, \mathcal{M}, \mu)$ be a measure space, $(Y, \mathcal{N})$ a measurable space and $f: X \to Y$ a measurable mapping. If h is a measurable function on Y, then the composite function $h \circ f$ is measurable on X and we have

$$\int_X (h \circ f) \, d\mu = \int_Y h \, d(f_* \mu),$$

in the sense that the existence of either side implies that of the other and they are equal. Here $f_* \mu$ is the direct image of μ under f.

Now we consider integration on product spaces. Let $(X, \mathcal{M}, \mu)$ and $(Y, \mathcal{N}, \nu)$ be σ-finite measure spaces, and $\mu \times \nu$ the product measure of μ and ν. If f is an $\mathcal{M} \times \mathcal{N}$-measurable function on $X \times Y$ such that its integral exists, then we customarily write for its integral

$$\iint_{X \times Y} f(x, y) \, d(\mu \times \nu)(x, y).$$

This integral is called the *double integral* of f. If it happens that the function

$$g(x) = \int_Y f(x, y)\, d\nu(y), \qquad x \in X,$$

is defined and also its integral exists, then we denote the integral $\int_X g\, d\mu$ by any one of the following:

$$\int_X \left(\int_Y f(x, y)\, d\nu(y) \right) d\mu(x), \qquad \int_X d\mu(x) \int_Y f(x, y)\, d\nu(y),$$

$$\iint_{X \times Y} f(x, y)\, d\nu(y)\, d\mu(x), \qquad \iint_{X \times Y} f\, d\nu\, d\mu.$$

Similarly we write:

$$\int_Y \left(\int_X f(x, y)\, d\mu(x) \right) d\nu(y), \qquad \int_Y d\nu(y) \int_X f(x, y)\, d\mu(x),$$

$$\iint_{X \times Y} f(x, y)\, d\mu(x)\, d\nu(y), \qquad \iint_{X \times Y} f\, d\mu\, d\nu.$$

These integrals are called the *iterated integrals* of f.

The next theorem describes the most important relation between double integrals and iterated integrals.

1.19.4 Theorem (Fubini)

(i) *If f is a $\mu \times \nu$-integrable function on $X \times Y$, then the function f_x on Y defined by $f_x(y) = f(x, y)$ is ν-integrable for μ-almost all $x \in X$, and the function f^y on X defined by $f^y(x) = f(x, y)$ is μ-integrable for ν-almost all $y \in Y$. Furthermore, the function defined by*

$$g(x) = \int_Y f_x(y)\, d\nu(y) = \int_Y f(x, y)\, d\nu(y)$$

for μ-almost all $x \in X$ is μ-integrable, and the function defined by

$$h(y) = \int_X f^y(x)\, d\mu(x) = \int_X f(x, y)\, d\mu(x)$$

for ν-almost all $y \in Y$ is ν-integrable; and we have

$$\iint_{X \times Y} f\, d(\mu \times \nu) = \int_X g\, d\mu = \int_Y h\, d\nu.$$

(ii) *Conversely, if f is an $\mathcal{M} \times \mathcal{N}$-measurable function on $X \times Y$, then the functions*

$$\varphi(x) = \int_Y |f(x, y)| \, d\nu(y), \qquad x \in X,$$

$$\psi(y) = \int_X |f(x, y)| \, d\mu(x), \qquad y \in Y,$$

are $\mathcal{M}$-measurable and $\mathcal{N}$-measurable, respectively; and we have

$$\iint_{X \times Y} |f| \, d(\mu \times \nu) = \int_X \varphi \, d\mu = \int_Y \psi \, d\nu.$$

Furthermore, if either φ or ψ is integrable, then f is integrable, and part (i) *applies.*

1.20. Probability Spaces

Let $(\Omega, \mathcal{F})$ be a measurable space. A non-negative measure P on $\mathcal{F}$ is called a *probability measure* if $P(\Omega) = 1$. The triple $(\Omega, \mathcal{F}, P)$ is called a *probability space.* The elements of Ω are known as sample points, those of $\mathcal{F}$ as events and the values $P(A)$ $(A \in \mathcal{F})$ are their probabilities.

An extended real-valued, $\mathcal{F}$-measurable function X on Ω is called a *random variable.* The integral

$$\int_\Omega X \, dP$$

(if it exists) is called the *expectation* of X, and is denoted by $E(X)$.

Let $(\Omega, \mathcal{F}, P)$ be a probability space, $\mathcal{G}$ a σ-algebra contained in $\mathcal{F}$ and X an integrable random variable. The *conditional expectation* of X for given $\mathcal{G}$ is any random variable Y such that:

(i) The function Y is $\mathcal{G}$-measurable;
(ii) $\int_A Y \, dP = \int_A X \, dP, \qquad A \in \mathcal{G}$.

One can verify that conditions (i) and (ii) determine Y up to a set in $\mathcal{G}$ of measure zero. We write

$$Y = E(X \mid \mathcal{G}).$$

When X is the characteristic function χ_B of a set $B \in \mathscr{F}$, we write

$$P(B|\mathscr{G})$$

instead of $E(\chi_B|\mathscr{G})$. The function $P(B|\mathscr{G})$ is called the *conditional probability* of B for given $\mathscr{G}$. This function can also be characterized as a $\mathscr{G}$-measurable function which satisfies

$$P(A \cap B) = \int_A P(B|\mathscr{G})\, dP, \qquad A \in \mathscr{G}.$$

Notes

For topological spaces, see Bourbaki [1] and Kelley [1]; or Jameson [1]. For the theory of measure and integration, see Bourbaki [3], Halmos [1], Lang [2] and Rudin [1]. The monotone class theorem, Theorem 1.16.1, was first proved by Dynkin [1]. The presentation here is due to Blumenthal–Getoor [1]. For probability spaces, see Lamperti [1].

2 Manifolds, Tensors and Densities

The purpose of this chapter is to summarize the basic facts about manifolds and mappings between them which are most frequently used in the theory of partial differential equations. Manifolds are an abstraction of the idea of a surface in Euclidean space. The virtue of manifold theory is that it provides the geometric insight into the study of partial differential equations, and intrinsic properties of partial differential equations may be revealed.

2.1. Manifolds

Let X be a set and $0 \leq r \leq \infty$. An *atlas* or *coordinate neighborhood system* of class C^r on X is a family of pairs $\mathscr{A} = \{(U_i, \varphi_i)\}_{i \in I}$ satisfying the following conditions:

(MA1) Each U_i is a subset of X and $X = \bigcup_{i \in I} U_i$.

(MA2) Each φ_i is a bijection of U_i onto an open subset of $\mathbf{R}^n$, and for every pair i, j of I with $U_i \cap U_j \neq \emptyset$, the set $\varphi_i(U_i \cap U_j)$ is open in $\mathbf{R}^n$.

(MA3) For each pair i, j of I with $U_i \cap U_j \neq \emptyset$ the mapping

$$\varphi_j \circ \varphi_i^{-1} : \varphi_i(U_i \cap U_j) \to \varphi_j(U_i \cap U_j)$$

is a C^r diffeomorphism. (Here a C^0 diffeomorphism means a homeomorphism.)

In other words, X is a set which can be covered by subsets U_i, each of which is parametrized by an open subset of $\mathbf{R}^n$. Each pair (U_i, φ_i) is called a *chart* or *coordinate neighborhood* of $\mathscr{A}$. The mappings $\varphi_j \circ \varphi_i^{-1}$ in condition (MA3) are called *transition maps* or *coordinate transformations.*

Let (U, φ) be a chart on X. If p is a point of U, then $\varphi(p)$ is a point of $\mathbf{R}^n$ and hence an n-tuple of real numbers. We let

$$\varphi(p) = (x^1(p), \ldots, x^n(p)), \qquad p \in U. \tag{1}$$

The n-tuple $(x^1(p), \ldots, x^n(p))$ of real numbers is called the *local coordinates* of p in the chart (U, φ), and the n-tuple $(x^1, \ldots, x^n)$ of real-valued functions on U is called the *local coordinate system* on (U, φ). Following standard notation, we shall write formula (1) as

$$\varphi(x) = (x^1, \ldots, x^n), \qquad x \in U. \tag{1'}$$

Two atlases $\mathscr{A}_1$ and $\mathscr{A}_2$ on X are said to be *compatible* if the union $\mathscr{A}_1 \cup \mathscr{A}_2$ is an atlas on X. It is easy to see that the relation of compatibility between atlases is an equivalence relation. An equivalence class of atlases on X is said to define a C^r *structure* $\mathscr{D}$ on X. The union $\mathscr{A}_{\mathscr{D}} = \cup\{\mathscr{A}; \mathscr{A} \in \mathscr{D}\}$ of the atlases in $\mathscr{D}$ is called the *maximal atlas* of $\mathscr{D}$, and a chart (U, φ) of $\mathscr{A}_{\mathscr{D}}$ is called an *admissible chart.*

An n-dimensional C^r *manifold* M is a pair consisting of a set X and a C^r structure $\mathscr{D}$ on X. We often identify M with the underlying set X for notational convenience.

Given an atlas $\mathscr{A}$ on X, we can obtain a maximal atlas just by including all charts whose transition maps with those in $\mathscr{A}$ are C^r diffeomorphisms. This maximal atlas is said to define the *C^r structure generated by $\mathscr{A}$.*

Topology on Manifolds

Now we will see how to define a topology on a manifold by means of atlases. Let M be an n-dimensional C^r manifold. A subset O of M is defined to be open if and only if, for each $x \in O$, there exists an admissible chart (U, φ) such that $x \in U$ and $U \subset O$. It is easy to verify that the open sets in M define a topology.

A C^r manifold is said to be *Hausdorff* if it is Hausdorff as a topological space.

From now on we assume that our manifolds are Hausdorff.

Let X be a topological space. A collection $\mathscr{C}$ of subsets of X is said to be *locally finite* if every point of X has a neighborhood which intersects only finitely many elements of $\mathscr{C}$. A covering $\{V_j\}$ of X is called a *refinement* of a covering $\{U_i\}$ of X if each V_j is contained in some U_i.

A topological space X is said to be *paracompact* if it is a Hausdorff space and every open covering of X has a locally finite refinement which is also an open covering of X.

The next theorem gives criteria for paracompactness.

2.1.1 Theorem. *If M is a C^0 manifold, then the following three conditions are equivalent:*

(i) *M satisfies the second axiom of countability.*
(ii) *M is a countable union of compact subsets.*
(iii) *M is paracompact and the number of connected components of M is at most countable.*

Submanifolds

Let M be a C^r manifold $(0 \le r \le \infty)$ and N a subset of M. We say that N is a *submanifold* of M if, at each point x of N, there exists an admissible chart (U, φ) on M such that:

(SM) $\varphi: U \to V_1 \times V_2$, where V_1 is open in $\mathbf{R}^m$ and V_2 is open in $\mathbf{R}^{n-m}$ $(1 \le m \le n)$, and we have

$$\varphi(U \cap N) = V_1 \times \{0\}.$$

The number $n - m$ is called the *codimension* of N in M.

An open subset of M is a submanifold if we take $m = n$, and is called an *open submanifold.* A submanifold of M is called a *closed submanifold* if it is a closed subset of M.

If N is a submanifold of M, then it is a C^r manifold in its own right with the C^r structure generated by the atlas

$$\{(U \cap N, \varphi|_{U \cap N}); (U, \varphi) \text{ is an admissible chart on } M \text{ having property (SM)}\}.$$

Furthermore, the topology on N defined by the above atlas is the relative topology.

2.2. C^∞ Mappings

Let M and N be two C^∞ manifolds. A mapping $f: M \to N$ is said to be *of class* C^∞ if, for each $x \in M$ and each admissible chart (V, ψ) on N with $f(x) \in V$, there exists a chart (U, φ) on M with $x \in U$ and $f(U) \subset V$ such that the mapping $\psi \circ f \circ \varphi^{-1}: \varphi(U) \to \psi(V)$ is of class C^∞. The mapping $\psi \circ f \circ \varphi^{-1}$ is called a *local representative* of f.

A mapping $f: M \to N$ is called a C^∞ *diffeomorphism* if it is a bijection and both f and f^{-1} are of class C^∞. Two C^∞ manifolds are said to be *diffeomorphic* if there exists a diffeomorphism between them.

Let M be an n-dimensional C^∞ manifold and $\{(U_\alpha, \varphi_\alpha)\}_{\alpha \in I}$ an atlas on M. Let U be an open set in M. A real-valued continuous function f defined on U is of class C^∞ if and only if, for each $\alpha \in I$, the local representative $f \circ \varphi_\alpha^{-1}$ of f is of class C^∞ on $\varphi_\alpha(U \cap U_\alpha)$.

Let $C^\infty(M)$ denote the space of real-valued C^∞ functions on M. The space $C^\infty(M)$ has an algebra structure. In fact, the product fg defined by $(fg)(x) = f(x)g(x)$, $x \in M$, enjoys the usual algebraic properties of a product.

Let $\varphi: M \to N$ be a C^∞ mapping of manifolds. If $g \in C^\infty(N)$, the *pull-back* φ^*g of g by φ is defined by

$$\varphi^*g = g \circ \varphi \in C^\infty(M).$$

If φ is a diffeomorphism, then $\varphi^*: C^\infty(N) \to C^\infty(M)$ is an isomorphism and $(\varphi^*)^{-1} = (\varphi^{-1})^*$. If $f \in C^\infty(M)$, the *push-forward* $\varphi_* f$ of f by φ is defined by

$$\varphi_* f = f \circ \varphi^{-1} \in C^\infty(N).$$

Note that

$$\varphi_* = (\varphi^{-1})^*, \qquad \varphi^* = (\varphi^{-1})_*.$$

Partitions of Unity

Let $\{U_i\}$ be an open covering of a C^∞ manifold M. A family $\{g_i\}$ of C^∞ functions on M is called a *partition of unity subordinate* to the covering $\{U_i\}$ if the following conditions are satisfied:

(PU1) $0 \leq g_i(x) \leq 1$ for all $x \in M$.

(PU2) $\operatorname{supp} g_i \subset U_i$ for each i.

(PU3) The collection $\{\operatorname{supp} g_i\}$ is locally finite and $\sum_i g_i(x) = 1$ for each $x \in M$.

Here supp g_i is the *support* of g_i, that is, the closure in M of the set $\{x \in M; g_i(x) \neq 0\}$.

We give a general theorem on the existence of partitions of unity.

2.2.1 Theorem. *Every paracompact C^∞ manifold has a partition of unity subordinate to any given open covering.*

2.3. Tangent Bundles

Let M be an n-dimensional C^∞ manifold. At each point x of M, we consider triples (U, φ, v) where (U, φ) is a chart at x and v is a vector in $\mathbf{R}^n$. We say that two such triples (U, φ, v) and (V, ψ, w) are equivalent if the derivative $(\psi \circ \varphi^{-1})'$ of $\psi \circ \varphi^{-1}$ at $\varphi(x)$ maps v on w, that is, if

$$(\psi \circ \varphi^{-1})'(\varphi(x))v = w.$$

It is easy to verify that this is an equivalence relation. An equivalence class of such triples is called a *tangent vector* of M at x.

The set of such tangent vectors is denoted by $T_x(M)$, and is called the *tangent space* of M at x. Each chart (U, φ) defines a bijection of $T_x(M)$ onto $\mathbf{R}^n$ in such a way that the equivalence class $\bar{v}$ of (U, φ, v) corresponds to the vector v. In the space $T_x(M)$ we can define addition and scalar multiplication as follows:

$$\begin{cases} \bar{v}_1 + \bar{v}_2 = \overline{v_1 + v_2}, & \\ c\bar{v} \qquad\;\; = \overline{cv}, & c \in \mathbf{R}. \end{cases}$$

Hence the tangent space $T_x(M)$ is a real linear space, and the mapping: $v \mapsto \bar{v}$ is an isomorphism of $\mathbf{R}^n$ onto $T_x(M)$.

We let

$$T(M) = \bigcup_{x \in M} T_x(M)$$

be the disjoint union of the tangent spaces $T_x(M)$, and define a mapping

$$\pi\colon T(M) \to M$$

by $\pi(\bar{v}) = x$ for $\bar{v} \in T_x(M)$.

Now we will make $T(M)$ into a $2n$-dimensional C^∞ manifold by giving natural charts for it.

Let (U, φ) be a chart on M. We define a mapping

$$\tau_\varphi\colon \pi^{-1}(U) \to \varphi(U) \times \mathbf{R}^n$$

by

$$\tau_\varphi(\bar{v}) = (\varphi(x), v),$$

if $\pi(\bar{v}) = x$ and $\bar{v}$ is a tangent vector at x represented by v in the chart (U, φ). Then the mapping τ_φ is a bijection. Further, if (U, φ) and (V, ψ) are two overlapping charts, that is, if $U \cap V \neq \varnothing$, then we have

$$\pi^{-1}(U) \cap \pi^{-1}(V) = \pi^{-1}(U \cap V),$$

and the transition map

$$\tau_\psi \circ \tau_\varphi^{-1}\colon \varphi(U \cap V) \times \mathbf{R}^n \to \psi(U \cap V) \times \mathbf{R}^n$$

is given by

$$(\varphi(x), v) \mapsto (\psi(x), (\psi \circ \varphi^{-1})'(\varphi(x))v), \qquad x \in U \cap V,\ v \in \mathbf{R}^n.$$

Since the derivative $(\psi \circ \varphi^{-1})'$ is of class C^∞ and is an isomorphism at $\varphi(x)$, we obtain that the family of pairs $\{(\pi^{-1}(U), \tau_\varphi)\}$, where (U, φ) ranges over all admissible charts, is an atlas on $T(M)$. This proves that $T(M)$ is a $2n$-dimensional C^∞ manifold.

We call $T(M)$ the *tangent bundle* of M and π the *tangent bundle projection* of M, respectively. Each chart $(\pi^{-1}(U), \tau_\varphi)$ is called a *trivializing chart* on $T(M)$ over U. Each such trivializing chart on $T(M)$ identifies the tangent bundle over U with the product $\varphi(U) \times \mathbf{R}^n$.

Let M, N be two C^∞ manifolds and $f\colon M \to N$ a C^∞ mapping. At each point x of M, we define a map

$$T_x f\colon T_x(M) \to T_{f(x)}(N)$$

as follows: if (U, φ) is a chart at x and (V, ψ) is a chart at $f(x)$ with $f(U) \subset V$, and if $\bar{v}$ is a tangent vector of M at x represented by $v \in \mathbf{R}^n$ in (U, φ), then we let

$$T_x f(\bar{v}) = \text{the tangent vector of } N \text{ at } f(x) \text{ represented by } f'_{\varphi\psi}(\varphi(x))v,$$

where $f_{\varphi\psi} = \psi \circ f \circ \varphi^{-1}$ is the local representative of f. It is easy to verify that the map $T_x f$ is independent of the charts used, and is linear. The map $T_x f$ is called the *tangent map* of f at x.

We define the tangent map

$$Tf\colon T(M) \to T(N)$$

to be the map equal to

$$T_x f : T_x(M) \to T_{f(x)}(N)$$

on each $T_x(M)$.

2.4. Vector Fields

Let M be an n-dimensional C^∞ manifold. A C^∞ *vector field* on M is a C^∞ mapping,

$$X : M \to T(M),$$

such that $X(x) \in T_x(M)$ for each $x \in M$. In other words, a vector field X assigns to each point x of M a tangent vector $X(x)$ of M at x. The set $\mathscr{X}(M)$ of all C^∞ vector fields on M is a real linear space with the obvious operations of addition and scalar multiplication.

If (U, φ) is a chart on M, then a C^∞ vector field X on M induces a C^∞ vector field $\tilde{X}$ on $\varphi(U)$ by defining

$$\tilde{X}(z) = \tau_\varphi \circ X(\varphi^{-1}(z)), \qquad z \in \varphi(U).$$

The vector field $\tilde{X}$ is called the *local representative* of X in the chart (U, φ). If we identify the tangent bundle over U with the product $U \times \mathbf{R}^n$, then X corresponds to a mapping

$$\begin{aligned} &U \to U \times \mathbf{R}^n, \\ &x \mapsto (x, X^1(x), \ldots, X^n(x)), \end{aligned} \tag{1}$$

where $X^1, \ldots, X^n$ are C^∞ functions on U. The n-component vector function $(X^1, \ldots, X^n)$ on U is called the *local components* of X relative to the chart (U, φ).

If $f \in C^\infty(M)$ and $X \in \mathscr{X}(M)$, then the mapping

$$M \ni x \mapsto f(x)X(x)$$

defines a C^∞ vector field on M. This is called the product of f and X. It is easy to verify that the space $\mathscr{X}(M)$ is a $C^\infty(M)$-module with respect to this operation of product.

Now we will define how vector fields operate on functions.

Let $f \in C^\infty(M)$. Since $Tf : T(M) \to T(\mathbf{R}) = \mathbf{R} \times \mathbf{R}$, we can write Tf acting on each $T_x(M)$ in the form

$$Tf(\bar{v}) = (f(x), df(x) \cdot \bar{v}), \qquad \bar{v} \in T_x(M).$$

Recall that $Tf = T_x f$ on $T_x(M)$ and that $T_x f\colon T_x(M) \to \mathbf{R}$ is linear. Hence $df(x)$ is an element of the dual space $T_x^*(M)$ of $T_x(M)$, and is called the *differential* of f at x. The dual space $T_x^*(M)$ is called the *space of differentials* or the *cotangent space* at x.

We work out the differential df in local charts. If (U, φ) is a chart on M, then the local representative of Tf is given by

$$(\tilde{f}(z), \tilde{f}'(z)v), \qquad z \in \varphi(U), \qquad v \in \mathbf{R}^n,$$

where $\tilde{f} = f \circ \varphi^{-1}$ is the local representative of f. Hence the local representative of df is the derivative of the local representative of f. That is, if $(x^1, \ldots, x^n)$ is a local coordinate system on (U, φ), then the local components of df are given by

$$(df)_i(x) = \frac{\partial f}{\partial x^i}(x) = \frac{\partial \tilde{f}}{\partial z^i}(\varphi(x)). \tag{2}$$

If $f \in C^\infty(M)$ and $X \in \mathscr{X}(M)$, we define the *derivative of f in the direction X* by

$$X[f](x) = df(x) \cdot X(x), \qquad x \in M.$$

The real-valued function $x \mapsto X[f](x)$ on M is denoted by $X[f]$ or $df(X)$. In view of formulas (1) and (2), it follows that

$$X[f] = \sum_{i=1}^{n} X^i \frac{\partial f}{\partial x^i}. \tag{3}$$

This proves that $X[f] \in C^\infty(M)$. The derivative $X[f]$ is also occasionally denoted by $\mathscr{L}_X f$, and is called the *Lie derivative of f along X*. It follows from formula (3) that the mapping $\mathscr{L}_X\colon C^\infty(M) \to C^\infty(M)$ satisfies

$$\mathscr{L}_X(fg) = \mathscr{L}_X f \cdot g + f \cdot \mathscr{L}_X g, \qquad f, g \in C^\infty(M). \tag{4}$$

A mapping $D\colon C^\infty(M) \to C^\infty(M)$ is called a *derivation* on $C^\infty(M)$ if it is linear and satisfies the following condition:

$$D(fg) = Df \cdot g + f \cdot Dg, \qquad f, g \in C^\infty(M).$$

The collection of all derivations on $C^\infty(M)$ is a real linear space with the obvious operations of addition and scalar multiplication.

Formula (4) tells us that for each $X \in \mathscr{X}(M)$ the Lie derivative $\mathscr{L}_X$ is a derivation. The next theorem shows the converse.

2.4.1 Theorem. *The collection of all derivations on $C^\infty(M)$ is a real linear space isomorphic to the space $\mathscr{X}(M)$. More precisely, for each derivation D on $C^\infty(M)$, there exists a unique C^∞ vector field X on M such that $\mathscr{L}_X = D$.*

This theorem allows us to use derivations to define vector fields (cf. Matsushima [1]). Further it provides a local basis for vector fields in the following way:

If (U, φ) is a chart on M with $\varphi(x) = (x^1, \dots, x^n)$, we define n derivations $\partial/\partial x^i$ on $C^\infty(U)$ by

$$\frac{\partial f}{\partial x^i}(x) = \frac{\partial \tilde{f}}{\partial z^i}(\varphi(x)), \qquad f \in C^\infty(U).$$

These derivations are linearly independent with coefficients in $C^\infty(U)$. In fact, since we have $(\partial/\partial x^i)(x^j) = \delta_i^j$, it follows that

$$\sum_{i=1}^n f^i \frac{\partial}{\partial x^i} = 0, \qquad f^i \in C^\infty(U) \Rightarrow f^j = \left(\sum_{i=1}^n f^i \frac{\partial}{\partial x^i}\right)(x^j) = 0.$$

Here the δ_i^j are the usual Kronecker symbols: $\delta_i^j = 1$ if $j = i$, and $\delta_i^j = 0$ otherwise. Theorem 2.4.1 tells us that the derivations $\partial/\partial x^1, \dots, \partial/\partial x^n$ may be identified with C^∞ vector fields on U. If $X \in \mathscr{X}(M)$ has the local components $(X^1, \dots, X^n)$ in (U, φ), then we have

$$\mathscr{L}_X = \sum_{i=1}^n X^i \frac{\partial}{\partial x^i},$$

and hence

$$X = \sum_{i=1}^n X^i \frac{\partial}{\partial x^i},$$

with the identification of vector fields with derivations. This proves that the vector fields $\partial/\partial x^1, \dots, \partial/\partial x^n$ form a local basis for the space $\mathscr{X}(M)$.

Moreover, since we have

$$dx^i\left(\frac{\partial}{\partial x^j}\right) = \frac{\partial x^i}{\partial x^j} = \delta_j^i,$$

we see that the differentials $dx^1, \dots, dx^n$ form a basis of $T_x^*(M)$ dual to the basis $\partial/\partial x^1, \dots, \partial/\partial x^n$ of $T_x(M)$ at each point x of U. Hence, if $f \in C^\infty(U)$, the differential df has the local expression

$$df = \sum_{i=1}^n \frac{\partial f}{\partial x^i} dx^i,$$

since we have

$$df\left(\frac{\partial}{\partial x^i}\right) = \frac{\partial f}{\partial x^i}.$$

If v is a C^∞ vector field on U, then it has the local expression

$$v = \sum_{i=1}^{n} \xi^i \frac{\partial}{\partial x^i},$$

where $\xi^1, \ldots, \xi^n$ are C^∞ functions on U. The functions $\xi^1, \ldots, \xi^n$ are called the *local components* of v relative to the chart (U, φ).

2.5. Integral Curves

Let M be a C^∞ manifold. A C^1 map c from an open interval I of $\mathbf{R}$ into M is called a *curve* of M. Let t be a point of I and (U, φ) a chart at $c(t)$. Shrinking the interval I to an open subinterval I_0 such that $c(I_0) \subset U$, we can take the derivative $(\varphi \circ c)'(t)$ as a vector in $\mathbf{R}^n$. This vector represents a tangent vector at $c(t)$, independently of the chart used. In this way we can define a mapping

$$\dot{c}: I \to T(M)$$

by

$$\dot{c}(t) = \text{the tangent vector of } M \text{ at } c(t) \text{ represented by } (\varphi \circ c)'(t).$$

Let X be a C^r vector field on M with $1 \leq r \leq \infty$. An *integral curve* of X is a C^1 map c from an open interval I into M such that

$$\dot{c}(t) = X(c(t)), \qquad t \in I. \tag{1}$$

The local representative of formula (1) is given by

$$\frac{d\tilde{c}}{dt}(t) = \tilde{X}(\tilde{c}(t)), \qquad t \in I, \tag{1'}$$

where $\tilde{c} = \varphi \circ c$ is the local representative of c and $\tilde{X}$ is the local representative of X in the chart (U, φ). If the interval I contains 0 and $c(0) = x_0$, we say that the map c is an integral curve of X at x_0.

We remark that Theorems 1.15.2 and 1.15.3 extend to this case.

2.6. Cotangent Bundles

Let M be an n-dimensional C^∞ manifold. We let

$$T^*(M) = \bigcup_{x \in M} T_x^*(M)$$

be the disjoint union of the cotangent spaces $T_x^*(M)$, and define a mapping

$$\pi^*: T^*(M) \to M$$

by $\pi^*(\omega) = x$ if $\omega \in T_x^*(M)$.

Now we will make $T^*(M)$ into a $2n$-dimensional C^∞ manifold by giving natural charts for it.

Let (U, φ) be a chart on M with $\varphi(x) = (x^1, \ldots, x^n)$. We define a mapping

$$\tau_\varphi^*: \pi^{*-1}(U) \to \varphi(U) \times \mathbf{R}^n$$

by

$$\tau_\varphi^*(\omega) = (\varphi(x), (\xi_1, \ldots, \xi_n)),$$

if $\pi^*(\omega) = x$ and $\omega = \sum_{i=1}^n \xi_i \, dx^i$. Then it follows that the mapping τ_φ^* is a bijection, since $(dx^1, \ldots, dx^n)$ is a basis of $T_x^*(M)$ at each point x of U. Further, it is easy to see that the family of pairs $\{(\pi^{*-1}(U), \tau_\varphi^*)\}$, where (U, φ) ranges over all admissible charts, is an atlas on $T^*(M)$. This shows that $T^*(M)$ is a $2n$-dimensional C^∞ manifold. We call $T^*(M)$ the *cotangent bundle* of M and π^* the *cotangent bundle projection* of M, respectively.

A C^∞ *covector field* or *differential one-form* on M is a C^∞ mapping,

$$\omega: M \to T^*(M),$$

such that $\omega(x) \in T_x^*(M)$ for each $x \in M$. In other words, a covector field ω assigns to each point x of M a cotangent vector $\omega(x)$ at x. The set $\mathscr{X}^*(M)$ of all C^∞ covector fields on M is a real linear space with the obvious operations of addition and scalar multiplication.

If $\omega \in \mathscr{X}^*(M)$ and (U, φ) is a chart with $\varphi(x) = (x^1, \ldots, x^n)$, then ω has the local expression

$$\omega = \sum_{i=1}^n \xi_i \, dx^i,$$

where $\xi_1, \ldots, \xi_n$ are C^∞ functions on U. The functions $\xi_1, \ldots, \xi_n$ are called the *local components* of ω relative to the chart (U, φ).

If $f \in C^\infty(M)$ and $\omega \in \mathscr{X}^*(M)$, then the mapping

$$M \ni x \mapsto f(x)\omega(x)$$

defines a C^∞ covector field on M. The space $\mathscr{X}^*(M)$ is a $C^\infty(M)$-module with respect to this operation of product.

2.7. Tensors

Let $\mathbf{K}$ be the real number field $\mathbf{R}$ or the complex number field $\mathbf{C}$, and let $E_1, \ldots, E_p$ be linear spaces over $\mathbf{K}$. A mapping

$$A: E_1 \times \cdots \times E_p \to \mathbf{K}$$

is said to be *p-multilinear* if $A(v_1, \ldots, v_p)$ is linear in each argument v_i separately, that is, if

$$\begin{aligned} A(v_1, \ldots, v_{i-1}, \lambda v_i + \mu w_i, v_{i+1}, \ldots, v_p) \\ = \lambda A(v_1, \ldots, v_i, \ldots, v_p) + \mu A(v_1, \ldots, w_i, \ldots, v_p). \end{aligned}$$

In the case $p = 2$, we say that A is *bilinear*. The set of all p-multilinear mappings of $E_1 \times \cdots \times E_p$ into $\mathbf{K}$ is a linear space over $\mathbf{K}$ with the obvious operations of addition and scalar multiplication. This linear space is denoted by $L(E_1, \ldots, E_p, \mathbf{K})$.

Let E be a finite dimensional linear space over $\mathbf{K}$. We write E^* for $L(E, \mathbf{K})$, the space of all linear functionals on E. The space E^* is called the *dual space* of E. We remark that E may be identified with its *bidual space* $E^{**} = L(E^*, \mathbf{K})$ by the isomorphism $e \mapsto e^{**}$ defined by $e^{**}(\alpha) = \alpha(e)$, $\alpha \in E^*$, $e \in E$.

We let

$T^r_s(E) = L(E^* \times \cdots \times E^* \times E \times \cdots \times E, \mathbf{K})$, r-copies of E^* and s-copies of E.

The elements of $T^r_s(E)$ are called *tensors on E, contravariant of order r and covariant of order s*, or simply *of type* $\binom{r}{s}$. In particular we have

$$T^1_0(E) = L(E^*, \mathbf{K}) = E,$$

$$T^0_1(E) = L(E, \mathbf{K}) = E^*.$$

If $t_1 \in T^{r_1}_{s_1}(E)$ and $t_2 \in T^{r_2}_{s_2}(E)$, we define the *tensor product* $t_1 \otimes t_2$ of t_1 and t_2 by the formula

$$\begin{aligned} (t_1 \otimes t_2)(\beta^1, \ldots, \beta^{r_1}, \gamma^1, \ldots, \gamma^{r_2}, v_1, \ldots, v_{s_1}, w_1, \ldots, w_{s_2}) \\ = t_1(\beta^1, \ldots, \beta^{r_1}, v_1, \ldots, v_{s_1}) t_2(\gamma^1, \ldots, \gamma^{r_2}, w_1, \ldots, w_{s_2}). \end{aligned}$$

Then we have $t_1 \otimes t_2 \in T^{r_1+r_2}_{s_1+s_2}(E)$. Also it is easy to see that the operation $\otimes$ is bilinear and associative.

Suppose that the linear space E has dimension n. Let $(e_1, \dots, e_n)$ be a basis of E and $(e^1, \dots, e^n)$ the corresponding dual basis of E^*, that is, $e^j(e_i) = \delta^j_i$. Then the n^{r+s} elements

$$\{e_{i_1} \otimes \cdots \otimes e_{i_r} \otimes e^{j_1} \otimes \cdots \otimes e^{j_s};\ 1 \le i_k \le n,\ 1 \le j_k \le n\}$$

form a basis of $T^r_s(E)$, so that the space $T^r_s(E)$ has dimension n^{r+s}. In fact, every element t of $T^r_s(E)$ can be written in the form

$$t = \sum_{\substack{i_1 \cdots i_r \\ j_1 \cdots j_s}} t(e^{i_1}, \dots, e^{i_r}, e_{j_1}, \dots, e_{j_s}) e_{i_1} \otimes \cdots \otimes e_{i_r} \otimes e^{j_1} \otimes \cdots \otimes e^{j_s},$$

The coefficients $t^{i_1 \cdots i_r}_{j_1 \cdots j_s} = t(e^{i_1}, \dots, e^{i_r}, e_{j_1}, \dots, e_{j_s})$ are called the *components* of t relative to the basis $(e_1, \dots, e_n)$.

2.8. Tensor Fields

Let M be an n-dimensional C^∞ manifold. We let

$$T^r_s(T(M)) = \bigcup_{x \in M} T^r_s(T_x(M))$$

be the disjoint union of the spaces $T^r_s(T_x(M))$ of tensors on $T_x(M)$, contravariant of order r and covariant of order s. This $T^r_s(T(M))$ carries a natural structure of a C^∞ manifold of dimension $n + n^{r+s}$, induced by the tangent bundle $T(M)$ and the cotangent bundle $T^*(M)$. (Cf. Abraham-Marsden-Ratiu [1], Theorem 5.2.6.) The manifold $T^r_s(T(M))$ is called the *vector bundle of tensors, contravariant of order r and covariant of order s*, or simply *of type* $\binom{r}{s}$.

Note that

$$T^1_0(T(M)) = T(M),$$
$$T^0_1(T(M)) = T^*(M).$$

A C^∞ *tensor field of type* $\binom{r}{s}$ on M is a C^∞ mapping,

$$t\colon M \to T^r_s(T(M)),$$

such that $t(x) \in T^r_s(T_x(M))$ for each $x \in M$. The set $\mathcal{T}^r_s(M)$ of all C^∞ tensor fields of type $\binom{r}{s}$ on M carries a real linear space structure, the addition and scalar multiplication of tensor fields being taken within each $T^r_s(T_x(M))$, $x \in M$.

Note that

$$\mathscr{T}^1_0(M) = \mathscr{X}(M),$$

$$\mathscr{T}^0_1(M) = \mathscr{X}^*(M).$$

We now give the expression of tensor fields in local charts. Recall that if (U, φ) is a chart on M with $\varphi(x) = (x^1, \ldots, x^n)$, then the vector fields $\partial/\partial x^1, \ldots, \partial/\partial x^n$ form a basis of the tangent space $T_x(M)$ and the differentials $dx^1, \ldots, dx^n$ form the corresponding dual basis of the cotangent space $T_x^*(M)$ at each point x of U. A tensor field $t \in \mathscr{T}^r_s(M)$ has the local expression

$$t = \sum_{\substack{i_1 \cdots i_r \\ j_1 \cdots j_s}} t^{i_1 \cdots i_r}_{j_1 \cdots j_s} \frac{\partial}{\partial x^{i_1}} \otimes \cdots \otimes \frac{\partial}{\partial x^{i_r}} \otimes dx^{j_1} \otimes \cdots \otimes dx^{j_s},$$

where $t^{i_1 \cdots i_r}_{j_1 \cdots j_s}$ are C^∞ functions on U. The functions $t^{i_1 \cdots i_r}_{j_1 \cdots j_s}$ are called the *local components* of t relative to the chart (U, φ).

A *Riemannian metric* on M is a C^∞ tensor field g of type $\binom{0}{2}$ on M such that $g(x) \in T^0_2(T_x(M))$ is an inner product on $T_x(M)$ for each $x \in M$. Hence, if (U, φ) is a chart on M with $\varphi(x) = (x^1, \ldots, x^n)$, then the local components

$$g_{ij}(x) = g(x)\left(\frac{\partial}{\partial x^i}, \frac{\partial}{\partial x^j}\right)$$

are C^∞ functions on U, and the matrix $(g_{ij}(x))$ is symmetric and positive definite at every point x of U.

A C^∞ manifold with a Riemannian metric is called a *Riemannian manifold.*

We give a general theorem on the existence of Riemannian metrics:

2.8.1 Theorem. *Every paracompact C^∞ manifold admits a Riemannian metric.*

2.9. Exterior Product

The permutation group S_k on k elements consists of all bijections $\sigma\colon \{1, \ldots, k\} \to \{1, \ldots, k\}$, usually given in the following form:

$$\sigma = \begin{pmatrix} 1 & \cdots & k \\ \sigma(1) & \cdots & \sigma(k) \end{pmatrix}.$$

A transposition is a permutation that swaps two elements of the set $\{1, \ldots, k\}$, leaving the remainder fixed. A permutation is said to be *even* (resp. *odd*) if it is written as the product of an even (resp. odd) number of transpositions. The expression of an even (resp. odd) permutation is not unique, but the number of transpositions is always even (resp. odd). We define the *signature*, sign σ, of a permutation σ by

$$\operatorname{sign} \sigma = \begin{cases} +1 & \text{if } \sigma \text{ is even,} \\ -1 & \text{if } \sigma \text{ is odd.} \end{cases}$$

Let $\mathbf{K}$ denote the real number field $\mathbf{R}$ or the complex number field $\mathbf{C}$. Throughout this section, let E be an n-dimensional linear space over $\mathbf{K}$.

Recall that

$$T_k^0(E) = \text{the space of } k\text{-multilinear mappings of } E \times \cdots \times E \text{ into } \mathbf{K}.$$

The group S_k acts on $T_k^0(E)$. In fact, each $\sigma \in S_k$ defines a mapping

$$\sigma\colon T_k^0(E) \to T_k^0(E)$$

by

$$(\sigma t)(e_1, \ldots, e_k) = t(e_{\sigma(1)}, \ldots, e_{\sigma(k)}), \qquad t \in T_k^0(E),$$

where $e_1, \ldots, e_k \in E$. A mapping $t \in T_k^0(E)$ is said to be *alternating* (resp. *symmetric*) if $\sigma t = (\operatorname{sign} \sigma)t$ (resp. $\sigma t = t$) for all $\sigma \in S_k$. It is easy to see that

$$\begin{aligned} &t \in T_k^0(E) \text{ is alternating if and only if} \\ &t(e_1, \ldots, e_k) = 0 \text{ when } e_i = e_j \text{ for some } i \neq j. \end{aligned} \tag{1}$$

The set of all alternating elements of $T_k^0(E)$ is a linear subspace of $T_k^0(E)$. This space is denoted by $\Lambda^k E^*$, and is called the k-th *exterior product* of E^*. The elements of $\Lambda^k E^*$ are called *exterior k-forms*. Note that by (1) we have

$$\Lambda^k E^* = \{0\} \qquad \text{if } k > n.$$

We define the *alternation mapping*

$$A\colon T_k^0(E) \to T_k^0(E)$$

by

$$At(e_1, \ldots, e_k) = \frac{1}{k!} \sum_{\sigma \in S_k} (\operatorname{sign} \sigma) t(e_{\sigma(1)}, \ldots, e_{\sigma(k)}).$$

Then we have:

2.9.1 Proposition. *The mapping A is a linear mapping onto $\Lambda^k E^*$, and is the identity map on $\Lambda^k E^*$.*

If $\alpha \in T_k^0(E)$ and $\beta \in T_l^0(E)$, we define the *exterior product* or *wedge product* $\alpha \wedge \beta$ of α and β by

$$\alpha \wedge \beta = \frac{(k+l)!}{k!l!} A(\alpha \otimes \beta).$$

Then we have $\alpha \wedge \beta \in \Lambda^{k+l}E^*$. Further, the following formula is a convenient way to compute exterior products:

$$(\alpha \wedge \beta)(e_1, \ldots, e_{k+l}) = \textstyle\sum'(\operatorname{sign} \sigma)\alpha(e_{\sigma(1)}, \ldots, e_{\sigma(k)})\beta(e_{\sigma(k+1)}, \ldots, e_{\sigma(k+l)}).$$

Here $\sum'$ denotes the sum over all (k, l) shuffles, that is, permutations σ of $\{1, 2, \ldots, k+l\}$ such that $\sigma(1) < \sigma(2) < \cdots < \sigma(k)$ and $\sigma(k+1) < \sigma(k+2) < \cdots < \sigma(k+l)$.

2.9.2 Example. If $\alpha^1, \ldots, \alpha^k \in E^*$, then we have

$$\begin{aligned}(\alpha^1 \wedge \cdots \wedge \alpha^k)(e_1, \ldots, e_k) &= \sum_{\sigma \in S_k} (\operatorname{sign} \sigma)\alpha^1(e_{\sigma(1)}) \cdots \alpha^k(e_{\sigma(k)}) \\ &= \det(\alpha^i(e_j)). \qquad (2)\end{aligned}$$

In particular, if $(e_1, \ldots, e_n)$ is a basis of E and $(e^1, \ldots, e^n)$ is the corresponding dual basis of E^*, then we have

$$(e^1 \wedge \cdots \wedge e^k)(e_1, \ldots, e_k) = 1.$$

The next proposition summarizes the basic properties of the operation $\wedge$.

2.9.3 Proposition. *Let $\alpha \in T_k^0(E)$, $\beta \in T_l^0(E)$ and $\gamma \in T_m^0(E)$. Then we have:*

(i) $\alpha \wedge \beta = A\alpha \wedge \beta = \alpha \wedge A\beta$.
(ii) *The operation $\wedge$ is bilinear.*
(iii) $\alpha \wedge \beta = (-1)^{kl}\beta \wedge \alpha$.
(iv) $\alpha \wedge (\beta \wedge \gamma) = (\alpha \wedge \beta) \wedge \gamma$.

The next proposition describes bases of $\Lambda^k E^*$.

2.9.4 Proposition. *For $2 \le k \le n$, the space $\Lambda^k E^*$ has dimension $\binom{n}{k}$. More precisely, if $(e_1, \ldots, e_n)$ is a basis of E and $(e^1, \ldots, e^n)$ is the corresponding dual basis of E^*, then the $\binom{n}{k}$ elements*

$$\{e^{i_1} \wedge \cdots \wedge e^{i_k};\ 1 \le i_1 < \cdots < i_k \le n\}$$

form a basis of $\Lambda^k E^*$.

2.10. Differential Forms

Let M be an n-dimensional C^∞ manifold. We let

$$\Lambda^k T^*(M) = \bigcup_{x \in M} \Lambda^k T_x^*(M)$$

be the disjoint union of the k-th exterior products of the cotangent spaces $T_x^*(M)$. The elements of $\Lambda^k T_x^*(M)$ are called *exterior k-forms* at x. This $\Lambda^k T^*(M)$ carries a natural structure of C^∞ manifold of dimension $n + \binom{n}{k}$, induced by the tangent bundle $T^*(M)$ (cf. Abraham-Marsden-Ratiu [1], Theorem 6.3.4). We call $\Lambda^k T^*(M)$ the *vector bundle of exterior k-forms* on the tangent spaces of M.

A *differential form of order k*, or simply a *k-form* on M, is a C^∞ mapping,

$$\omega: M \to \Lambda^k T^*(M),$$

such that $\omega(x) \in \Lambda^k T_x^*(M)$ for each $x \in M$. The set $\Omega^k(M)$ of all k-forms on M is a real linear space with the obvious operations of addition and scalar multiplication.

We now give the expression of differential forms in local charts. We remark that if (U, φ) is a chart on M with $\varphi(x) = (x^1, \ldots, x^n)$, then the $\binom{n}{k}$ elements

$$\{dx^{i_1} \wedge \cdots \wedge dx^{i_k};\ 1 \leq i_1 < \cdots < i_k \leq n\}$$

form a basis of $\Lambda^k T_x^*(M)$ at each point x of U. A differential form $\omega \in \Omega^k(M)$ has the local expression

$$\omega = \sum_{1 \leq i_1 < \cdots < i_k \leq n} \xi_{i_1 \cdots i_k}\, dx^{i_1} \wedge \cdots \wedge dx^{i_k},$$

where $\xi_{i_1 \cdots i_k}$ are C^∞ functions on U. The functions $\xi_{i_1 \cdots i_k}$ are called the *local components* of ω relative to the chart (U, φ).

If $f \in C^\infty(M)$ and $\omega \in \Omega^k(M)$, then the mapping

$$M \ni x \mapsto f(x)\omega(x)$$

defines a k-form on M. The space $\Omega^k(M)$ is a $C^\infty(M)$-module with respect to this operation of product.

2.11. Densities

Let E be an n-dimensional linear space over $\mathbf{R}$ and $E^* = L(E, \mathbf{R})$ its dual space. Let $\Lambda^n E$ be the n-th exterior product of E and $\Lambda^n E^*$ the n-th exterior product of E^*. Proposition 2.9.4 shows that $\Lambda^n E$ and $\Lambda^n E^*$ are both one-dimensional. Further, the spaces $\Lambda^n E$ and $\Lambda^n E^*$ are dual to each other. The non-zero elements of $\Lambda^n E^*$ are called *volume elements* on E.

A complex-valued *density* on E is a mapping

$$\rho: \Lambda^n E \to \mathbf{C}$$

such that

$$\rho(\lambda\sigma) = |\lambda| \rho(\sigma), \qquad \lambda \in \mathbf{R}.$$

The set of all densities on E is a complex linear space with the obvious operations of addition and scalar multiplication. This linear space is denoted by $\Omega(E^*)$, and is called the *space of densities* on E.

Densities can be constructed from volume elements in the following way: If $\omega \in \Lambda^n E^*$, we define a mapping

$$|\omega|: \Lambda^n E \to \mathbf{C}$$

by

$$|\omega|(\sigma) = |\langle \sigma, \omega \rangle|, \qquad \sigma \in \Lambda^n E,$$

where $\langle\ ,\ \rangle$ is the pairing of $\Lambda^n E$ and $\Lambda^n E^*$. Then we have $|\omega| \in \Omega(E^*)$.

The space $\Omega(E^*)$ is one-dimensional. In fact, if $(e_1, \ldots, e_n)$ is a basis of E and $(e^1, \ldots, e^n)$ is the corresponding dual basis of E^*, then every ρ of $\Omega(E^*)$ can be written in the form

$$\rho = \rho(e_1 \wedge \cdots \wedge e_n) |e^1 \wedge \cdots \wedge e^n|.$$

Let M be an n-dimensional C^∞ manifold. We remark that if (U, φ) is a chart on M with $\varphi(x) = (x^1, \ldots, x^n)$, then the density $|dx^1 \wedge \cdots \wedge dx^n|$ is a basis of the space $\Omega(T_x^*(M))$ of densities on $T_x(M)$ at each point x of U.

We let

$$\Omega(T^*(M)) = \bigcup_{x \in M} \Omega(T_x^*(M))$$

be the disjoint union of the spaces $\Omega(T_x^*(M))$, and define a mapping

$$|\pi|: \Omega(T^*(M)) \to M$$

by $|\pi|(\rho) = x$ if $\rho \in \Omega(T_x^*(M))$.

Now we will make $\Omega(T^*(M))$ into an $(n + 2)$-dimensional C^∞ manifold by giving natural charts for it.

Let (U, φ) be a chart on M with $\varphi(x) = (x^1, \ldots, x^n)$. We define a mapping

$$|\varphi|: |\pi|^{-1}(U) \to \varphi(U) \times \mathbf{R}^2$$

by

$$|\varphi|(\rho) = \left(\varphi(x), \rho\left(\frac{\partial}{\partial x^1} \wedge \cdots \wedge \frac{\partial}{\partial x^n}\right)\right),$$

where $|\pi|(\rho) = x$. Here we identify $\mathbf{C}$ with $\mathbf{R}^2$. If (V, ψ) is another chart with $\psi(y) = (y^1, \ldots, y^n)$ such that $U \cap V \neq \varnothing$, then we have

$$\begin{aligned}\rho &= \rho\left(\frac{\partial}{\partial x^1} \wedge \cdots \wedge \frac{\partial}{\partial x^n}\right)|dx^1 \wedge \cdots \wedge dx^n| \\ &= \rho\left(\frac{\partial}{\partial x^1} \wedge \cdots \wedge \frac{\partial}{\partial x^n}\right)\left|\det\left(\frac{\partial x^i}{\partial y^j}\right)\right| |dy^1 \wedge \cdots \wedge dy^n|,\end{aligned}$$

so that

$$\rho\left(\frac{\partial}{\partial y^1} \wedge \cdots \wedge \frac{\partial}{\partial y^n}\right) = \left|\det\left(\frac{\partial x^i}{\partial y^j}\right)\right| \rho\left(\frac{\partial}{\partial x^1} \wedge \cdots \wedge \frac{\partial}{\partial x^n}\right). \tag{1}$$

We remark that

$$\det\left(\frac{\partial x^i}{\partial y^j}\right) = \det(J(\varphi \circ \psi^{-1})),$$

the Jacobian determinant of $\varphi \circ \psi^{-1}$. Thus, the transition map

$$|\psi| \circ |\varphi|^{-1}: \varphi(U \cap V) \times \mathbf{R}^2 \to \psi(U \cap V) \times \mathbf{R}^2$$

is given by

$$(\varphi(x), \zeta) \mapsto (\psi(x), |\det(J(\varphi \circ \psi^{-1})(\psi(x)))|\zeta),$$

for $x \in U \cap V$ and $\zeta \in \mathbf{R}^2$. Since $\det(J(\varphi \circ \psi^{-1}))$ is of class C^∞ and is non-zero at $\psi(x)$, we obtain that the family of pairs $\{(|\pi|^{-1}(U), |\varphi|)\}$, where (U, φ) ranges over all admissible charts, is an atlas on $\Omega(T^*(M))$. This proves that $\Omega(T^*(M))$ is an $(n + 2)$-dimensional C^∞ manifold. We call $\Omega(T^*(M))$ the *fiber bundle of densities* on the tangent spaces of M.

A C^∞ *density* on M is a C^∞ mapping,

$$\rho: M \to \Omega(T^*(M)),$$

such that $\rho(x) \in \Omega(T_x^*(M))$ for each $x \in M$. The set $C^\infty(|M|)$ of all C^∞ densities on M is a complex linear space with the obvious operations of addition and scalar multiplication.

If $\rho \in C^\infty(|M|)$, and (U, φ) is a chart on M with $\varphi(x) = (x^1, \ldots, x^n)$, then ρ has the local expression

$$\rho(x) = f(x)\,|dx^1 \wedge \cdots \wedge dx^n|,$$

where f is a C^∞ function on U given by

$$f(x) = \rho(x)\left(\frac{\partial}{\partial x^1} \wedge \cdots \wedge \frac{\partial}{\partial x^n}\right).$$

If $h \in C^\infty(M)$ and $\rho \in C^\infty(|M|)$, then the mapping

$$M \ni x \mapsto h(x)\rho(x)$$

defines a C^∞ density on M. The space $C^\infty(|M|)$ is a $C^\infty(M)$-module with respect to this operation of product.

We now discuss the behavior of densities relative to a change of coordinates.

If (U, φ) and (V, ψ) are two overlapping charts on M with $\varphi(x) = (x^1, \ldots, x^n)$ and $\psi(y) = (y^1, \ldots, y^n)$, then ρ has the local expressions

$$\begin{cases} \rho(x) = f(x)\,|dx^1 \wedge \cdots \wedge dx^n|, \\ f(x) = \rho(x)\left(\dfrac{\partial}{\partial x^1} \wedge \cdots \wedge \dfrac{\partial}{\partial x^n}\right), \end{cases} \qquad (x \in U),$$

and

$$\begin{cases} \rho(y) = g(y)\,|dy^1 \wedge \cdots \wedge dy^n|, \\ g(y) = \rho(y)\left(\dfrac{\partial}{\partial y^1} \wedge \cdots \wedge \dfrac{\partial}{\partial y^n}\right), \end{cases} \qquad (y \in V).$$

Hence formula (1) tells us that

$$\tilde{g}(\psi(x)) = \tilde{f}(\varphi(x))|\det(J(\varphi \circ \psi^{-1})(\psi(x)))|, \qquad x \in U \cap V, \tag{2}$$

where $\tilde{f} = f \circ \varphi^{-1}$ and $\tilde{g} = g \circ \psi^{-1}$.

By virtue of formula (2), we can speak of real-valued densities and also of strictly positive densities.

By smoothly patching together densities on the tangent spaces, we obtain the following theorem on the existence of densities:

2.11.1 Theorem. *Let (M, g) be an n-dimensional, Riemannian C^∞ manifold. Then there exists a strictly positive C^∞ density μ on M such that μ equals 1 on all orthonormal bases of the tangent spaces of M. If $(X_1, \ldots, X_n)$ is such a basis in an open subset U of M with dual basis $(\xi^1, \ldots, \xi^n)$, then $\mu = |\xi^1 \wedge \cdots \wedge \xi^n|$.*

More generally, if $v_1, \ldots, v_n \in T_x(M)$, *then we have*

$$\mu(x)(v_1, \ldots, v_n) = (\det(g(x)(v_i, v_j)))^{1/2}. \tag{3}$$

2.11.2 Remark. Formula (3) shows that if (U, φ) is a chart on M with $\varphi(x) = (x^1, \ldots, x^n)$, then μ has the local expression

$$\mu(x) = (\det(g_{ij}(x)))^{1/2} \, |dx^1 \wedge \cdots \wedge dx^n|,$$

where

$$g_{ij}(x) = g(x)\left(\frac{\partial}{\partial x^i}, \frac{\partial}{\partial x^j}\right).$$

2.12. Integration on Manifolds

We define the integral of a density on an n-dimensional C^∞ manifold in terms of integrals over sets in $\mathbf{R}^n$ by means of a partition of unity subordinate to an atlas.

The next theorem guarantees that the integral is well-defined, independent of the choice of atlas and partition of unity.

2.12.1 Theorem (the change of variable formula). *Let U, V be two open subsets of $\mathbf{R}^n$ and $\chi\colon U \to V$ a C^1 diffeomorphism. If v is a Lebesgue integrable function on V, then the function $v \circ \chi |\det(J\chi)|$ is Lebesgue integrable on U, and we have*

$$\int_V v(y)\, dy = \int_U v(\chi(x)) |\det(J\chi(x))| \, dx, \tag{1}$$

where $J\chi$ is the Jacobian matrix of χ.

Let M be an n-dimensional C^∞ manifold equipped with a strictly positive density μ. If f is a non-negative, Borel measurable function on M, then we can define the integral

$$\int_M f \cdot \mu$$

in the following manner: let $\{(U_i, \chi_i)\}_{i \in I}$ be an atlas on M, and $\{\eta_i\}_{i \in I}$ a partition of unity subordinate to the covering $\{U_i\}_{i \in I}$. Suppose that μ has the local expression

$$\mu = h_i \, |dx^1 \wedge \cdots \wedge dx^n|$$

relative to the chart (U_i, χ_i), with $\chi_i(x) = (x^1, \ldots, x^n)$. Here note that $h_i > 0$ on U_i. Then we define

$$\int_M f \cdot \mu = \sum_{i \in I} \int_{\chi_i(U_i)} \eta_i(\chi_i^{-1}(z)) f(\chi_i^{-1}(z)) h_i(\chi_i^{-1}(z))\, dz, \tag{2}$$

where $dz = dz^1 \cdots dz^n$ is the Lebesgue measure on $\mathbf{R}^n$. By virtue of Theorem 2.12.1, one can verify that this integral is independent of the atlas $\{(U_i, \chi_i)\}$ and the partition of unity $\{\eta_i\}$ used. We say that f is *integrable* if $\int_M f \cdot \mu < \infty$.

A Borel measurable function f on M is said to be *integrable* if $|f|$ is integrable in the above sense. Then we define $\int_M f \cdot \mu$ by formula (2). Note that

$$\left| \int_M f \cdot \mu \right| \leq \int_M |f| \cdot \mu.$$

We call $\int_M f \cdot \mu$ the *integral* of f with respect to the density μ.

2.13. Manifolds with Boundary

We denote by $\mathbf{R}^n_+$ the open half space

$$\mathbf{R}^n_+ = \{x = (x^1, \ldots, x^n) \in \mathbf{R}^n;\ x^n > 0\}.$$

We let

$$\overline{\mathbf{R}^n_+} = \{x \in \mathbf{R}^n;\ x^n \geq 0\},$$

and let U be an open set in $\overline{\mathbf{R}^n_+}$ in the topology induced on $\overline{\mathbf{R}^n_+}$ from $\mathbf{R}^n$. We define the *boundary* ∂U of U to be the intersection of U with $\mathbf{R}^{n-1} \times \{0\}$ and the *interior* Int U of U to be the complement of ∂U in U, that is,

$$\partial U = U \cap \{x \in \mathbf{R}^n;\ x^n = 0\},$$

$$\operatorname{Int} U = U \cap \{x \in \mathbf{R}^n;\ x^n > 0\}.$$

It is clear that Int U is open in U and that ∂U is closed in U, but not in $\mathbf{R}^n$. This inconsistent use of the notation ∂U is temporary.

Let U and V be two open sets in $\overline{\mathbf{R}^n_+}$. We say that a mapping $f\colon U \to V$ is *of class* C^r $(0 \leq r \leq \infty)$ if, for each point x of U, there exist a neighborhood U_1 of x in $\mathbf{R}^n$ and a neighborhood V_1 of $f(x)$ in $\mathbf{R}^n$, and a C^r mapping $f_1\colon U_1 \to V_1$ such that $f_1|_{U \cap U_1} = f|_{U \cap U_1}$.

Then we have:

2.13.1 Lemma. *Let U, V be open sets in $\overline{\mathbf{R}^n_+}$, and $f: U \to V$ a C^r diffeomorphism with $1 \le r \le \infty$. Then the mapping f induces two C^r diffeomorphisms* $\operatorname{Int} f: \operatorname{Int} U \to \operatorname{Int} V$ *and* $\partial f: \partial U \to \partial V$.

Now we can define a C^r manifold ($1 \le r \le \infty$) with boundary in the following way:

Let M be a set. An *atlas of charts with boundary* on M is a family of pairs $\{(U_i, \varphi_i)\}_{i \in I}$ satisfying the following conditions:

(MB1) Each U_i is a subset of M and $M = \bigcup_{i \in I} U_i$.

(MB2) Each φ_i is a bijection of U_i onto an open subset of $\overline{\mathbf{R}^n_+}$, and for every pair i, j of I with $U_i \cap U_j \neq \varnothing$ the set $\varphi_i(U_i \cap U_j)$ is open in $\overline{\mathbf{R}^n_+}$.

(MB3) For each pair i, j of I with $U_i \cap U_j \neq \varnothing$ the mapping

$$\varphi_j \circ \varphi_i^{-1}: \varphi_i(U_i \cap U_j) \to \varphi_j(U_i \cap U_j)$$

is a C^r diffeomorphism.

Each pair (U_i, φ_i) is called a *chart with boundary* of the atlas.

An n-dimensional C^r *manifold with boundary* is a set M together with an atlas of charts with boundary on M. By virtue of Lemma 2.13.1, we can define

$$\operatorname{Int} M = \bigcup_{i \in I} (\operatorname{Int} \varphi_i)^{-1}(\operatorname{Int}(\varphi_i(U_i))),$$

$$\partial M = \bigcup_{i \in I} (\partial \varphi_i)^{-1}(\partial(\varphi_i(U_i))).$$

We call $\operatorname{Int} M$ the *interior* of M and ∂M the *boundary* of M, respectively. The set $\operatorname{Int} M$ is an n-dimensional C^r manifold (without boundary) with atlas obtained from (U_i, φ_i) by replacing $\varphi_i(U_i)$ by $\operatorname{Int}(\varphi_i(U_i))$, and the set ∂M is an $(n-1)$-dimensional C^r manifold (without boundary) with atlas obtained from (U_i, φ_i) by replacing $\varphi_i(U_i)$ by $\partial(\varphi_i(U_i))$.

The Double of a Manifold

We give two fundamental theorems on C^∞ manifolds with boundary. In what follows, let M be an n-dimensional paracompact C^∞ manifold with boundary ∂M.

The first theorem states that ∂M has an open neighborhood in M which is diffeomorphic to $\partial M \times [0, 1)$.

2.13.2 Theorem (the product neighborhood theorem). *There exists a C^∞ diffeomorphism φ of $\partial M \times [0, 1)$ onto an open neighborhood W of ∂M in M which is the identity map on ∂M.*

The diffeomorphism φ is called a *collar* for M and the neighborhood W is called a *product neighborhood* of ∂M.

The second theorem states that M is a submanifold of some n-dimensional C^∞ manifold without boundary. Let $M_0 = M \times \{0\}$ and $M_1 = M \times \{1\}$ be two copies of M. The *double* $\hat{M}$ of M is the topological space obtained from the union $M_0 \cup M_1$ by identifying $(x, 0)$ with $(x, 1)$ for each x in ∂M.

Using the product neighborhood theorem, we have:

2.13.3 Theorem. *The double $\hat{M}$ of M is an n-dimensional C^∞ manifold without boundary, and is uniquely determined up to C^∞ diffeomorphisms.*

Notes

The material in this chapter is adapted from Abraham–Marsden–Ratiu [1], Lang [1] and Matsushima [1], while Theorems 2.13.2 and 2.13.3 are taken from Munkres [1].

3 Functional Analysis

This chapter is devoted to a review of standard topics from functional analysis such as Banach and Hilbert spaces, the Riesz representation theorem relating linear functionals on spaces of continuous functions to integrals, compact and Fredholm operators, and the theory of semigroups. These topics form a necessary background for what follows.

3.1. Quasinormed Linear Spaces

Let X be a linear space over the real or complex number field $\mathbf{K}$. A real-valued function p defined on X is called a *seminorm* on X if it satisfies the following conditions:

(S1) $0 \leq p(x) < \infty$, $x \in X$.
(S2) $p(\alpha x) = |\alpha| p(x)$, $\alpha \in \mathbf{K}$, $x \in X$.
(S3) $p(x + y) \leq p(x) + p(y)$, $x, y \in X$.

Let $\{p_i\}$ be a countable family of seminorms on X such that

$$p_1(x) \leq p_2(x) \leq \cdots \leq p_i(x) \leq \cdots, \qquad x \in X, \tag{1}$$

and define

$$V_{ij} = \left\{x \in X;\, p_i(x) < \frac{1}{j}\right\}, \qquad i, j = 1, 2, \dots .$$

Then it is easy to verify that a countable family of the sets

$$x + V_{ij} = \{x + y;\, y \in V_{ij}\}$$

satisfies axioms $(V1)$, $(V2)$ and $(V3)$ of a fundamental neighborhood system of x; hence X is a topological space which satisfies the first axiom of countability.

Furthermore we have:

3.1.1 Theorem. *Let $\{p_i\}$ be a countable family of seminorms on a linear space X which satisfies condition* (1). *Suppose that*:

For every non-zero $x \in X$, there exists a seminorm p_i such that $p_i(x) > 0$. (2)

Then the space X is metrizable by the metric:

$$\rho(x, y) = \sum_{i=1}^{\infty} \frac{1}{2^i} \frac{p_i(x - y)}{1 + p_i(x - y)}, \qquad x, y \in X.$$

If we let

$$|x| = \rho(x, 0) = \sum_{i=1}^{\infty} \frac{1}{2^i} \frac{p_i(x)}{1 + p_i(x)}, \qquad x \in X, \tag{3}$$

then the quantity $|x|$ enjoys the following properties:

($Q1$) $|x| \geq 0$; $|x| = 0$ if and only if $x = 0$.
($Q2$) $|x + y| \leq |x| + |y|$ (triangle inequality).
($Q3$) $\alpha_n \to 0$ in $\mathbf{K} \Rightarrow |\alpha_n x| \to 0$, $x \in X$.
($Q4$) $|x_n| \to 0 \Rightarrow |\alpha x_n| \to 0$, $\alpha \in \mathbf{K}$.

This $|x|$ is called a *quasinorm* of x, and the space X is called a *quasinormed linear space*.

Theorem 3.1.1 may be restated as follows:

3.1.2 Theorem. *A linear space X, topologized by a countable family $\{p_i\}$ of seminorms satisfying conditions* (1) *and* (2), *is a quasinormed linear space with respect to the quasinorm $|x|$ defined by formula* (3).

Let X be a quasinormed linear space. The convergence

$$\lim_{n\to\infty} |x_n - x| = 0$$

in X is denoted by $s\text{-}\lim_{n\to\infty} x_n = x$, or simply by $x_n \to x$, and we say that the sequence $\{x_n\}$ *converges strongly* to x. A sequence $\{x_n\}$ is called a *Cauchy sequence* if it satisfies Cauchy's condition

$$\lim_{m,n\to\infty} |x_m - x_n| = 0.$$

A quasinormed linear space X is called a *Fréchet space* if it is complete, that is, if every Cauchy sequence in X converges strongly to a point in X.

If a quasinormed linear space X is topologized by a countable family $\{p_i\}$ of seminorms which satisfies conditions (1) and (2), then the above definitions may be reformulated in terms of seminorms as follows:

1) A sequence $\{x_n\}$ in X converges strongly to a point x in X if and only if, for every seminorm p_i and every $\varepsilon > 0$, there exists a positive integer $N = N(i, \varepsilon)$ such that

$$n \geq N \Rightarrow p_i(x_n - x) < \varepsilon.$$

2) A sequence $\{x_n\}$ in X is a Cauchy sequence if and only if, for every seminorm p_i and every $\varepsilon > 0$, there exists a positive integer $N = N(i, \varepsilon)$ such that

$$m, n \geq N \Rightarrow p_i(x_m - x_n) < \varepsilon.$$

Let X be a quasinormed linear space. A linear subspace of X is called a *closed subspace* if it is a closed subset of X. For example, the closure $\bar{M}$ of a linear subspace M is a closed subspace. In fact, the elements of $\bar{M}$ are limits of sequences in M; thus if $x = \lim_n x_n$, $x_n \in M$ and $y = \lim_n y_n$, $y_n \in M$, then we have

$$\begin{cases} x + y = \lim\limits_n (x_n + y_n), \\ \alpha x = \lim\limits_n \alpha x_n, \qquad \alpha \in \mathbf{K}, \end{cases}$$

so that $x + y \in \bar{M}$ and $\alpha x \in \bar{M}$.

Bounded Sets

Let X be a quasinormed linear space, topologized by a countable family $\{p_i\}$ of seminorms which satisfies conditions (1) and (2). A set B in X is said to be *bounded* if, for every seminorm p_i, we have

$$\sup_{x \in B} p_i(x) < +\infty.$$

We remark that every compact set is bounded.

Throughout the rest of this section, let X and Y be quasinormed linear spaces over the same scalar field, topologized respectively by countable families $\{p_i\}$ and $\{q_i\}$ of seminorms which satisfy conditions (1) and (2).

Continuity of Linear Operators

Let T be a linear operator from X into Y with domain $D(T)$. By virtue of the linearity of T, it follows that T is continuous everywhere on $D(T)$ if and only if it is continuous at one point of $D(T)$.

Furthermore we have:

3.1.3 Theorem. *A linear operator T from X into Y with domain $D(T)$ is continuous everywhere on $D(T)$ if and only if, for every seminorm q_j on Y, there exist a seminorm p_i on X and a constant $C > 0$ such that*

$$q_j(Tx) \leq Cp_i(x), \qquad x \in D(T).$$

Topologies of Linear Operators

We let

$L(X, Y)$ = the collection of continuous linear operators on X into Y.

We define in the set $L(X, Y)$ addition and scalar multiplication of operators in the usual way:

$$\begin{cases} (T + S)x = Tx + Sx, & x \in X, \\ (\alpha T)x = \alpha(Tx), & \alpha \in \mathbf{K}. \end{cases}$$

Then $L(X, Y)$ is a linear space.

We introduce three different topologies on the space $L(X, Y)$:

1) *Simple convergence topology*: This is the topology of convergence at each point of X; a sequence $\{T_n\}$ in $L(X, Y)$ converges to an element T of

$L(X, Y)$ in the simple convergence topology if and only if $T_n x \to Tx$ in Y for each $x \in X$.

2) *Compact convergence topology*: This is the topology of uniform convergence on compact sets in X; $T_n \to T$ in the compact convergence topology if and only if $T_n x \to Tx$ in Y uniformly for x ranging over compact sets in X.

3) *Bounded convergence topology*: This is the topology of uniform convergence on bounded sets in X; $T_n \to T$ in the bounded convergence topology if and only if $T_n x \to Tx$ in Y uniformly for x ranging over bounded sets in X.

The simple convergence topology is weaker than the compact convergence topology, and the compact convergence topology is weaker than the bounded convergence topology.

The Banach–Steinhaus Theorem

We introduce three different definitions of boundedness for sets in the space $L(X, Y)$:

1) A set H in $L(X, Y)$ is said to be bounded in the simple convergence topology if, for each $x \in X$, the set $\{Tx; T \in H\}$ is bounded in Y.

2) A set H in $L(X, Y)$ is said to be bounded in the compact convergence topology if, for every compact set K in X, the set $\bigcup_{T \in H} T(K)$ is bounded in Y.

3) A set H in $L(X, Y)$ is said to be bounded in the bounded convergence topology if, for every bounded set B in X, the set $\bigcup_{T \in H} T(B)$ is bounded in Y.

Further, a set H in $L(X, Y)$ is said to be *equicontinuous* if, for every seminorm q_j on Y, there exist a seminorm p_i on X and a constant $C > 0$ such that

$$\sup_{T \in H} q_j(Tx) \leq C p_i(x), \qquad x \in X.$$

The next theorem states one of the fundamental properties of Fréchet spaces.

3.1.4 Theorem (Banach–Steinhaus). *Let X be a Fréchet space and Y a quasinormed linear space. Then the following four conditions are equivalent:*

(i) *A set H in $L(X, Y)$ is bounded in the simple convergence topology.*
(ii) *A set H in $L(X, Y)$ is bounded in the compact convergence topology.*
(iii) *A set H in $L(X, Y)$ is bounded in the bounded convergence topology.*
(iv) *A set H in $L(X, Y)$ is equicontinuous.*

Product Spaces

Let X and Y be quasinormed linear spaces over the same scalar field $\mathbf{K}$. Then the Cartesian product $X \times Y$ becomes a linear space over $\mathbf{K}$ if we define the algebraic operations coordinatewise:

$$\begin{cases} \{x_1, y_1\} + \{x_2, y_2\} = \{x_1 + x_2, y_1 + y_2\}, \\ \alpha\{x, y\} = \{\alpha x, \alpha y\}, \qquad \alpha \in \mathbf{K}. \end{cases}$$

It is easy to verify that the quantity

$$|\{x, y\}| = (|x|^2 + |y|^2)^{1/2} \tag{4}$$

satisfies axioms ($Q1$) through ($Q4$) of a quasinorm; hence the product space $X \times Y$ is a quasinormed linear space with respect to the quasinorm defined by (4). Furthermore, if X and Y are Fréchet spaces, then so is $X \times Y$. In other words, the completeness is inherited by the product space.

3.2. Normed Linear Spaces

A quasinormed linear space is called a *normed linear space* if it is topologized by just one seminorm which satisfies condition (3.1.2). We give the precise definition of a normed linear space:

Let X be a linear space over the real or complex number field $\mathbf{K}$. A real-valued function $\|\cdot\|$ defined on X is called a *norm* on X if it satisfies the following conditions:

($N1$) $\|x\| \geq 0$; $\|x\| = 0$ if and only if $x = 0$.
($N2$) $\|\alpha x\| = |\alpha| \cdot \|x\|$, $\alpha \in \mathbf{K}$, $x \in X$.
($N3$) $\|x + y\| \leq \|x\| + \|y\|$, $x, y \in X$ (triangle inequality).

A linear space X equipped with a norm $\|\cdot\|$ is called a normed linear space. The topology on X is defined by the metric

$$\rho(x, y) = \|x - y\|.$$

The convergence

$$\lim_{n \to \infty} \|x_n - x\| = 0$$

in X is denoted by s-$\lim_{n\to\infty} x_n = x$, or simply $x_n \to x$, and we say that the sequence $\{x_n\}$ converges strongly to x. A sequence $\{x_n\}$ in X is called a Cauchy sequence if it satisfies Cauchy's condition

$$\lim_{n,\, m\to\infty} \|x_n - x_m\| = 0.$$

A normed linear space X is called a *Banach space* if it is complete, that is, if every Cauchy sequence in X converges strongly to a point in X.

Two norms $\|\cdot\|_1$ and $\|\cdot\|_2$ defined on the same linear space X are said to be equivalent if there exist constants $c > 0$ and $C > 0$ such that

$$c\|x\|_1 \le \|x\|_2 \le C\|x\|_1, \qquad x \in X.$$

Equivalent norms induce the same topology.

If X and Y are normed linear spaces over the same scalar field, then the product space $X \times Y$ is a normed linear space with the norm

$$\|\{x, y\} = (\|x\|_X^2 + \|y\|_Y^2)^{1/2}.$$

If X and Y are Banach spaces, then so is $X \times Y$.

Let X be a normed linear space. If Y is a closed linear subspace of X, then the factor space X/Y is a normed linear space with the norm

$$\|\tilde{x}\| = \inf_{z \in \tilde{x}} \|z\|. \tag{1}$$

If X is a Banach space, then so is X/Y. The space X/Y, normed by (1), is called a *normed factor space.*

Throughout the rest of this section, the letters X, Y, Z denote normed linear spaces over the same scalar field.

The next theorem is a normed linear space version of Theorem 3.1.3.

3.2.1 Theorem. *Let T be a linear operator from X into Y with domain $D(T)$. Then T is continuous everywhere on $D(T)$ if and only if there exists a constant $C > 0$ such that*

$$\|Tx\| \le C\|x\|, \qquad x \in D(T). \tag{2}$$

3.2.2 Remark. In inequality (2), the quantity $\|x\|$ is the norm of x in X and the quantity $\|Tx\|$ is the norm of Tx in Y. Frequently several norms appear together, but it is clear from the context which is which.

One of the consequences of Theorem 3.2.1 is the following extension theorem for a continuous linear operator:

3.2.3 Theorem. *If T is a continuous linear operator from X into Y with domain $D(T)$, and if Y is a Banach space, then T has a unique continuous extension $\tilde{T}$ whose domain is the closure $\overline{D(T)}$ of $D(T)$.*

As another consequence of Theorem 3.2.1, we give a necessary and sufficient condition for the existence of the continuous inverse of a linear operator:

3.2.4 Theorem. *Let T be a linear operator from X into Y with domain $D(T)$. Then T admits a continuous inverse T^{-1} if and only if there exists a constant $c > 0$ such that*

$$\|Tx\| \geq c\|x\|, \qquad x \in D(T).$$

A linear operator T from X into Y with domain $D(T)$ is called an *isometry* if it is norm-preserving, that is, if we have

$$\|Tx\| = \|x\|, \qquad x \in D(T).$$

It is clear that if T is an isometry, then it is injective and both T and T^{-1} are continuous.

If T is a continuous, one-to-one linear mapping of X onto Y, and if its inverse T^{-1} is also a continuous mapping, then it is called an *isomorphism* of X onto Y. Two normed linear spaces are said to be *isomorphic* if there is an isomorphism between them.

Combining Theorems 3.2.1 and 3.2.4, we obtain:

3.2.5 Theorem. *Let T be a linear operator on X onto Y. Then T is an isomorphism if and only if there exist constants $c > 0$ and $C > 0$ such that*

$$c\|x\| \leq \|Tx\| \leq C\|x\|, \qquad x \in X.$$

If T is a continuous linear operator from X into Y with domain $D(T)$, we let

$$\|T\| = \inf\{C;\ \|Tx\| \leq C\|x\|,\ x \in D(T)\}.$$

Then, in view of the linearity of T, we have

$$\|T\| = \sup_{\substack{x \in D(T) \\ x \neq 0}} \frac{\|Tx\|}{\|x\|} = \sup_{\substack{x \in D(T) \\ \|x\| = 1}} \|Tx\| = \sup_{\substack{x \in D(T) \\ \|x\| \leq 1}} \|Tx\|. \tag{3}$$

This proves that $\|T\|$ is the smallest non-negative number such that

$$\|Tx\| \leq \|T\| \cdot \|x\|, \qquad x \in D(T). \tag{4}$$

Theorem 3.2.1 tells us that a linear operator T on X into Y is continuous if and only if it maps bounded sets in X into bounded sets in Y. Thus a continuous linear operator on X into Y is usually called a *bounded linear operator* on X into Y.

We let

$L(X, Y) =$ the space of bounded (continuous) linear operators on X into Y.

In the case of normed linear spaces, the simple convergence topology on $L(X, Y)$ is usually called the *strong topology* of operators, and the bounded convergence topology on $L(X, Y)$ is called the *uniform topology* of operators.

In view of (3) and (4), it follows that the quantity $\|T\|$ satisfies axioms ($N1$), ($N2$) and ($N3$) of a norm; hence the space $L(X, Y)$ is a normed linear space with the norm $\|T\|$ given by (3). The topology on $L(X, Y)$ induced by $\|T\|$ is just the uniform topology of operators.

We give a sufficient condition for the space $L(X, Y)$ to be complete:

3.2.6 Theorem. *If Y is a Banach space, then so is $L(X, Y)$.*

If T is a linear operator from X into Y with domain $D(T)$ and S is a linear operator from Y into Z with domain $D(S)$, then we define the *product* ST as follows:

(a) $D(ST) = \{x \in D(T);\ Tx \in D(S)\}$,
(b) $(ST)(x) = S(Tx)$, $x \in D(ST)$.

As for the product of linear operators, we have:

3.2.7 Proposition. *If $T \in L(X, Y)$ and $S \in L(Y, Z)$, then $ST \in L(X, Z)$ and we have*

$$\|ST\| \leq \|S\| \cdot \|T\|.$$

We often make use of the following theorem in constructing the bounded inverse of a bounded linear operator.

3.2.8 Theorem. *If T is a bounded linear operator on a Banach space X into itself and satisfies*

$$\|T\| < 1,$$

then the operator $I - T$ has a unique bounded linear inverse $(I - T)^{-1}$ which is given by C. Neumann's series

$$(I - T)^{-1} = \sum_{n=0}^{\infty} T^n.$$

Here I is the identity operator $Ix = x$, $x \in X$, and $T^0 = I$.

The next theorem is a normed linear space version of the Banach–Steinhaus theorem (Theorem 3.1.4).

3.2.9 Theorem (the resonance theorem). *Let X be a Banach space, Y a normed linear space and H a subset of $L(X, Y)$. Then the boundedness of $\{\|Tx\|; T \in H\}$ at each $x \in X$ implies the boundedness of $\{\|T\|; T \in H\}$.*

3.2.10 Corollary. *Let X be a Banach space, Y a normed linear space and $\{T_n\}$ a sequence in $L(X, Y)$. If*

$$\operatorname*{s-lim}_{n\to\infty} T_n x = Tx \tag{5}$$

exists for each $x \in X$, then we have $T \in L(X, Y)$ and

$$\|T\| \leq \liminf_{n\to\infty} \|T_n\|.$$

The operator T obtained above is called the *strong limit* of the sequence $\{T_n\}$, since the convergence in (5) is the strong topology of operators. We then write

$$T = \operatorname*{s-lim}_{n\to\infty} T_n.$$

Finite Dimensional Spaces

The next theorem tells us that there is no point in studying abstract finite dimensional normed linear spaces.

3.2.11 Theorem. *All n-dimensional normed linear spaces over the same scalar field* $\mathbf{K}$ *are isomorphic to* $\mathbf{K}^n$ *with the maximum norm*

$$\|\alpha\| = \max_{1 \leq i \leq n} |\alpha_i|, \qquad \alpha = (\alpha_1, \ldots, \alpha_n) \in \mathbf{K}^n.$$

Topological properties of the space $\mathbf{K}^n$ apply to all finite dimensional normed linear spaces.

3.2.12 Corollary. *All finite dimensional normed linear spaces are complete.*

3.2.13 Corollary. *Every finite dimensional linear subspace of a normed linear space is closed.*

3.2.14 Corollary. *A subset of a finite dimensional normed linear space is compact if and only if it is closed and bounded.*

By Corollary 3.2.14, the closed unit ball in a finite dimensional normed linear space is compact. Conversely, this property characterizes finite dimensional spaces:

3.2.15 Theorem. *If the closed unit ball in a normed linear space* X *is compact, then* X *is finite dimensional.*

The Hahn–Banach Extension Theorem

3.2.16 Theorem (Hahn–Banach). *Let* X *be a normed linear space over the real or complex number field,* M *a linear subspace of* X *and* f *a continuous linear functional defined on* M*. Then* f *can be extended to a continuous linear functional* $\tilde{f}$ *on* X *so that*

$$\|\tilde{f}\| = \|f\|.$$

3.2.17 Corollary. *Let* X *be a normed linear space. For each non-zero element* x_0 *of* X*, there exists a continuous linear functional* f *on* X *such that*

$$\begin{cases} f(x_0) = \|x_0\|, \\ \|f\| = 1. \end{cases}$$

A continuous linear functional on X is usually called a *bounded linear functional* on X.

Dual Spaces

Let X be a normed linear space over the real or complex number field $\mathbf{K}$. Then the space $L(X, \mathbf{K})$ of all bounded linear functionals on X is called the *dual space* of X, and is denoted by X'. The bounded (resp. simple) convergence topology on X' is called the *strong* (resp. *weak**) *topology* on X', and the dual space X' equipped with this topology is called the *strong* (resp. *weak**) *dual space* of X.

It follows from an application of Theorem 3.2.6 with $Y = \mathbf{K}$ that the strong dual space X' is a Banach space with the norm

$$\|f\| = \sup_{\substack{x \in X \\ \|x\| \leq 1}} |f(x)|.$$

Corollary 3.2.17 tells us that the dual space X' separates points of X, that is, for two arbitrary distinct points x_1, x_2 of X there exists a functional $f \in X'$ such that $f(x_1) \neq f(x_2)$.

Annihilators

Let A be a subset of a normed linear space X. An element f of the dual space X' is called an *annihilator* of A if it satisfies

$$f(x) = 0 \qquad \text{for all } x \in A.$$

We let

$$A^0 = \{f \in X';\ f(x) = 0 \text{ for all } x \in A\}$$

be the set of all annihilators of A. This is not a one way proposition. If B is a subset of X', we let

$$^0B = \{x \in X;\ f(x) = 0 \text{ for all } f \in B\}$$

be the set of all annihilators of B.

Here are some basic properties of annihilators:

1. The sets A^0 and ^0B are closed linear subspaces of X and X', respectively.
2. If M is a closed linear subspace of X, then $^0(M^0) = M$.
3. If A is a subset of X and M is the closure of the subspace spanned by A, then $M^0 = A^0$ and $M = {}^0(A^0)$.

Dual Spaces of Normed Factor Spaces

Let M be a closed linear subspace of a normed linear space X. Then each element f of M^0 defines a bounded linear functional $\tilde{f}$ on the normed factor space X/M by

$$\tilde{f}(\tilde{x}) = f(x), \qquad \tilde{x} \in X/M.$$

In fact, the value $f(x)$ on the right-hand side does not depend on the choice of a representative x of the equivalence class $\tilde{x}$, and we have

$$\|\tilde{f}\| = \|f\|.$$

Further it is easy to see that the mapping $\pi\colon f \mapsto \tilde{f}$ of M^0 into $(X/M)'$ is linear and surjective; hence we have:

3.2.18 Theorem. *The strong dual space $(X/M)'$ of the factor space X/M can be identified with the space M^0 of all annihilators of M by the linear isometry π.*

Bidual Spaces

Each element x of a normed linear space X defines a bounded linear functional Jx on the strong dual space X' by

$$Jx(f) = f(x), \qquad f \in X'. \tag{6}$$

Then Corollary 3.2.17 tells us that

$$\|Jx\| = \sup_{\substack{f \in X' \\ \|f\| \leq 1}} |Jx(f)| = \|x\|,$$

so that the mapping J is a linear isometry of X into the strong dual space $(X')'$ of X'. The space $(X')'$ is called the *strong bidual* (or *second dual*) *space of* X.

Summing up, we have:

3.2.19 Theorem. *A normed linear space X can be embedded into its strong bidual space $(X')'$ by the linear isometry J defined by formula* (6).

If the mapping J is surjective, that is, if $X = (X')'$, then we say that X is *reflexive*.

Weak Convergence

A sequence $\{x_n\}$ in a normed linear space X is said to be *weakly convergent* if the limit $\lim_{n\to\infty} f(x_n)$ exists and is finite for each f in the dual space X' of X. A sequence $\{x_n\}$ in X is said to *converge weakly* to an element x of X if $\lim_{n\to\infty} f(x_n) = f(x)$ for every $f \in X'$; we then write $\text{w-}\lim_{n\to\infty} x_n = x$, or simply $x_n \to x$ weakly. Since the space X' separates points of X, the limit x is uniquely determined. Theorem 3.2.19 tells us that X may be considered as a linear subspace of its bidual space $(X')'$; hence the weak topology on X is just the simple convergence topology on the bidual space $(X')' = L(X', \mathbf{K})$.

For weakly convergent sequences, we have:

3.2.20 Theorem

(i) $\text{s-}\lim_{n\to\infty} x_n = x$ *implies* $\text{w-}\lim_{n\to\infty} x_n = x$.

(ii) *A weakly convergent sequence* $\{x_n\}$ *is bounded*

$$\sup_n \|x_n\| < +\infty.$$

Furthermore, if $\text{w-}\lim_{n\to\infty} x_n = x$, *then the sequence* $\{x_n\}$ *is bounded and we have*

$$\|x\| \leq \liminf_{n\to\infty} \|x_n\|.$$

Part (ii) of Theorem 3.2.20 has a converse:

3.2.21 Theorem. *A sequence* $\{x_n\}$ *in* X *converges weakly to an element* x *of* X *if the following two conditions are satisfied*:

(a) *The sequence* $\{x_n\}$ *is bounded.*

(b) $\lim_{n\to\infty} f(x_n) = f(x)$ *for every* f *in some strongly dense subset of* X'.

Weak Convergence*

A sequence $\{f_n\}$ in the dual space X' is said to be *weakly* convergent* if the limit $\lim_{n\to\infty} f_n(x)$ exists and is finite for every $x \in X$. A sequence $\{f_n\}$ in X' is said to *converge weakly** to an element f of X' if $\lim_{n\to\infty} f_n(x) = f(x)$ for every $x \in X$; we then write $\text{w*-}\lim_{n\to\infty} f_n = f$, or simply $f_n \to f$ weakly*. The weak* topology on X' is just the simple topology on the space $X' = L(X, \mathbf{K})$.

We have the following analogue of Theorem 3.2.20:

3.2.22 Theorem

(i) s-$\lim_{n\to\infty} f_n = f$ *implies* w*-$\lim_{n\to\infty} f_n = f$.

(ii) *If X is a Banach space, then a weakly* convergent sequence $\{f_n\}$ in X' converges weakly* to an element f of X' and we have*

$$\|f\| \leq \liminf_{n\to\infty} \|f_n\|.$$

One of the important consequences of Theorem 3.2.22 is the *sequential weak* compactness* of bounded sets:

3.2.23 Theorem. *Let X be a separable Banach space. Then every bounded sequence in the strong dual space X' has a subsequence which converges weakly* to an element of X'.*

Transposes

Let T be a linear operator from X into Y with domain $D(T)$ everywhere dense in X. Such operators are called *densely defined operators.*

Each element g of the dual space Y' of Y defines a linear functional G on $D(T)$ by the formula

$$G(x) = g(Tx), \qquad x \in D(T).$$

If this functional G is continuous everywhere on $D(T)$, it follows from an application of Theorem 3.2.3 that G can be extended uniquely to a continuous linear functional g' on $\overline{D(T)} = X$, that is, there exists a unique element g' of the dual space X' of X which is an extension of G. So we let

$D(T')$ = the totality of those $g \in Y'$ such that the mapping $x \mapsto g(Tx)$ is continuous everywhere on $D(T)$,

and define

$$T'g = g'.$$

In other words, the mapping T' is a linear operator from Y' into X', with domain $D(T')$, such that

$$g(Tx) = T'g(x), \qquad x \in D(T), \qquad g \in D(T'). \tag{7}$$

The operator T' is called the *transpose* of T.

Frequently we write $\langle f, x\rangle$ or $\langle x, f\rangle$ for the value $f(x)$ of a functional f at a point x; for example, we write formula (7) as

$$\langle Tx, g\rangle = \langle x, T'g\rangle, \qquad x \in D(T),\ g \in D(T'). \tag{7'}$$

The next theorem states that the continuity of operators is inherited by the transposes.

3.2.24 Theorem. *Let X, Y be normed linear spaces and X', Y' their strong dual spaces, respectively. If T is a bounded linear operator on X into Y, then its transpose T' is a bounded linear operator on Y' into X', and we have $\|T'\| = \|T\|$.*

3.3. The Riesz Representation Theorem

One of the fundamental theorems in analysis is the Riesz representation theorem which describes an intimate relationship between measures and linear functionals.

The Space of Continuous Functions

Let K be a *compact* metric space, and $C(K)$ the collection of real-valued continuous functions on K. We define in the set $C(K)$ addition and scalar multiplication of functions in the usual way:

$$\begin{cases} (f+g)(x) = f(x) + g(x), & x \in K, \\ (\alpha f)(x) = \alpha f(x), & \alpha \in \mathbf{R},\ x \in K. \end{cases}$$

Then $C(K)$ is a real linear space. Further, it is a Banach space with the supremum (maximum) norm

$$\|f\| = \sup_{x \in K} |f(x)|.$$

The Space of Signed Measures

Let $(X, \mathscr{M})$ be a measurable space. If μ and λ are signed measures on $\mathscr{M}$, we define the sum $\mu + \lambda$ and the scalar multiple $\alpha\mu$ ($\alpha \in \mathbf{R}$) as follows:

$$\begin{cases} (\mu + \lambda)(A) = \mu(A) + \lambda(A), & A \in \mathscr{M}, \\ (\alpha\mu)(A) = \alpha\mu(A), & A \in \mathscr{M}. \end{cases}$$

Then it is clear that $\mu + \lambda$ and $\alpha\mu$ are signed measures. Further, one can verify that the quantity

$$\|\mu\| = \text{the total variation } |\mu|(X) \text{ of } \mu \tag{1}$$

satisfies axioms $(N1)$, $(N2)$ and $(N3)$ of a norm. Thus the totality of signed measures on $\mathscr{M}$ is a normed linear space with the norm $\|\mu\|$ defined by (1).

Now we define

$$\begin{cases} \mu^+ = \dfrac{1}{2}(|\mu| + \mu), \\ \mu^- = \dfrac{1}{2}(|\mu| - \mu). \end{cases}$$

Then it follows from inequality (1.18.1) that both μ^+ and μ^- are finite non-negative measures on $\mathscr{M}$. Also we have the *Jordan decomposition* of μ:

$$\mu = \mu^+ - \mu^-.$$

The measures μ^+ and μ^- are called the *positive* and *negative variation measures* of μ respectively.

The Riesz Representation Theorem

First we characterize the non-negative linear functionals on $C(K)$. A linear functional T on $C(K)$ is said to be *non-negative* if it satisfies

$$f \in C(K), \qquad f \geq 0 \text{ on } K \Rightarrow Tf \geq 0.$$

We remark that a non-negative linear functional T is bounded on $C(K)$ with norm

$$\|T\| = \sup_{\substack{f \in C(K) \\ \|f\| \leq 1}} |Tf| = T1.$$

Then we have:

3.3.1 Theorem. *To each non-negative linear functional T on $C(K)$, there corresponds a unique finite non-negative Borel measure μ on K such that*

$$Tf = \int_K f(x)\, d\mu(x), \qquad f \in C(K), \tag{2}$$

and we have

$$\|T\| = \mu(K). \tag{3}$$

Conversely, every finite non-negative Borel measure μ *on* K *defines a non-negative linear functional* T *on* $C(K)$ *through formula* (2), *and relation* (3) *holds.*

3.3.2 Remark. It is easy to see that every open set in a compact metric space is a σ-compact. Thus we find from Theorem 1.18.2 that every finite non-negative Borel measure μ on K is regular.

Now we characterize the space of all bounded linear functionals on $C(K)$, that is, the dual space $C(K)'$ of $C(K)$. Recall that the dual space $C(K)'$ is a Banach space with the norm

$$\|T\| = \sup_{\substack{f \in C(K) \\ \|f\| \leq 1}} |Tf|.$$

The Riesz representation theorem reads as follows:

3.3.3 Theorem (Riesz). *To each* $T \in C(K)'$, *there corresponds a unique real Borel measure* μ *on* K *such that*

$$Tf = \int_K f(x)\, d\mu(x), \qquad f \in C(K), \tag{2}$$

and we have

$$\|T\| = \text{the total variation } |\mu|(K) \text{ of } \mu. \tag{4}$$

Conversely, every real Borel measure μ *on* K *defines a bounded linear functional* $T \in C(K)'$ *through formula* (2), *and relation* (4) *holds.*

3.3.4 Remark. In view of Theorem 1.18.2, we obtain that the positive and negative variation measures μ^+, μ^- of a real Borel measure μ on K are both regular.

Note that the space $\mu(K)$ of all real Borel measures μ on K is a normed linear space with the norm

$$\|\mu\| = \text{the total variation } |\mu|(K) \text{ of } \mu. \tag{5}$$

Therefore we can restate Theorem 3.3.3 as follows:

3.3.5 Theorem. *The dual space* $C(K)'$ *of* $C(K)$ *can be identified with the space* $\mu(K)$ *of all real Borel measures on* K *normed by formula* (5).

Weak Convergence of Measures

Let K be a compact metric space, and let $C(K)$ be the Banach space of real-valued continuous functions on K with the supremum (maximum) norm

$$\|f\| = \sup_{x \in K} |f(x)|.$$

A sequence $\{\mu_n\}$ of real Borel measures on K is said to *converge weakly* to a real Borel measure μ on K if we have

$$\lim_{n \to \infty} \int_K f(x)\, d\mu_n(x) = \int_K f(x)\, d\mu(x), \qquad f \in C(K). \tag{6}$$

Theorem 3.3.5 tells us that the space $\mu(K)$ of all real Borel measures on K normed by formula (5) can be identified with the strong dual space $C(K)'$ of $C(K)$. Thus the weak convergence (6) of real Borel measures is just the weak* convergence of $C(K)'$.

One more result is important when studying the weak* convergence of measures:

3.3.6 Theorem. *The Banach space $C(K)$ is separable, that is, it contains a countable, dense subset.*

The next theorem is one of the fundamental theorems in measure theory.

3.3.7 Theorem. *Every sequence $\{\mu_n\}$ of real Borel measures on K satisfying*

$$\sup_n |\mu_n|(K) < +\infty \tag{7}$$

has a subsequence which converges weakly to a real Borel measure μ on K.

Furthermore, if the measures μ_n are all non-negative, then the measure μ is also non-negative.

Proof. By virtue of Theorem 3.3.6, we can apply Theorem 3.2.23 with $X = C(K)$ to obtain the first assertion, since condition (7) implies the boundedness of $\{\|\mu_n\|\}$. The second assertion is an immediate consequence of the first assertion of Theorem 3.3.1. ■

3.4. Closed Operators

Let X and Y be normed linear spaces over the same scalar field. Let T be a linear operator from X into Y with domain $D(T)$. The *graph* $G(T)$ of T is the set

$$G(T) = \{\{x, Tx\}; x \in D(T)\}$$

in the product space $X \times Y$. Note that $G(T)$ is a linear subspace of $X \times Y$. We say that T is *closed* if its graph $G(T)$ is closed in $X \times Y$. This is equivalent to saying that:

$$\{x_n\} \subset D(T), \qquad x_n \to x \text{ in } X, \qquad Tx_n \to y \text{ in } Y \Rightarrow x \in D(T), \qquad Tx = y.$$

In particular, if T is continuous and its domain $D(T)$ is closed in X, then T is a closed linear operator.

We remark that if T is a closed linear operator which is also injective, then its inverse T^{-1} is a closed linear operator. In fact, this follows from the fact that the mapping $\{x, y\} \mapsto \{y, x\}$ is a homeomorphism of $X \times Y$ onto $Y \times X$.

A linear operator T is said to be *closable* if the closure $\overline{G(T)}$ in $X \times Y$ of $G(T)$ is the graph of a linear operator, say $\bar{T}$, that is, if

$$\overline{G(T)} = G(\bar{T}).$$

A linear operator is called a *closed extension* of T if it is a closed linear operator which is also an extension of T. It is easy to see that if T is closable, then every closed extension of T is an extension of $\bar{T}$. Thus the operator $\bar{T}$ is called the *minimal closed extension* of T.

The next theorem gives a necessary and sufficient condition for a linear operator to be closable.

3.4.1 Theorem. *A linear operator T from X into Y with domain $D(T)$ is closable if and only if the following condition is satisfied:*

$$\{x_n\} \subset D(T), \qquad x_n \to 0 \textit{ in } X, \qquad Tx_n \to y \textit{ in } Y \Rightarrow y = 0.$$

Now we state two important theorems concerning closed linear operators:

3.4.2 Theorem (Banach's open mapping theorem). *Let X and Y be Banach spaces. Then every closed linear operator on X onto Y is open, that is, it maps every open set in X onto an open set in Y.*

3.4.3 Theorem (Banach's closed graph theorem). *Let X and Y be Banach spaces. Then every closed linear operator on X into Y is continuous.*

3.4.4 Corollary. *Let X and Y be Banach spaces. If T is a continuous, one-to-one linear operator on X onto Y, then its inverse T^{-1} is also continuous; hence T is an isomorphism.*

In fact, the inverse T^{-1} is a closed linear operator, so that Theorem 3.4.3 applies.

We give useful characterizations of closed linear operators with closed range:

3.4.5 Theorem. *Let X and Y be Banach spaces, and T a closed linear operator from X into Y with domain $D(T)$. Then the range $R(T)$ of T is closed in Y if and only if there exists a constant $C > 0$ such that*

$$\operatorname{dist}(x, N(T)) \leq C\|Tx\|, \qquad x \in D(T).$$

Here $\operatorname{dist}(x, N(T)) = \inf_{z \in N(T)} \|x - z\|$ is the distance from x to the null space $N(T)$ of T.

3.4.6 Theorem (Banach's closed range theorem). *Let X and Y be Banach spaces, and T a densely defined, closed linear operator from X into Y. Then the following four conditions are equivalent:*

(i) *The range $R(T)$ of T is closed in Y.*
(ii) *The range $R(T')$ of the transpose T' is closed in X'.*
(iii) $R(T) = {}^0N(T') = \{y \in Y; \langle y, y' \rangle = 0$ *for all* $y' \in N(T')\}$.
(iv) $R(T') = N(T)^0 = \{x' \in X'; \langle x', x \rangle = 0$ *for all* $x \in N(T)\}$.

3.5. Complemented Subspaces

Let X be a linear space. Two linear subspaces M and N of X are said to be *algebraic complements* in X if X is the direct sum of M and N, that is, if $X = M \dotplus N$.

Algebraic complements M and N in a normed linear space X are said to be *topological complements* in X if the addition mapping $\{y, z\} \mapsto y + z$ is an isomorphism of $M \times N$ onto X. We then write

$$X = M \oplus N.$$

As an application of Corollary 3.4.4, we obtain:

3.5.1 Theorem. *Let X be a Banach space. If M and N are closed algebraic complements in X, then they are topological complements.*

A closed linear subspace of a normed linear space X is said to be *complemented* in X if it has a topological complement. By Theorem 3.5.1, this is equivalent in Banach spaces to the existence of a closed algebraic complement.

The next theorem gives two criteria for a closed subspace to be complemented.

3.5.2 Theorem. *Let X be a Banach space and M a closed subspace of X. If M has either finite dimension or finite codimension, then it is complemented in X.*

3.6. Compact Operators

Let X and Y be normed linear spaces over the same scalar field $\mathbf{K}$. A linear operator T on X into Y is said to be *compact* or *completely continuous* if it maps every bounded subset of X onto a relatively compact subset of Y, that is, if the closure of $T(B)$ is compact in Y for every bounded subset B of X. This is equivalent to saying that, for every bounded sequence $\{x_n\}$ in X, the sequence $\{Tx_n\}$ has a subsequence which converges in Y.

We list some facts which follow at once:

1. Every compact operator is bounded.

In fact, a compact operator maps the unit sphere onto a bounded set.

2. Every bounded linear operator with finite dimensional range is compact.

This is an immediate consequence of Corollary 3.2.14.

3. No isomorphism between infinite dimensional spaces is compact.

This follows from Theorem 3.2.15.

4. A linear combination of compact operators is compact.

5. The product of a compact operator with a bounded operator is compact.

The next theorem states that if Y is a Banach space, then the compact operators on X into Y form a closed subspace of $L(X, Y)$.

3.6.1 Theorem. *Let X be a normed linear space and Y a Banach space. If $\{T_n\}$ is a sequence of compact linear operators which converges to an operator T in the space $L(X, Y)$ with the uniform topology, then T is compact.*

As for the transposes of compact operators, we have:

3.6.2 Theorem. *Let X and Y be normed linear spaces. If T is a compact linear operator on X into Y, then its transpose T' is a compact linear operator on Y' into X'.*

The Riesz–Schauder Theory

Now we state the most interesting results on compact linear operators, which are essentially due to F. Riesz in the Hilbert space setting. The results were extended to Banach spaces by Schauder.

3.6.3 Theorem. *Let X be a Banach space and T a compact linear operator on X into itself. Set*

$$S = I - T.$$

Then we have:

(i) *The null space $N(S)$ of S is finite dimensional and the range $R(S)$ of S is closed in X.*
(ii) *The null space $N(S')$ of the transpose S' is finite dimensional and the range $R(S')$ of S' is closed in X'.*
(iii) $\dim N(S) = \dim N(S')$.

The next result is an extension of the theory of linear mappings in finite dimensional linear spaces.

3.6.4 Corollary (the Fredholm alternative). *Let T be a compact linear operator on a Banach space X into itself. If $S = I - T$ is either one-to-one or onto, then it is an isomorphism of X onto itself.*

Let T be a bounded linear operator on X into itself. The *resolvent set* of T, denoted $\rho(T)$, is defined to be the set of scalars $\lambda \in \mathbf{K}$ such that $\lambda I - T$ is an isomorphism of X onto itself. In this case, the inverse $(\lambda I - T)^{-1}$ is called the *resolvent* of T. The complement of $\rho(T)$, that is, the set of scalars $\lambda \in \mathbf{K}$ such that $\lambda I - T$ is not an isomorphism of X onto itself, is called the *spectrum* of T and is denoted by $\sigma(T)$. The set $\sigma_p(T)$ of scalars $\lambda \in \mathbf{K}$ such that $\lambda I - T$ is not one-to-one forms a subset of $\sigma(T)$, and is called the *point spectrum* of T. A scalar $\lambda \in \mathbf{K}$ belongs to $\sigma_p(T)$ if and only if there exists a non-zero element $x \in X$ such that $Tx = \lambda x$. In this case, λ is called an *eigenvalue* of T and x an *eigenvector* of T corresponding to λ. Also the null space $N(\lambda I - T)$ of $\lambda I - T$ is called the *eigenspace* of T corresponding to λ, and the dimension of $N(\lambda I - T)$ is called the *multiplicity* of λ.

Using C. Neumann's series (Theorem 3.2.8), we find that the resolvent set $\rho(T)$ is open in $\mathbf{K}$ and

$$\{\lambda \in \mathbf{K};\, |\lambda| > \|T\|\} \subset \rho(T).$$

Hence the spectrum $\sigma(T) = \mathbf{K} \setminus \rho(T)$ is closed and bounded in $\mathbf{K}$.

If T is a compact operator and λ is a non-zero element of $\sigma(T)$, then applying Corollary 3.6.4 to the operator $\lambda^{-1}T$, we obtain that $\lambda I - T$ is not one-to-one, that is, $\lambda \in \sigma_p(T)$. Also note that if X is infinite dimensional, then T is not an isomorphism of X onto itself; hence $0 \in \sigma(T)$. Therefore the scalar field $\mathbf{K}$ can be decomposed as follows:

$$\mathbf{K} = (\sigma_p(T) \cup \{0\}) \cup \rho(T).$$

We can say rather more about the spectrum $\sigma(T)$:

3.6.5 Theorem (Riesz–Schauder). *Let T be a compact linear operator on a Banach space X into itself. Then we have:*

(i) *The spectrum $\sigma(T)$ of T is either a finite set or a countable set accumulating only at 0; and every non-zero element of $\sigma(T)$ is an eigenvalue of T.*
(ii) $\dim N(\lambda I - T) = \dim N(\bar{\lambda} I - T') < +\infty$ *for all* $\lambda \neq 0$.
(iii) *Let $\lambda \neq 0$. The equation $(\lambda I - T)x = y$ has a solution if and only if y is orthogonal to the space $N(\bar{\lambda} I - T')$. Similarly, the equation $(\bar{\lambda} I - T')z = w$ has a solution if and only if w is orthogonal to the space $N(\lambda I - T)$. Moreover, the operator $\lambda I - T$ is onto if and only if it is one-to-one.*

3.7. Fredholm Operators

Throughout this section, the letters X, Y, Z denote Banach spaces over the same scalar field.

A linear operator T from X into Y is called a *Fredholm operator* if the following five conditions are satisfied:

(i) The domain $D(T)$ of T is everywhere dense in X.
(ii) T is a closed operator.
(iii) The null space $N(T)$ of T has finite dimension.
(iv) The range $R(T)$ of T is closed in Y.
(v) The range $R(T)$ has finite codimension.

Then the *index* of T is defined by

$$\text{ind } T = \dim N(T) - \text{codim } R(T).$$

Combining Theorem 3.6.3 and Theorem 3.4.6, we obtain that if $X = Y$ and T is compact, then the operator $I - T$ is a Fredholm operator and $\text{ind}(I - T) = 0$.

We give a characterization of Fredholm operators. First we have:

3.7.1 Theorem. *If T is a Fredholm operator from X into Y with domain $D(T)$, then there exist a bounded linear operator $S: Y \to X$ and compact linear operators $P: X \to X$, $Q: Y \to Y$ such that:*

(a) *$ST = I - P$ on $D(T)$,*
(b) *$TS = I - Q$ on Y.*

Furthermore, we have $R(P) = N(T)$ and $\dim R(Q) = \text{codim } R(T)$.

Theorem 3.7.1 has a converse:

3.7.2 Theorem. *Let T be a closed linear operator from X into Y, with domain $D(T)$ everywhere dense in X. Suppose that there exist bounded linear operators $S_1: Y \to X$, $S_2: Y \to X$ and compact linear operators $K_1: X \to X$, $K_2: Y \to Y$ such that:*

(a) *$S_1 T = I - K_1$ on $D(T)$,*
(b) *$TS_2 = I - K_2$ on Y.*

Then T is a Fredholm operator.

Now we state important properties of Fredholm operators:

3.7.3 Theorem. *If T is a Fredholm operator from X into Y and S is a Fredholm operator from Y into Z, then the product ST is a Fredholm operator from X into Z and we have*

$$\text{ind}(ST) = \text{ind } S + \text{ind } T.$$

3.7.4 Theorem. *If T is a Fredholm operator from X into Y and K is a compact linear operator on X into Y, then the sum $T + K$ is a Fredholm operator and we have*

$$\text{ind}(T + K) = \text{ind } T.$$

As for the transposes of Fredholm operators, we have:

3.7.5 Theorem. *If T is a Fredholm operator from X into Y and Y is reflexive, then the transpose T' of T is a Fredholm operator from Y' into X' and we have*

$$\text{ind } T' = -\text{ ind } T.$$

We give a very useful criterion for conditions (iii) and (iv) made for Fredholm operators:

3.7.6 Theorem (Peetre). *Let X, Y, Z be Banach spaces such that $X \subset Z$ is a compact injection, and let T be a closed linear operator from X into Y with domain $D(T)$. Then the following two conditions are equivalent:*

(i) *The null space $N(T)$ of T has finite dimension and the range $R(T)$ of T is closed in Y.*
(ii) *There is a constant $C > 0$ such that*

$$\|x\|_X \leq C(\|Tx\|_Y + \|x\|_Z), \qquad x \in D(T). \tag{1}$$

Proof. (i) $\Rightarrow$ (ii): By Theorem 3.5.2, the null space $N(T)$ has a closed topological complement X_0:

$$X = N(T) \oplus X_0. \tag{2}$$

This gives that

$$D(T) = N(T) \oplus (D(T) \cap X_0).$$

Hence every element x of $D(T)$ can be written in the form

$$x = x_0 + x_1, \qquad x_0 \in D(T) \cap X_0, \qquad x_1 \in N(T).$$

Since the range $R(T)$ is closed in Y, it then follows from an application of Theorem 3.4.5 that

$$\|x_0\|_X \leq C\|Tx_0\|_Y. \tag{3}$$

Here and in the following the letter C denotes a generic positive constant independent of x.

On the other hand, Theorem 3.2.11 tells us that all norms on a finite dimensional linear space are equivalent. This gives that

$$\|x_1\|_X \leq C\|x_1\|_Z. \tag{4}$$

But, since the injection $X \to Z$ is compact and hence is continuous, we have

$$\begin{aligned}\|x_1\|_Z &\leq \|x\|_Z + \|x_0\|_Z \\ &\leq \|x\|_Z + C\|x_0\|_X. \end{aligned} \tag{5}$$

Thus it follows from inequalities (4) and (5) that

$$\|x_1\|_X \leq C(\|x\|_Z + \|x_0\|_X). \tag{6}$$

Therefore, combining inequalities (3) and (6), we obtain inequality (1),

$$\begin{aligned}\|x\|_X &\leq \|x_0\|_X + \|x_1\|_X \\ &\leq C(\|Tx\|_Y + \|x\|_Z),\end{aligned}$$

since $Tx_0 = Tx$.

(ii) $\Rightarrow$ (i): By inequality (1), we have

$$\|x\|_X \leq C\|x\|_Z, \qquad x \in N(T). \tag{7}$$

But, the null space $N(T)$ is closed in X, and so it is a Banach space. Since the injection $X \to Z$ is compact, it follows from inequality (7) that the closed unit ball $\{x \in N(T); \|x\|_X \leq 1\}$ of $N(T)$ is compact. Therefore we obtain from Theorem 3.2.15 that

$$\dim N(T) < +\infty.$$

Let X_0 be a closed topological complement of $N(T)$ as in the decomposition (2).

To prove the closedness of $R(T)$, by virtue of Theorem 3.4.5, it suffices to show that

$$\|x\|_X \leq C\|Tx\|_Y, \qquad x \in D(T) \cap X_0.$$

Assume to the contrary that:

For every $n \in \mathbf{N}$ there is an element x_n of $D(T) \cap X_0$ such that

$$\|x_n\|_X > n\|Tx_n\|_Y.$$

We let

$$x_n' = \frac{x_n}{\|x_n\|_X}.$$

Then we have

$$\begin{cases} x'_n \in D(T) \cap X_0, \qquad \|x'_n\|_X = 1, & (8) \\ \|Tx'_n\|_Y < \dfrac{1}{n}. & (9) \end{cases}$$

Since the injection $X \to Z$ is compact, by passing to a subsequence one may assume that the sequence $\{x'_n\}$ is a Cauchy sequence in Z. Then, in view of (9), it follows from inequality (1) that the sequence $\{x'_n\}$ is a Cauchy sequence in X, and hence converges to an element x' of X. Since the operator T is closed, we obtain that

$$x' \in D(T), \qquad Tx' = 0,$$

so that

$$x' \in N(T).$$

On the other hand, in view of (8), it follows that

$$x' \in X_0,$$

and further that

$$\|x'\|_X = \lim_n \|x'_n\|_X = 1.$$

This is a contradiction, since we have

$$x' \in N(T) \cap X_0 = \{0\}. \qquad \blacksquare$$

3.8. Hilbert Spaces

A complex (or real) linear space X is called a *pre-Hilbert space* or *inner product space* if, to each ordered pair of elements x and y of X, there is associated a complex (or real) number (x, y) in such a way that:

(*I*1) $(y, x) = \overline{(x, y)}$.
(*I*2) $(\alpha x, y) = \alpha(x, y)$, $\alpha \in \mathbf{C}$ (or $\alpha \in \mathbf{R}$).
(*I*3) $(x + y, z) = (x, z) + (y, z)$.
(*I*4) $(x, x) \geq 0$; $(x, x) = 0$ if and only if $x = 0$.

Here $\overline{(x, y)}$ denotes the complex conjugate of (x, y). In the real case condition (*I*1) becomes simply $(y, x) = (x, y)$. The number (x, y) is called the *inner product* or *scalar product* of x and y.

The following are immediate consequences of conditions ($I1$), ($I2$) and ($I3$):

1. $(\alpha x + \beta y, z) = \alpha(x, z) + \beta(y, z)$, $\alpha, \beta \in \mathbf{C}$.
2. $(x, \alpha y + \beta z) = \bar{\alpha}(x, y) + \bar{\beta}(x, z)$, $\alpha, \beta \in \mathbf{C}$.

These properties 1 and 2 are frequently called *sesquilinearity*. In the real case they reduce to bilinearity.

We list some basic properties of the inner product:

1) The *Schwarz inequality* holds:

$$|(x, y)|^2 \leq (x, x)(y, y).$$

Here the equality holds if and only if x and y are linearly dependent.

2) The quantity

$$\|x\| = (x, x)^{1/2} \qquad \text{(the non-negative square root)}$$

satisfies axioms ($N1$), ($N2$) and ($N3$) of a norm; hence a pre-Hilbert space is a normed linear space with the norm $\|x\| = (x, x)^{1/2}$.

3) The inner product (x, y) is a continuous function of x and y:

$$\|x_n - x\| \to 0, \qquad \|y_n - y\| \to 0 \Rightarrow (x_n, y_n) \to (x, y).$$

4) The *parallelogram law* holds:

$$\|x + y\|^2 + \|x - y\|^2 = 2(\|x\|^2 + \|y\|^2). \tag{1}$$

Conversely, suppose that X is a normed linear space whose norm satisfies condition (1). We let

$$(x, y) = \frac{1}{4}(\|x + y\|^2 - \|x - y\|^2)$$

if X is a real normed linear space, and let

$$(x, y) = \frac{1}{4}(\|x + y\|^2 - \|x - y\|^2 + i\|x + iy\|^2 - i\|x - iy\|^2)$$

if X is a complex normed linear space. Then it is easy to verify that the number (x, y) satisfies axioms ($I1$) through ($I4$) of an inner product; hence X is a pre-Hilbert space.

A pre-Hilbert space is called a *Hilbert space* if it is complete with respect to the norm derived from the inner product.

If X and Y are pre-Hilbert spaces over the same scalar field, then the product space $X \times Y$ is made into a pre-Hilbert space by the inner product

$$(\{x_1, y_1\}, \{x_2, y_2\}) = (x_1, x_2) + (y_1, y_2).$$

Further, if X and Y are Hilbert spaces, then so is $X \times Y$.

Orthogonality

Let X be a pre-Hilbert space. Two elements x, y of X are said to be *orthogonal* if $(x, y) = 0$; we then write $x \perp y$. We remark that

$$\begin{cases} x \perp y \Leftrightarrow y \perp x. \\ x \perp x \Leftrightarrow x = 0. \end{cases}$$

If A is a subset of X, we let

$$A^{\perp} = \{x \in X; (x, y) = 0 \text{ for all } y \in A\}.$$

In other words, $A^{\perp}$ is the set of all those elements of X which are orthogonal to every element of A.

We list some facts which follow at once:

1. The set $A^{\perp}$ is a linear subspace of X.
2. $A \subset B \Rightarrow B^{\perp} \subset A^{\perp}$.
3. $A \cap A^{\perp} = \{0\}$.
4. The set $A^{\perp}$ is closed.
5. $A^{\perp} = \bar{A}^{\perp} = \overline{[A]}^{\perp}$ where $\bar{A}$ is the closure of A and $[A]$ is the space spanned by A, that is, the space of finite linear combinations of elements of A.

Facts 4 and 5 follow from the continuity of the inner product.

The Closest-Point Theorem and Applications

3.8.1 Theorem (the closest-point theorem). *Let X be a Hilbert space and A a closed, convex subset of X. If x is a point not in A, then there is a unique point a in A such that*

$$\|x - a\| = \operatorname{dist}(x, A).$$

Theorem 3.8.1 is proved by using the parallelogram law.

One of the consequences of Theorem 3.8.1 is that every closed linear subspace of a Hilbert space is complemented:

3.8.2 Theorem. *Let M be a closed linear subspace of a Hilbert space X. Then every element x of X can be decomposed uniquely in the form*

$$x = y + z, \qquad y \in M, \qquad z \in M^{\perp}, \tag{2}$$

Moreover, the mapping $x \mapsto \{y, z\}$ is an isomorphism of X onto $M \times M^{\perp}$.

We shall write the decomposition (2) as

$$X = M \oplus M^{\perp}, \tag{2'}$$

emphasizing that the mapping $x \mapsto \{y, z\}$ is an isomorphism of X onto $M \times M^{\perp}$. The space $M^{\perp}$ is called the *orthogonal complement* of M.

3.8.3 Corollary. *If M is a closed linear subspace of a Hilbert space X, then $M^{\perp\perp} = (M^{\perp})^{\perp} = M$. Furthermore, if A is a subset of X, then $A^{\perp\perp} = \overline{[A]}$.*

With the above notation (2), we define a mapping P_M of X into M by

$$P_M x = y.$$

Since the decomposition (2) is unique, it follows that P_M is linear.

Further we easily obtain:

3.8.4 Theorem. *The linear operator P_M enjoys the following properties:*

(i) $P_M^2 = P_M$ (*idempotent property*).
(ii) $(P_M x, x') = (x, P_M x')$ (*symmetric property*).
(iii) $\|P_M\| \leq 1$.

The operator P_M is called the *orthogonal projection* onto M.

Similarly, we define a mapping $P_{M^{\perp}}$ of X into $M^{\perp}$ by

$$P_{M^{\perp}} x = z.$$

Then Corollary 3.8.3 tells us that $P_{M^{\perp}}$ is the orthogonal projection onto $M^{\perp}$. It is clear that

$$P_M + P_{M^{\perp}} = I.$$

Now we give an important characterization of bounded linear functionals on a Hilbert space:

3.8.5 Theorem (Riesz). *Every element y of a Hilbert space X defines a bounded linear functional $J_X y$ on X by*

$$J_X y(x) = (x, y), \qquad x \in X, \tag{3}$$

and we have

$$\|J_X y\| = \sup_{\substack{x \in X \\ \|x\| \leq 1}} |J_X y(x)| = \|y\|.$$

Conversely, for every bounded linear functional f on X, there exists a unique element y of X such that $f = J_X y$, that is,

$$f(x) = (x, y), \qquad x \in X,$$

and so

$$\|f\| = \|y\|.$$

In view of formula (3), it follows that the mapping J_X enjoys the following property:

$$J_X(\alpha y + \beta z) = \bar{\alpha} J_X y + \bar{\beta} J_X z, \qquad y, z \in X, \alpha, \beta \in \mathbf{C}.$$

We express this by saying that J_X is *conjugate linear* or *antilinear*. In the real case, J_X is linear.

Let X' be the strong dual space of a Hilbert space X, that is, the space of bounded linear functionals on X with the norm

$$\|f\| = \sup_{\substack{x \in X \\ \|x\| \leq 1}} |f(x)|.$$

Then Theorem 3.8.5 may be restated as follows:

There is a conjugate linear, norm-preserving isomorphism J_X of X onto X'. (4)

In this case, we say that X' is *antidual* to X.

Recall that a sequence $\{x_n\}$ in a normed linear space X is said to converge weakly to an element x of X if $f(x_n) \to f(x)$ for every $f \in X'$. Assertion (4) tells us that a sequence $\{x_n\}$ in a Hilbert space X converges weakly to an element x of X if and only if $(x_n, y) \to (x, y)$ for every $y \in X$.

Another important consequence of Theorem 3.8.5 is the *reflexivity* of Hilbert spaces:

3.8.6 Corollary. *Every Hilbert space can be identified with its strong bidual space.*

Orthonormal Sets

Let X be a pre-Hilbert space. A subset S of X is said to be *orthogonal* if every pair of distinct elements of S is orthogonal. Further, if each element of S has norm one, then S is said to be *orthonormal*. We remark that if S is an orthogonal set of non-zero elements, one can construct an orthonormal set from S by normalizing each element of S. If $\{x_1, \ldots, x_n\}$ is an orthonormal set and if $x = \sum_{i=i}^{n} \alpha_i x_i$, then we have

$$\|x\|^2 = \sum_{i=1}^{n} |\alpha_i|^2, \qquad \alpha_i = (x, x_i).$$

Therefore, every orthonormal set is linearly independent.

Now we state the Gram–Schmidt orthogonalization theorem:

3.8.7 Theorem (Gram–Schmidt). *Let $\{x_i\}_{i \in I}$ be a finite or countable infinite set of linearly independent vectors of X. Then we can construct an orthonormal set $\{u_i\}_{i \in I}$ such that for each $i \in I$:*

(a) *u_i is a linear combination of $\{x_1, \ldots, x_i\}$;*
(b) *x_i is a linear combination of $\{u_1, \ldots, u_i\}$.*

3.8.8 Corollary. *Every n-dimensional pre-Hilbert space over the scalar field* **K** *is isomorphic to the space* $\mathbf{K}^n$ *with the usual inner product.*

Let $\{u_\lambda\}_{\lambda \in \Lambda}$ be an orthonormal set of a pre-Hilbert space X. For each $x \in X$, we let

$$\hat{x}_\lambda = (x, u_\lambda), \qquad \lambda \in \Lambda.$$

The scalars $\hat{x}_\lambda$ are called the *Fourier coefficients* of x with respect to $\{u_\lambda\}$.

Then we have:

3.8.9 Theorem. *For each $x \in X$, the set of those $\lambda \in \Lambda$ such that $\hat{x}_\lambda \neq 0$ is at most countable. Further, the* Bessel inequality *holds*:

$$\sum_{\lambda \in \Lambda} |\hat{x}_\lambda|^2 \leq \|x\|^2.$$

An orthonormal set S of X is called a *complete orthonormal system* if it is not contained in a larger orthonormal set of X.

As for the existence of such systems, we have:

3.8.10 Theorem. *Let X be a Hilbert space having a non-zero element. Then, for every orthonormal set S in X, there exists a complete orthonormal system which contains S.*

The next theorem gives useful criteria for the completeness of orthonormal sets.

3.8.11 Theorem. *Let $S = \{u_\lambda\}_{\lambda \in \Lambda}$ be an orthonormal set in a Hilbert space X. Then the following five conditions are equivalent:*

(i) *The set S is complete.*

(ii) $S^\perp = \{0\}$.

(iii) *The space $[S]$ spanned by S is dense in X: $\overline{[S]} = X$.*

(iv) *For every $x \in X$, we have*

$$\|x\|^2 = \sum_{\lambda \in \Lambda} |\hat{x}_\lambda|^2. \tag{5}$$

(v) *For every $x \in X$, we have*

$$x = \sum_{\lambda \in \Lambda} \hat{x}_\lambda u_\lambda \text{ in } X. \tag{6}$$

Formula (5) is called the *Parseval identity* and formula (6) is called the *Fourier expansion* of x with respect to $\{u_\lambda\}$.

Adjoints

Throughout this subsection, the letters X, Y, Z denote Hilbert spaces over the same scalar field.

Let T be a linear operator from X into Y, with domain $D(T)$ everywhere dense in X. Each element y of Y defines a linear functional f on $D(T)$ by the formula

$$f(x) = (Tx, y), \qquad x \in D(T).$$

If this functional f is continuous everywhere on $D(T)$, applying Theorem 3.2.3, we obtain that f can be extended uniquely to a continuous linear functional $\tilde{f}$ on $\overline{D(T)} = X$. Therefore the Riesz theorem (Theorem 3.8.5) tells us that there exists a unique element $y*$ of X such that

$$\tilde{f}(x) = (x, y*), \qquad x \in X.$$

In particular, we have

$$(Tx, y) = f(x) = (x, y*), \qquad x \in D(T).$$

So we let

$D(T^*)$ = the totality of those $y \in Y$ such that the mapping $x \mapsto (Tx, y)$ is continuous everywhere on $D(T)$,

and define

$$T^*y = y*.$$

In other words, the mapping T^* is a linear operator from Y into X with domain $D(T^*)$, such that

$$(Tx, y) = (x, T^*y), \qquad x \in D(T), \qquad y \in D(T^*).$$

The operator T^* is called the *adjoint operator* or simply the *adjoint* of T.

We list some basic properties of adjoints:

1. The operator T^* is closed.
2. If $T \in L(X, Y)$, then $T^* \in L(Y, X)$ and $\|T^*\| = \|T\|$.
3. If $T, S \in L(X, Y)$, then $(\alpha T + \beta S)^* = \bar{\alpha}T^* + \bar{\beta}S^*$, $\alpha, \beta \in \mathbf{C}$.
4. If $T \in L(X, Y)$ and $S \in L(Y, Z)$, then $(ST)^* = T^*S^*$.

A densely defined linear operator T from X into itself is said to be *self-adjoint* if $T = T^*$. Note that every self-adjoint operator is closed.

As for the adjoints of closed operators, we have:

3.8.12 Theorem. *If T is a densely defined, closed linear operator from X into Y, then the adjoint T^* is a densely defined, closed linear operator from Y into X and we have $T^{**} = (T^*)^* = T$.*

3.8.13 Corollary. *If T is a densely defined, closable linear operator, then the adjoint T^* is densely defined and the operator T^{**} coincides with the minimal closed extension $\bar{T}$ of T.*

The Hilbert–Schmidt Theory

In the finite dimensional case, the spectral theorem for self-adjoint linear operators states that there exists an orthonormal basis consisting of eigenvectors. We shall generalize this theorem to the Hilbert space case.

Let T be a self-adjoint linear operator on a Hilbert space X into itself. We remark that any eigenvalue λ of T is real. In fact, if x is an eigenvector of T corresponding to λ, then $\lambda(x, x) = (Tx, x) = (x, Tx) = \bar{\lambda}(x, x)$, so that $\lambda = \bar{\lambda}$.

Furthermore, Theorem 3.6.5 tells us that if T is compact, then the non-zero eigenvalues of T forms a countable set accumulating only at 0; hence one can order them in a sequence $\{\lambda_j\}$ such that

$$|\lambda_1| \geq |\lambda_2| \geq \cdots \geq |\lambda_j| \geq |\lambda_{j+1}| \geq \cdots \to 0,$$

where each λ_j is repeated according to its multiplicity. For each λ_j, we let

$$V_{\lambda_j} = \text{the eigenspace } N(\lambda_j I - T) \text{ of } T \text{ corresponding to the eigenvalue } \lambda_j.$$

The eigenspaces V_{λ_j} are mutually orthogonal. In fact, if $x \in V_{\lambda_i}$ and $y \in V_{\lambda_j}$, then $\lambda_i(x, y) = (Tx, y) = (x, Ty) = \lambda_j(x, y)$, so that $(x, y) = 0$ if $\lambda_i \neq \lambda_j$. Therefore we can choose an orthonormal basis of V_{λ_j}, and combine these into an orthonormal set $\{x_j\}$ of eigenvectors of T such that $Tx_j = \lambda_j x_j$.

The spectral theorem extends to the Hilbert space case as follows:

3.8.14 (Hilbert–Schmidt). *Let T be a self-adjoint, compact linear operator on a Hilbert space X into itself. Then, for any $x \in X$, we have*

$$Tx = \sum_{j=1}^{\infty} \lambda_j(x, x_j)x_j = \text{s-}\lim_{n\to\infty} \sum_{j=1}^{n} \lambda_j(x, x_j)x_j.$$

In particular, if T is one-to-one, then we have

$$x = \sum_{j=1}^{\infty} (x, x_j)x_j = \text{s-}\lim_{n\to\infty} \sum_{j=1}^{n} (x, x_j)x_j,$$

that is, the family $\{x_j\}$ of eigenvectors is a complete orthonormal system of X.

3.9. Theory of Semigroups

Banach Space Valued Functions

Let X be a Banach space. A function $u(t)$, defined on an interval I with values in X, is said to be *strongly continuous* at a point t_0 of I if it satisfies

$$\lim_{t\to t_0} \|u(t) - u(t_0)\| = 0.$$

If $u(t)$ is strongly continuous at every point of I, then it is said to be strongly continuous on I. If $u(t)$ is strongly continuous on I, then the function $\|u(t)\|$ is continuous on I and also, for any f in the dual space X' of X, the function $f(u(t))$ is continuous on I.

If $u(t)$ is a strongly continuous function on I such that

$$\int_I \|u(t)\| \, dt < +\infty,$$

then the Riemann integral

$$\int_I u(t) \, dt$$

can be defined just as in the case of scalar valued functions; we then say that the function $u(t)$ is *strongly integrable* on I. By the triangle inequality, we have

$$\left\| \int_I u(t) \, dt \right\| \leq \int_I \|u(t)\| \, dt.$$

A function $u(t)$ defined on an open interval I is said to be *strongly differentiable* at a point t_0 of I if the limit

$$\lim_{h \to 0} \frac{u(t_0 + h) - u(t_0)}{h} \tag{1}$$

exists in X. The value of (1) is denoted by

$$\frac{du}{dt}(t_0) \qquad \text{or} \qquad u'(t_0).$$

If $u(t)$ is strongly differentiable at every point of I, then it is said to be strongly differentiable on I. A strongly differentiable function is strongly continuous.

Operator Valued Functions

Let X be a Banach space, and $L(X, X)$ the space of all bounded linear operators on X into itself. Theorem 3.2.6 tells us that $L(X, X)$ is a Banach space with the norm

$$\|T\| = \sup_{\substack{x \in X \\ \|x\| \leq 1}} \|Tx\|.$$

A function $T(t)$, defined on an interval I with values in the space $L(X, X)$, is said to be *strongly continuous* at a point t_0 of I if it satisfies

$$\lim_{t \to t_0} \|T(t)x - T(t_0)x\| = 0, \qquad x \in X.$$

We say that $T(t)$ is *norm continuous* at t_0 if it satisfies

$$\lim_{t \to t_0} \| T(t) - T(t_0) \| = 0.$$

If $T(t)$ is strongly (resp. norm) continuous at every point of I, then it is said to be strongly (resp. norm) continuous on I. A norm continuous function is strongly continuous.

A function $T(t)$ defined on an open interval I is said to be *strongly differentiable* at a point t_0 of I if there exists $S(t_0)$ in $L(X, X)$ such that

$$\lim_{h \to 0} \left\| \left(\frac{T(t_0 + h) - T(t_0)}{h} \right) x - S(t_0) x \right\| = 0, \qquad x \in X.$$

We say that $T(t)$ is *norm differentiable* at t_0 if it satisfies

$$\lim_{h \to 0} \left\| \frac{T(t_0 + h) - T(t_0)}{h} - S(T_0) \right\| = 0.$$

The operator $S(t_0)$ is denoted by

$$\frac{dT}{dt}(t_0) \qquad \text{or} \qquad T'(t_0).$$

If $T(t)$ is strongly (resp. norm) differentiable at every point of I, then it is said to be strongly (resp. norm) differentiable on I. A norm differentiable function is strongly differentiable.

We remark that the Leibniz formula can be extended to strongly or norm differentiable functions.

Exponential Functions

Just as in the case of numerical series, we have:

3.9.1 Theorem. *If A is a bounded linear operator on a Banach space X into itself, then the series*

$$e^{tA} = \sum_{m=0}^{\infty} \frac{t^m}{m!} A^m \qquad (-\infty < t < \infty)$$

converges in the space $L(X, X)$, and enjoys the following properties:

(a) $\|e^{tA}\| \leq e^{|t| \cdot \|A\|}$.
(b) $e^{tA} e^{sA} = e^{(t+s)A} (-\infty < t, s < \infty)$.

(c) *The function* e^{tA} *is norm differentiable on* $\mathbf{R}$, *and we have*:

$$\frac{d}{dt}(e^{tA}) = Ae^{tA} = e^{tA}A.$$

Contraction Semigroups

Let X be a Banach space. A one-parameter family $\{T_t\}_{t\geq 0}$ of bounded linear operators on X into itself is called a *contraction semigroup of class* (C_0) or simply a *contraction semigroup* if it satisfies the following conditions:

(i) $T_{t+s} = T_t \cdot T_s$, $t, s \geq 0$.
(ii) $\lim_{t\downarrow 0} \|T_t x - x\| = 0$, $x \in X$.
(iii) $\|T_t\| \leq 1$, $t \geq 0$.

Condition (i) is called the *semigroup property*.

3.9.2 Remark. In view of conditions (i) and (ii), it follows that

$$T_0 = I.$$

Hence condition (ii) is equivalent to the strong continuity of $\{T_t\}_{t\geq 0}$ at $t = 0$. Moreover it is easy to verify that a contraction semigroup $\{T_t\}_{t\geq 0}$ is strongly continuous on $[0, \infty)$.

Let $\{T_t\}_{t\geq 0}$ be a contraction semigroup of class (C_0) defined on a Banach space X. We let

$$\mathscr{D} = \text{the set of all } x \in X \text{ such that the limit } \lim_{h\downarrow 0} \frac{T_h x - x}{h} \text{ exists in } X.$$

Then we define a linear operator $\mathfrak{A}$ from X into itself as follows:

(a) The domain $D(\mathfrak{A})$ of $\mathfrak{A}$ is the set $\mathscr{D}$.
(b) $\mathfrak{A}x = \lim_{h\downarrow 0}(T_h x - x)/h$, $x \in D(\mathfrak{A})$.

The operator $\mathfrak{A}$ is called the *infinitesimal generator* of $\{T_t\}_{t\geq 0}$.

Now we give a differential equation associated with the semigroup.

3.9.3 Proposition. *Let* $\mathfrak{A}$ *be the infinitesimal generator of a contraction semigroup* $\{T_t\}_{t\geq 0}$. *If* $x \in D(\mathfrak{A})$, *then we have* $T_t x \in D(\mathfrak{A})$ *for all* $t > 0$, *and the function* $T_t x$ *is strongly differentiable on* $(0, \infty)$ *and satisfies the equation*

$$\frac{d}{dt}(T_t x) = \mathfrak{A}(T_t x) = T_t(\mathfrak{A}x), \qquad t > 0.$$

Let $\{T_t\}_{t\geq 0}$ be a contraction semigroup. Then the integral

$$\int_0^s e^{-\alpha t} T_t x \, dt, \qquad x \in X, \tag{2}$$

is strongly integrable for all $s > 0$, since the integrand is strongly continuous on $[0, \infty)$. Further, if $\alpha > 0$, then the limit $G_\alpha x$ of integral (2) exists in X as $s \to \infty$:

$$G_\alpha x = \int_0^\infty e^{-\alpha t} T_t x \, dt, \qquad x \in X, \alpha > 0. \tag{3}$$

Thus $G_\alpha x$ is defined for all $x \in X$ if $\alpha > 0$. It is easy to see that the operator G_α is a bounded linear operator on X into itself with norm $1/\alpha$:

$$\|G_\alpha x\| \leq \frac{1}{\alpha} \|x\|, \qquad x \in X. \tag{4}$$

The family $\{G_\alpha\}_{\alpha > 0}$ of bounded linear operators is called the *resolvent* of the semigroup $\{T_t\}_{t\geq 0}$.

The next theorem characterizes the resolvent.

3.9.4 Theorem. *Let $\{T_t\}_{t\geq 0}$ be a contraction semigroup defined on a Banach space X and $\mathfrak{A}$ the infinitesimal generator of $\{T_t\}$. For each $\alpha > 0$, the operator $(\alpha I - \mathfrak{A})$ is a bijection of $D(\mathfrak{A})$ onto X, and its inverse $(\alpha I - \mathfrak{A})^{-1}$ is the resolvent G_α.*

3.9.5 Corollary. *For all $x \in X$, we have*

$$x = \lim_{\alpha \to \infty} \alpha G_\alpha x.$$

Thus the domain $D(\mathfrak{A})$ is everywhere dense in X.

Now we consider when a linear operator is the infinitesimal generator of some contraction semigroup. This question is answered by the following:

3.9.6 Theorem (Hille–Yosida). *Let $\mathfrak{A}$ be a linear operator from a Banach space X into itself, with domain $D(\mathfrak{A})$. In order that $\mathfrak{A}$ be the infinitesimal generator of some contraction semigroup, it is necessary and sufficient that $\mathfrak{A}$ satisfies the following three conditions:*

(i) *The domain $D(\mathfrak{A})$ is everywhere dense in X.*

(ii) *For every $\alpha > 0$, the equation $(\alpha I - \mathfrak{A})x = y$ has a unique solution $x \in D(\mathfrak{A})$ for any $y \in X$; we then write*

$$x = (\alpha I - \mathfrak{A})^{-1} y.$$

(iii) *For any $\alpha > 0$, we have*

$$\|(\alpha I - \mathfrak{A})^{-1}\| \leq \frac{1}{\alpha}.$$

Proof. The necessity of conditions (i) through (iii) follows from Theorem 3.9.4, Corollary 3.9.5 and inequality (4).

For the sufficiency, we only give a sketch of the proof.

1) If $\alpha > 0$, we let

$$J_\alpha = \alpha(\alpha I - \mathfrak{A})^{-1},$$

and

$$\mathfrak{A}_\alpha = \mathfrak{A} J_\alpha.$$

Then we have

$$\begin{cases} \|J_\alpha\| \leq 1, \\ \operatorname*{s-lim}_{\alpha \to \infty} J_\alpha = I, \end{cases}$$

and

$$\begin{cases} \|\mathfrak{A}_\alpha\| \leq 2\alpha, \\ \lim_{\alpha \to \infty} \mathfrak{A}_\alpha x = \mathfrak{A} x, \qquad x \in D(\mathfrak{A}). \end{cases}$$

The operators $\mathfrak{A}_\alpha$ are called the *Yosida approximations* to $\mathfrak{A}$.

2) We define

$$T_t(\alpha) = e^{t\mathfrak{A}_\alpha}, \qquad \alpha > 0.$$

Since we have $\mathfrak{A}_\alpha = \alpha\mathfrak{A}(\alpha I - \mathfrak{A})^{-1} = \alpha(J_\alpha - I)$, it follows from an application of Theorem 3.9.1 that the operators

$$T_t(\alpha) = e^{-\alpha t} e^{\alpha t J_\alpha}, \qquad t \geq 0, \tag{5}$$

form a contraction semigroup for each $\alpha > 0$.

3) The operator $T_t(\alpha)$ has a strong limit T_t as $\alpha \to \infty$, and this convergence is uniform in t over bounded intervals contained in $[0, \infty)$.

4) Now we can define a family $\{T_t\}_{t \geq 0}$ of linear operators as follows:

$$T_t x = \lim_{\alpha \to \infty} T_t(\alpha) x, \qquad x \in X. \tag{6}$$

Then it follows from an application of the resonance theorem (Theorem 3.2.9) that the operator T_t is bounded and satisfies

$$\|T_t\| \leq \liminf_{n\to\infty} \|T_t(\alpha)\| \leq 1, \qquad t \geq 0.$$

The semigroup property of $\{T_t\}$ follows from that of $\{T_t(\alpha)\}$. Further, since the convergence of (6) is uniform in t over bounded intervals contained in $[0, \infty)$, it follows that the function $T_t x$, $x \in X$, is strongly continuous on $[0, \infty)$. Consequently the family $\{T_t\}_{t\geq 0}$ forms a contraction semigroup.

5) It is easy to verify that the infinitesimal generator of the semigroup $\{T_t\}_{t\geq 0}$ thus obtained is precisely the operator $\mathfrak{A}$.

6) Finally we can prove that if $\{U_t\}_{t\geq 0}$ is another contraction semigroup which has $\mathfrak{A}$ as its infinitesimal generator, then we have $U_t = T_t$ for all $t \geq 0$. This implies the uniqueness of the semigroup. ■

Notes

The material in this chapter is adapted from the book of Yosida [1] and also part of Schechter's [1].

Section 3.1: For more leisurely treatments of linear topological spaces, the reader is referred to Bourbaki [2], Köthe [1] and Treves [1].

Section 3.3: The Riesz representation theorem, Theorem 3.3.3, is adapted from Rudin [1]. For a proof of Theorem 3.3.6, see Jameson [1].

Section 3.7: For further material on Fredholm operators, see Gohberg–Kreĭn [1]. Theorem 3.7.6, first proved by Peetre [2] for bounded operators, is taken from Taira [5].

4 Distributions, Operators and Kernels

In this chapter we present a brief description of the basic concepts and results of the theory of distributions, or generalized functions, which will be used in subsequent chapters. Distribution theory has become a convenient tool in the study of partial differential equations. Many problems in partial differential equations can be formulated in terms of abstract operators acting between suitable spaces of distributions, and these operators are then analyzed by the methods of functional analysis. The virtue of this approach is that a given problem is stripped of extraneous data, so that the analytic core of the problem is revealed.

4.1. Notation

Points in Euclidean Spaces

Let $\mathbf{R}^n$ be the n-dimensional Euclidean space. In Chapter 2, points in $\mathbf{R}^n$ or the standard coordinate system of $\mathbf{R}^n$ is denoted by $(x^1, \ldots, x^n)$; however, in this chapter, we shall use instead the conventional notation

$$x = (x_1, \ldots, x_n).$$

If $x = (x_1, \dots, x_n)$ and $y = (y_1, \dots, y_n)$ are points in $\mathbf{R}^n$, we set

$$x \cdot y = \sum_{j=1}^{n} x_j y_j,$$

$$|x| = \left(\sum_{j=1}^{n} x_j^2 \right)^{1/2}.$$

Multi-Indices and Derivations

Let $\alpha = (\alpha_1, \dots, \alpha_n)$ be an n-tuple of non-negative integers. Such an n-tuple α is called a *multi-index*. We let

$$|\alpha| = \alpha_1 + \cdots + \alpha_n,$$

$$\alpha! = \alpha_1! \dots \alpha_n!.$$

If $\alpha = (\alpha_1, \dots, \alpha_n)$ and $\beta = (\beta_1, \dots, \beta_n)$ are multi-indices, we define

$$\alpha + \beta = (\alpha_1 + \beta_1, \dots, \alpha_n + \beta_n).$$

The notation $\alpha \le \beta$ means that $\alpha_j \le \beta_j$ for $1 \le j \le n$. Then we let

$$\binom{\beta}{\alpha} = \frac{\beta!}{(\beta - \alpha)!\,\alpha!} = \binom{\beta_1}{\alpha_1} \cdots \binom{\beta_n}{\alpha_n}.$$

We use the shorthand

$$\partial_j = \frac{\partial}{\partial x_j},$$

$$D_j = \frac{1}{i} \frac{\partial}{\partial x_j} \qquad (i = \sqrt{-1}),$$

for derivatives on $\mathbf{R}^n$. Higher-order derivatives are expressed by multi-indices as follows:

$$\partial^\alpha = \partial_1^{\alpha_1} \dots \partial_n^{\alpha_n},$$

$$D^\alpha = D_1^{\alpha_1} \dots D_n^{\alpha_n}.$$

Similarly, if $x = (x_1, \dots, x_n) \in \mathbf{R}^n$, we write

$$x^\alpha = x_1^{\alpha_1} \dots x_n^{\alpha_n}.$$

4.2. Function Spaces

L^p-Spaces

Let Ω be an open subset of $\mathbf{R}^n$. Two Lebesgue measurable functions f, g on Ω are said to be equivalent if they are equal almost everywhere in Ω with respect to the Lebesgue measure dx, that is, if $f(x) = g(x)$ for all x outside a set of Lebesgue measure zero. This is obviously an equivalence relation.

If $1 \leq p < \infty$, we let

$L^p(\Omega)$ = the space of equivalence classes of Lebesgue measurable functions f on Ω such that $|f|^p$ is integrable on Ω.

The space $L^p(\Omega)$ is a Banach space with the norm

$$\|f\|_p = \left(\int_\Omega |f(x)|^p \, dx\right)^{1/p}.$$

Furthermore, the space $L^2(\Omega)$ is a Hilbert space with the inner product

$$(f, g) = \int_\Omega f(x)\overline{g(x)} \, dx.$$

A Lebesgue measurable function f on Ω is said to be *essentially bounded* if there exists a constant $C > 0$ such that $|f(x)| \leq C$ almost everywhere (a.e.) in Ω. We define

$$\operatorname*{ess\,sup}_{x \in \Omega} |f(x)| = \inf\{C; |f(x)| \leq C \text{ a.e. in } \Omega\}.$$

For $p = \infty$, we let

$L^\infty(\Omega)$ = the space of equivalence classes of essentially bounded, Lebesgue measurable functions on Ω.

The space $L^\infty(\Omega)$ is a Banach space with the norm

$$\|f\|_\infty = \operatorname*{ess\,sup}_{x \in \Omega} |f(x)|.$$

If $1 < p < \infty$, we let $q = p/(p - 1)$, so that $1 < q < \infty$ and

$$\frac{1}{p} + \frac{1}{q} = 1.$$

The number q is called the *exponent conjugate* to p.

Recall that the most basic inequality for L^p-functions is the following:

4.2.1 Theorem (Hölder's inequality). *If $1 < p < \infty$ and $f \in L^p(\Omega)$, $g \in L^q(\Omega)$, then we have $fg \in L^1(\Omega)$ and*

$$\|fg\|_1 \leq \|f\|_p \|g\|_q. \tag{1}$$

We remark that inequality (1) is true for the two cases $p = 1$, $q = \infty$ and $p = \infty$, $q = 1$. Inequality (1) in the case $p = q = 2$ is referred to as *Schwarz's inequality*.

Convolutions

We give a general theorem about integral operators on a measure space:

4.2.2 Theorem (the generalized Young inequality). *Let $(X, \mathcal{M}, \mu)$ be a measure space. Suppose that K is a measurable function on the product space $X \times X$ such that*

$$\sup_{x \in X} \int_X |K(x, y)| \, d\mu(y) \leq C$$

and

$$\sup_{y \in X} \int_X |K(x, y)| \, d\mu(x) \leq C,$$

where $C > 0$ is a constant. If $f \in L^p(X)$ with $1 \leq p \leq \infty$, then the function Tf, defined by

$$Tf(x) = \int_X K(x, y) f(y) \, d\mu(y),$$

is well-defined for almost all $x \in X$, and is in $L^p(X)$. Furthermore, we have

$$\|Tf\|_p \leq C \|f\|_p.$$

Theorem 4.2.2 is an immediate consequence of Fubini's theorem (Theorem 1.19.4) and Theorem 4.2.1.

4.2.3 Corollary (the Young inequality). *If $f \in L^1(\mathbf{R}^n)$ and $g \in L^p(\mathbf{R}^n)$ with $1 \leq p \leq \infty$, then the function $f * g$, defined by*

$$(f * g)(x) = \int_{R^n} f(x - y) g(y) \, dy,$$

is well-defined for almost all $x \in \mathbf{R}^n$, *and is in* $L^p(\mathbf{R}^n)$. *Furthermore we have*

$$\|f * g\|_p \leq \|f\|_1 \|g\|_p.$$

The function $f * g$ is called the *convolution* of f and g.

Spaces of C^k Functions

Let Ω be an open subset of $\mathbf{R}^n$. We let

$$C(\Omega) = \text{the space of continuous functions on } \Omega.$$

If K is a compact subset of Ω, we define a seminorm p_K on $C(\Omega)$ by

$$C(\Omega) \ni \varphi \mapsto p_K(\varphi) = \sup_{x \in K} |\varphi(x)|.$$

We equip the space $C(\Omega)$ with the topology defined by the family $\{p_K\}$ of seminorms where K ranges over all compact subsets of Ω.

If k is a positive integer, we let

$$C^k(\Omega) = \text{the space of } C^k \text{ functions on } \Omega.$$

We define a seminorm $p_{K,k}$ on $C^k(\Omega)$ by

$$C^k(\Omega) \ni \varphi \mapsto p_{K,k}(\varphi) = \sup_{\substack{x \in K \\ |\alpha| \leq k}} |\partial^\alpha \varphi(x)|. \tag{2}$$

We equip the space $C^k(\Omega)$ with the topology defined by the family $\{p_{K,k}\}$ of seminorms where K ranges over all compact subsets of Ω. This is the topology of uniform convergence on compact subsets of Ω of the functions and their derivatives of order $\leq k$.

We set

$$C^\infty(\Omega) = \bigcap_{k=1}^{\infty} C^k(\Omega),$$

and

$$C^0(\Omega) = C(\Omega).$$

Let m be a non-negative integer or $m = \infty$. Let $\{K_\ell\}$ be a sequence of compact subsets of Ω such that K_ℓ is contained in the interior of $K_{\ell+1}$ for each ℓ and $\Omega = \bigcup_{\ell=1}^{\infty} K_\ell$. For example, one may take

$$K_\ell = \left\{x \in \Omega; |x| \leq \ell, \operatorname{dist}(x, \partial\Omega) \geq \frac{1}{\ell}\right\}.$$

Such a sequence $\{K_\ell\}$ is called an exhaustive sequence of compact subsets of Ω. It is easy to see that the countable family $\{p_{K_\ell, j}\}_{\substack{\ell=1,2,\dots \\ 0 \leq j \leq m}}$ of seminorms suffices to define the topology on $C^m(\Omega)$, and further that $C^m(\Omega)$ is complete. Hence the space $C^m(\Omega)$ is a Fréchet space.

Further we let

$C(\bar{\Omega})$ = the space of functions in $C(\Omega)$ having continuous extensions to the closure $\bar{\Omega}$ of Ω.

If k is a positive integer, we let

$C^k(\bar{\Omega})$ = the space of functions in $C^k(\Omega)$ all of whose derivatives of order $\leq k$ have continuous extensions to $\bar{\Omega}$.

We set

$$C^\infty(\bar{\Omega}) = \bigcap_{k=1}^{\infty} C^k(\bar{\Omega}),$$

and

$$C^0(\bar{\Omega}) = C(\bar{\Omega}).$$

Let m be a non-negative integer or $m = \infty$. We equip the space $C^m(\bar{\Omega})$ with the topology defined by the family $\{p_{K,j}\}$ of seminorms where K ranges over all compact subsets of $\bar{\Omega}$ and $0 \leq j \leq m$.

Let $\{F_\ell\}$ be an increasing sequence of compact subsets of $\bar{\Omega}$ such that $\bigcup_{\ell=1}^{\infty} F_\ell = \bar{\Omega}$. For example, one may take

$$F_\ell = \{x \in \bar{\Omega};\ |x| \leq \ell\}.$$

Such a sequence $\{F_\ell\}$ is called an exhaustive sequence of compact subsets of $\bar{\Omega}$. It is easy to see that the countable family $\{p_{F_\ell, j}\}_{\substack{\ell=1,2,\dots \\ 0 \leq j \leq m}}$ of seminorms suffices to define the topology on $C^m(\bar{\Omega})$, and further that $C^m(\Omega)$ is complete. Hence the space $C^m(\Omega)$ is a Fréchet space.

If Ω is bounded and $0 \leq m < \infty$, then the space $C^m(\bar{\Omega})$ is a Banach space with the norm

$$\|\varphi\|_{C^m(\bar{\Omega})} = \sup_{\substack{x \in \bar{\Omega} \\ |\alpha| \leq m}} |\partial^\alpha \varphi(x)|.$$

Space of Test Functions

Let Ω be an open subset of $\mathbf{R}^n$ and let u be a continuous function on Ω. The *support* of u, denoted supp u, is the closure in Ω of the set $\{x \in \Omega; u(x) \neq 0\}$. In other words, the support of u is the smallest closed subset of Ω outside of which u vanishes.

Let m be a non-negative integer or $m = \infty$. If K is a compact subset of Ω, we let

$$C_K^m(\Omega) = \text{the space of functions in } C^m(\Omega) \text{ with support in } K.$$

The space $C_K^m(\Omega)$ is a closed subspace of $C^m(\Omega)$. Further we let

$$C_0^m(\Omega) = \bigcup_{K \subset \Omega} C_K^m(\Omega),$$

where K ranges over all compact subsets of Ω, so that $C_0^m(\Omega)$ is the space of functions in $C^m(\Omega)$ with compact support in Ω. We remark that the space $C_0^m(\Omega)$ can be identified with the space of functions in $C_0^m(\mathbf{R}^n)$ with support in Ω. If $\{K_\ell\}$ is an exhaustive sequence of compact subsets of Ω, we equip the space $C_0^m(\Omega)$ with the *inductive limit topology* of the spaces $C_{K_\ell}^m(\Omega)$, that is, the strongest locally convex linear space topology such that each injection $C_{K_\ell}^m(\Omega) \to C_0^m(\Omega)$ is continuous. One can verify that this topology on $C_0^m(\Omega)$ is independent of the sequence $\{K_\ell\}$ used.

We list some basic properties of the topology on $C_0^m(\Omega)$:

1) A sequence $\{\varphi_j\}$ in $C_0^m(\Omega)$ converges to an element φ in $C_0^m(\Omega)$ if and only if the functions φ_j, φ are supported in a common compact subset K of Ω and $\varphi_j \to \varphi$ in $C_K^m(\Omega)$.

2) A subset of $C_0^m(\Omega)$ is bounded if and only if it is bounded in $C_K^m(\Omega)$ for some compact $K \subset \Omega$.

3) A linear mapping of $C_0^m(\Omega)$ into a linear topological space is continuous if and only if its restriction to $C_K^m(\Omega)$ for every compact $K \subset \Omega$ is continuous.

The elements of $C_0^\infty(\Omega)$ are often called *test functions*.

If K is a compact subset of $\bar{\Omega}$, we let

$$C_K^m(\bar{\Omega}) = \text{the space of functions in } C^m(\bar{\Omega}) \text{ which vanish in } \Omega \setminus K.$$

The space $C_K^m(\bar{\Omega})$ is a closed subspace of $C^m(\bar{\Omega})$. Further we let

$$C_0^m(\bar{\Omega}) = \bigcup_{K \subset \Omega} C_K^m(\bar{\Omega}),$$

where K ranges over all compact subsets of $\bar{\Omega}$. If $\{F_\ell\}$ is an exhaustive sequence of compact subsets of $\bar{\Omega}$, we equip the space $C_0^m(\bar{\Omega})$ with the inductive limit topology of the spaces $C^m_{F_\ell}(\bar{\Omega})$. This topology on $C^m_0(\bar{\Omega})$ is independent of the sequence $\{F_\ell\}$ is used.

Hölder Spaces

Let D be a subset of $\mathbf{R}^n$ and $0 < \theta < 1$. A function φ defined on D is said to be *Hölder continuous* with exponent θ on D if the quantity

$$[\varphi]_{\theta;D} = \sup_{\substack{x,y \in D \\ x \neq y}} \frac{|\varphi(x) - \varphi(y)|}{|x-y|^\theta}$$

is finite. We say that φ is *locally Hölder continuous* with exponent θ on D if it is Hölder continuous with exponent θ on compact subsets of D. Hölder continuity may be viewed as a fractional differentiability.

Let Ω be an open subset of $\mathbf{R}^n$. We let

$C^\theta(\Omega)$ = the space of functions in $C(\Omega)$ which are locally Hölder continuous with exponent θ on Ω.

If k is a positive integer, we let

$C^{k+\theta}(\Omega)$ = the space of functions in $C^k(\Omega)$ whose k-th order derivatives are locally Hölder continuous with exponent θ on Ω.

If K is a compact subset of Ω, we define a seminorm $q_{K,k}$ on $C^{k+\theta}(\Omega)$ by

$$C^{k+\theta}(\Omega) \ni \varphi \mapsto q_{K,k}(\varphi) = \sup_{\substack{x \in K \\ |\alpha| \leq k}} |\partial^\alpha \varphi(x)| + \sup_{|\alpha| = k} [\partial^\alpha \varphi]_{\theta;K}.$$

We equip the space $C^{k+\theta}(\Omega)$ with the topology defined by the family $\{q_{K,k}\}$ of seminorms where K ranges over all compact subsets of Ω. It is easy to see that the space $C^{k+\theta}(\Omega)$ is a Fréchet space.

Further we let

$C^\theta(\bar{\Omega})$ = the space of functions in $C(\bar{\Omega})$ which are Hölder continuous with exponent θ on $\bar{\Omega}$.

If k is a positive integer, we let

$C^{k+\theta}(\bar{\Omega})$ = the space of functions in $C^k(\bar{\Omega})$ whose k-th order derivatives are Hölder continuous with exponent θ on $\bar{\Omega}$.

Let m be a non-negative integer. We equip the space $C^{m+\theta}(\bar{\Omega})$ with the topology defined by the family $\{q_{K,k}\}$ of seminorms where K ranges over all compact subsets of $\bar{\Omega}$. It is easy to see that the space $C^{m+\theta}(\bar{\Omega})$ is a Fréchet space.

If Ω is bounded, then $C^{m+\theta}(\bar{\Omega})$ is a Banach space with the norm

$$\|\varphi\|_{C^{m+\theta}(\bar{\Omega})} = \|\varphi\|_{C^m(\bar{\Omega})} + \sup_{|\alpha|=m} [\partial^\alpha \varphi]_{\theta;\bar{\Omega}}.$$

Mollifiers

Let ρ be a non-negative C^∞ function on $\mathbf{R}^n$ satisfying the following conditions:

(a) $\operatorname{supp} \rho = \{x \in \mathbf{R}^n; |x| \leq 1\}$.

(b) $\displaystyle\int_{\mathbf{R}^n} \rho(x)\, dx = 1.$

For example, one may take

$$\rho(x) = \begin{cases} k \exp[-1/(1-|x|^2)] & \text{if } |x| < 1, \\ 0 & \text{if } |x| \geq 1, \end{cases}$$

where the constant factor k is so chosen that condition (b) is satisfied.

For each $\varepsilon > 0$, we define

$$\rho_\varepsilon(x) = \frac{1}{\varepsilon^n} \rho\left(\frac{x}{\varepsilon}\right).$$

Then ρ_ε is a non-negative C^∞ function on $\mathbf{R}^n$, and satisfies:

(c) $\operatorname{supp} \rho_\varepsilon = \{x \in \mathbf{R}^n; |x| \leq \varepsilon\}$.

(d) $\displaystyle\int_{\mathbf{R}^n} \rho_\varepsilon(x)\, dx = 1.$

The functions ρ_ε are called *mollifiers*.

The next theorem shows how mollifiers can be used to approximate a function by C^∞ functions.

4.2.4 Theorem. *Let Ω be an open subset of $\mathbf{R}^n$. Then we have:*

(i) *If $u \in L^p(\Omega)$ with $1 \leq p < \infty$ and vanishes outside a compact subset K of Ω, then $\rho_\varepsilon * u \in C_0^\infty(\Omega)$ provided $\varepsilon <$ dist$(K, \partial\Omega)$, and $\rho_\varepsilon * u \to u$ in $L^p(\Omega)$ as $\varepsilon \downarrow 0$.*

(ii) *If $u \in C_0^m(\Omega)$ with $0 \leq m < \infty$, then $\rho_\varepsilon * u \in C_0^\infty(\Omega)$ provided $\varepsilon <$ dist(supp u, $\partial\Omega$), and $\rho_\varepsilon * u \to u$ in $C_0^m(\mu)$ as $\varepsilon \downarrow 0$.*

The functions $\rho_\varepsilon * u$ are called *regularizations* of u.

4.2.5 Corollary. *The space $C_0^\infty(\Omega)$ is dense in $L^p(\Omega)$ for $1 \leq p < \infty$.*

Corollary 4.2.5 is an immediate consequence of part (i) of Theorem 4.2.4, since L^p-functions with compact support are dense in $L^p(\Omega)$.

4.3. Differential Operators

Let Ω be an open subset of $\mathbf{R}^n$. If m is a non-negative integer, we let

$$P(x, D) = \sum_{|\alpha| \leq m} a_\alpha(x) D^\alpha, \qquad a_\alpha \in C^\infty(\Omega).$$

It is clear that $P(x, D)$ is a continuous linear mapping of $C^\infty(\Omega)$ into itself. Such mappings are called *differential operators* of order m on Ω. We remark that $P = P(x, D)$ satisfies

$$\operatorname{supp} Pu \subset \operatorname{supp} u, \qquad u \in C^\infty(\Omega), \tag{1}$$

since differentiation is a purely local process. We express this by saying that differential operators are *local.*

The next theorem states that the converse is also true.

4.3.1 Theorem (Peetre). *Suppose that P is a linear mapping of $C^\infty(\Omega)$ into itself which satisfies condition* (1). *Then, for every relatively compact subset Ω' of Ω, there exist a non-negative integer m and C^∞ functions a_α on Ω' such that*

$$Pu(x) = \sum_{|\alpha| \leq m} a_\alpha(x) D^\alpha u(x), \qquad u \in C^\infty(\Omega'),\ x \in \Omega'.$$

4.4. Distributions

Definitions and Basic Properties

Let Ω be an open subset of $\mathbf{R}^n$. A *distribution* on Ω is a continuous linear functional on $C_0^\infty(\Omega)$. The space of distributions on Ω is denoted by $\mathscr{D}'(\Omega)$. In other words, the space $\mathscr{D}'(\Omega)$ is the dual space of $C_0^\infty(\Omega)$. If $u \in \mathscr{D}'(\Omega)$ and $\varphi \in C_0^\infty(\Omega)$, we denote the action of u on φ by $\langle u, \varphi \rangle$ or sometimes by $\langle \varphi, u \rangle$.

We state useful characterizations of distributions:

4.4.1 Theorem. *Let u be a linear functional on $C_0^\infty(\Omega)$. Then the following three conditions are equivalent*:

(i) *The functional u is a distribution.*
(ii) *For any compact subset K of Ω, there exist a constant $C > 0$ and a non-negative integer m such that*

$$|\langle u, \varphi\rangle| \le C p_{K,m}(\varphi), \qquad \varphi \in C_K^\infty(\Omega),$$

where

$$p_{K,m}(\varphi) = \sup_{\substack{x \in K \\ |\alpha| \le m}} |\partial^\alpha \varphi(x)|.$$

(iii) $\langle u, \varphi_j\rangle \to 0$ *whenever* $\varphi_j \to 0$ *in* $C_0^\infty(\Omega)$.

Theorem 4.2.4 shows that the space $C_0^\infty(\Omega)$ is a dense subspace of $C_0^m(\Omega)$ for $0 \le m < \infty$. Also it is clear that the injection of $C_0^\infty(\Omega)$ into $C_0^m(\Omega)$ is continuous. Hence the dual space $\mathscr{D}'^m(\Omega)$ of $C_0^m(\Omega)$ can be identified with a linear subspace of $\mathscr{D}'(\Omega)$, by the identification of a continuous linear functional on $C_0^m(\Omega)$ with its restriction to $C_0^\infty(\Omega)$. The elements of $\mathscr{D}'^m(\Omega)$ are called distributions of *order* $\le m$ on Ω. In other words, the distributions of order $\le m$ on Ω are precisely those distributions on Ω that have continuous extensions to $C_0^m(\Omega)$.

Now we give some examples of distributions:

1) We let

$L^1_{\text{loc}}(\Omega)$ = the space of equivalence classes of Lebesgue measurable functions on Ω which are integrable on every compact subset of Ω.

The elements of $L^1_{\text{loc}}(\Omega)$ are called *locally integrable functions* on Ω.

Every element f of $L^1_{\text{loc}}(\Omega)$ defines a distribution T_f of order zero on Ω by the formula

$$\langle T_f, \varphi\rangle = \int_\Omega f(x)\varphi(x)\,dx, \qquad \varphi \in C_0^\infty(\Omega).$$

In fact we have, for all $\varphi \in C_K^\infty(\Omega)$,

$$|\langle T_f, \varphi\rangle| \le \left(\int_K |f(x)|\,dx\right) p_{K,0}(\varphi).$$

Since the mapping $f \mapsto T_f$ induces an injection of $L^1_{\text{loc}}(\Omega)$ into $\mathscr{D}'(\Omega)$, we can regard locally integrable functions as distributions. We say that such distributions "are" functions. In particular, the functions in $C^m(\Omega)$ ($0 \leq m \leq \infty$) and in $L^p(\Omega)$ ($1 \leq p \leq \infty$) are distributions on Ω.

2) More generally, every complex Borel measure μ on Ω defines a distribution of order zero on Ω by the formula

$$\langle \mu, \varphi \rangle = \int_\Omega \varphi(x)\, d\mu(x), \qquad \varphi \in C_0^\infty(\Omega).$$

In particular, if we take μ to be the point mass at a point x_0 of Ω, we obtain the *Dirac measure* δ_{x_0} defined by

$$\langle \delta_{x_0}, \varphi \rangle = \varphi(x_0), \qquad \varphi \in C_0^\infty(\Omega).$$

We define various operations on distributions:

(a) *Restriction*: If $u \in \mathscr{D}'(\Omega)$ and V is an open subset of Ω, we define the restriction $u|_V$ of u to V by the formula

$$\langle u|_V, \varphi \rangle = \langle u, \varphi \rangle, \qquad \varphi \in C_0^\infty(V).$$

Then we have $u|_V \in \mathscr{D}'(V)$.

(b) *Differentiation*: The derivative $\partial^\alpha u$ of a distribution $u \in \mathscr{D}'(\Omega)$ is the distribution on Ω defined by

$$\langle \partial^\alpha u, \varphi \rangle = (-1)^{|\alpha|} \langle u, \partial^\alpha \varphi \rangle, \qquad \varphi \in C_0^\infty(\Omega).$$

(c) *Multiplication by functions*: The product au of a function $a \in C^\infty(\Omega)$ and a distribution $u \in \mathscr{D}'(\Omega)$ is the distribution on Ω defined by

$$\langle au, \varphi \rangle = \langle u, a\varphi \rangle, \qquad \varphi \in C_0^\infty(\Omega).$$

The *Leibniz formula* for the differentiation of a product remains valid:

$$D^\beta(au) = \sum_{\alpha \leq \beta} \binom{\beta}{\alpha} D^{\beta-\alpha} a \cdot D^\alpha u. \tag{1}$$

(d) We can combine (b) and (c). We let

$$P(x, D) = \sum_{|\alpha| \leq m} a_\alpha(x) D^\alpha, \qquad a_\alpha \in C^\infty(\Omega),$$

be a differential operator of order m on Ω. If $u \in \mathscr{D}'(\Omega)$, we define $P(x, D)u$ by the formula

$$\langle P(x, D)u, \varphi \rangle = \left\langle u, \sum_{|\alpha| \leq m} (-1)^{|\alpha|} D^\alpha (a_\alpha \varphi) \right\rangle, \qquad \varphi \in C_0^\infty(\Omega).$$

Then we have $P(x, D)u \in \mathscr{D}'(\Omega)$.

The function $p(x, \xi) = \sum_{|\alpha| \leq m} a_\alpha(x)\xi^\alpha$ is called the *complete symbol* of $P(x, D)$ and the function $p_m(x, \xi) = \sum_{|\alpha| = m} a_\alpha(x)\xi^\alpha$ is called the *principal symbol* of $P(x, D)$. We denote $P^{(\alpha)}(x, D)$ the differential operator of order $m - |\alpha|$ having complete symbol $\partial_\xi^\alpha p(x, \xi)$. Then we have the following generalization of formula (1):

$$P(x, D)(au) = \sum_{|\alpha| \leq m} \frac{1}{\alpha!} (P^{(\alpha)}(x, D)a) D^\alpha u. \tag{1'}$$

This is referred to as the *Leibniz-Hörmander formula*.

(e) *Conjugation*: The conjugate $\bar{u}$ of a distribution $u \in \mathscr{D}'(\Omega)$ is the distribution on Ω defined by

$$\langle \bar{u}, \varphi \rangle = \overline{\langle u, \bar{\varphi} \rangle}, \qquad \varphi \in C_0^\infty(\Omega),$$

where $\overline{}$ denotes complex conjugation.

Topologies on $\mathscr{D}'(\Omega)$

There are two natural topologies on the space $\mathscr{D}'(\Omega)$:

1) *Weak* topology* τ_s: This is the topology of convergence at each element of $C_0^\infty(\Omega)$. The space $\mathscr{D}'(\Omega)$ endowed with this topology is denoted by $\mathscr{D}'_s(\Omega)$. A sequence $\{u_j\}$ of distributions converges to a distribution u in $\mathscr{D}'_s(\Omega)$ if and only if the sequence $\{\langle u_j, \varphi \rangle\}$ converges to $\langle u, \varphi \rangle$ for every $\varphi \in C_0^\infty(\Omega)$.

2) *Strong topology* τ_b: This is the topology of uniform convergence on bounded subsets of $C_0^\infty(\Omega)$. The space $\mathscr{D}'(\Omega)$ endowed with this topology is denoted by $\mathscr{D}'_b(\Omega)$. A sequence $\{u_j\}$ of distributions converges to a distribution u in $\mathscr{D}'_b(\Omega)$ if and only if the sequence $\{\langle u_j, \varphi \rangle\}$ converges to $\langle u, \varphi \rangle$ uniformly in φ over bounded subsets of $C_0^\infty(\Omega)$.

We list some basic topological properties of $\mathscr{D}'(\Omega)$:

1) In the case of a sequence of distributions, the two notions of convergence coincide, that is, $u_j \to u$ in $\mathscr{D}'_s(\Omega)$ if and only if $u_j \to u$ in $\mathscr{D}'_b(\Omega)$.

Let Ω_1 and Ω_2 be open subsets of $\mathbf{R}^{n_1}$ and $\mathbf{R}^{n_2}$, respectively, and let A be a linear operator on $C_0^\infty(\Omega_2)$ into $\mathscr{D}'(\Omega_1)$. Then the continuity of A does not depend on the topology (τ_s or τ_b) on $\mathscr{D}'(\Omega_1)$. In fact, $A: C_0^\infty(\Omega_2) \to \mathscr{D}'(\Omega_1)$ is continuous if and only if its restriction to $C_{K_2}^\infty(\Omega_2)$ for every compact $K_2 \subset \Omega_2$ is continuous; so it suffices to base our reasoning on sequences.

2) If $\{u_j\}$ is a sequence in $\mathscr{D}'(\Omega)$ and

$$\langle u, \varphi \rangle = \lim_{j \to \infty} \langle u_j, \varphi \rangle$$

exists for every $\varphi \in C_0^\infty(\Omega)$, then we have $u \in \mathscr{D}'(\Omega)$. Thus, $u_j \to u$ in $\mathscr{D}'_s(\Omega)$ and hence in $\mathscr{D}'_b(\Omega)$. This is one of the important consequences of the Banach-Steinhaus theorem (Theorem 3.1.4).

3) The strong dual space of $\mathscr{D}'_b(\Omega)$ can be identified with $C_0^\infty(\Omega)$. This fact is referred to as the *reflexivity* of $C_0^\infty(\Omega)$.

Support of a Distribution

Two distributions u_1 and u_2 on Ω are said to be equal in an open subset V of Ω if the restrictions $u_1|_V$ and $u_2|_V$ are equal. In particular, $u = 0$ in V if and only if $\langle u, \varphi \rangle = 0$ for all $\varphi \in C_0^\infty(V)$.

The local behavior of a distribution determines it completely. More precisely, we have:

4.4.2 Theorem. *The space $\mathscr{D}'(\Omega)$ has the* sheaf property; *this means the following*:

(S1) *If $\{V_j\}$ is an open covering of Ω and a distribution $u \in \mathscr{D}'(\Omega)$ is zero in every V_j, then $u = 0$ in Ω.*

(S2) *Given an open covering $\{V_j\}$ of Ω, and a family of distributions $u_j \in \mathscr{D}'(V_j)$ such that $u_j = u_k$ in $V_j \cap V_k$, there exists a distribution $u \in \mathscr{D}'(\Omega)$ such that $u = u_j$ in each V_j.*

If $u \in \mathscr{D}'(\Omega)$, the *support* of u is the smallest closed subset of Ω outside of which u is zero. The support of u is denoted by supp u. We remark that if $\varphi \in C_0^\infty(\Omega)$ such that supp $\varphi \cap$ supp $u = \varnothing$, then $\langle u, \varphi \rangle = 0$. It is clear that the present definition of support coincides with the previous one if u is a continuous function on Ω.

Dual Space of $C^\infty(\Omega)$

The injection of $C_0^\infty(\Omega)$ into $C^\infty(\Omega)$ is continuous and the space $C_0^\infty(\Omega)$ is a dense subspace of $C^\infty(\Omega)$. Hence the dual space $\mathscr{E}'(\Omega)$ of $C^\infty(\Omega)$ can be identified with a linear subspace of $\mathscr{D}'(\Omega)$, by the identification of a continuous linear functional on $C^\infty(\Omega)$ with its restriction to $C_0^\infty(\Omega)$. In other words, the elements of $\mathscr{E}'(\Omega)$ are precisely those distributions that have continuous extensions to $C^\infty(\Omega)$.

More precisely, we have:

4.4.3 Theorem

(i) *The dual space* $\mathscr{E}'(\Omega)$ *of* $C^\infty(\Omega)$ *consists of those elements of* $\mathscr{D}'(\Omega)$ *with compact support.*

(ii) *The dual space* $\mathscr{E}'^m(\Omega)$ *of* $C_0^m(\Omega)$ $(0 \le m < \infty)$ *consists of those elements of* $\mathscr{D}'^m(\Omega)$ *with compact support, and* $\mathscr{E}'(\Omega) = \bigcup_{m=0}^{\infty} \mathscr{E}'^m(\Omega)$.

As in the case of $\mathscr{D}'(\Omega)$, we equip the space $\mathscr{E}'(\Omega)$ with two natural topologies τ_s and τ_b, and denote $(\mathscr{E}'(\Omega), \tau_s)$ and $(\mathscr{E}'(\Omega), \tau_b)$ by $\mathscr{E}'_s(\Omega)$ and $\mathscr{E}'_b(\Omega)$, respectively. We have the same topological properties of $\mathscr{E}'(\Omega)$ as those of $\mathscr{D}'(\Omega)$.

Tensor Product of Distributions

Let Ω_1 and Ω_2 be open subsets of $\mathbf{R}^{n_1}$ and $\mathbf{R}^{n_2}$, respectively. If $\varphi \in C_0^\infty(\Omega_1)$ and $\psi \in C_0^\infty(\Omega_2)$, we define the *tensor product* $\varphi \otimes \psi$ of φ and ψ by

$$(\varphi \otimes \psi)(x_1, x_2) = \varphi(x_1)\psi(x_2).$$

It is clear that $\varphi \otimes \psi \in C_0^\infty(\Omega_1 \times \Omega_2)$. We let

$C_0^\infty(\Omega_1) \otimes C_0^\infty(\Omega_2) =$ the space of finite linear combinations of the form $\varphi \otimes \psi$, where $\varphi \in C_0^\infty(\Omega_1)$ and $\psi \in C_0^\infty(\Omega_2)$.

The space $C_0^\infty(\Omega_1) \otimes C_0^\infty(\Omega_2)$ is a linear subspace of $C_0^\infty(\Omega_1 \times \Omega_2)$. Further, it is sequentially dense in $C_0^\infty(\Omega_1 \times \Omega_2)$, that is, for every $\Phi \in C_0^\infty(\Omega_1 \times \Omega_2)$, there exists a sequence $\{\Phi_j\}$ in $C_0^\infty(\Omega_1) \otimes C_0^\infty(\Omega_2)$ such that $\Phi_j \to \Phi$ in $C_0^\infty(\Omega_1 \times \Omega_2)$.

The sequential density of $C_0^\infty(\Omega_1) \otimes C_0^\infty(\Omega_2)$ in $C_0^\infty(\Omega_1 \times \Omega_2)$ allows us to obtain the following:

4.4.4 Theorem. *If* $u \in \mathscr{D}'(\Omega_1)$ *and* $v \in \mathscr{D}'(\Omega_2)$, *there exists a unique distribution* $u \otimes v \in \mathscr{D}'(\Omega_1 \times \Omega_2)$ *such that*

$$\langle u \otimes v, \Phi \rangle = \langle u, \varphi \rangle = \langle v, \psi \rangle, \qquad \Phi \in C_0^\infty(\Omega_1 \times \Omega_2),$$

where $\varphi(x_1) = \langle v, \Phi(x_1, \cdot) \rangle$ *and* $\psi(x_2) = \langle u, \Phi(\cdot, x_2) \rangle$.

The distribution $u \otimes v$ is called the *tensor product* of u and v.

We list some basic properties of the tensor product:

1. $\langle u \otimes v, \varphi \otimes \psi \rangle = \langle u, \varphi \rangle \langle v, \psi \rangle$, $\varphi \in C_0^\infty(\Omega_1)$, $\psi \in C_0^\infty(\Omega_2)$.
2. $\operatorname{supp}(u \otimes v) = \operatorname{supp} u \times \operatorname{supp} v$.
3. $D_{x_1}^\alpha D_{x_2}^\beta (u \otimes v) = D^\alpha u \otimes D^\beta v$.

Convolution of Distributions

Corollary 4.2.3 tells us that if $u \in L^1(\mathbf{R}^n)$ and $v \in L^p(\mathbf{R}^n)$ with $1 \leq p \leq \infty$, then the convolution

$$(u * v)(x) = \int_{\mathbf{R}^n} u(x - y)v(y)\, dy$$

is well-defined for almost all $x \in \mathbf{R}^n$, and is in $L^p(\mathbf{R}^n)$. Further it follows from Fubini's theorem (Theorem 1.19.4) that

$$\langle u * v, \varphi \rangle = \iint_{\mathbf{R}^n \times \mathbf{R}^n} u(x)v(y)\varphi(x + y)\, dx\, dy, \qquad \varphi \in C_0^\infty(\mathbf{R}^n).$$

We use this formula to extend the definition of convolution to the case of distributions.

Let u, $v \in \mathscr{D}'(\mathbf{R}^n)$ and suppose that one of them has compact support. If $\varphi \in C_0^\infty(\mathbf{R}^n)$, then the support of the function $\tilde{\varphi}\colon (x, y) \mapsto \varphi(x + y)$ is contained in the strip $\{(x, y) \in \mathbf{R}^n \times \mathbf{R}^n;\ x + y \in \operatorname{supp} \varphi\}$. Thus it is easy to see that the intersection $\operatorname{supp}(u \otimes v) \cap \operatorname{supp} \tilde{\varphi}$ is a compact subset of $\mathbf{R}^n \times \mathbf{R}^n$. We choose a function θ in $C_0^\infty(\mathbf{R}^n \times \mathbf{R}^n)$ such that $\theta = 1$ in a neighborhood of $\operatorname{supp}(u \otimes v) \cap \operatorname{supp} \tilde{\varphi}$, and define

$$\langle u \otimes v, \tilde{\varphi} \rangle = \langle u \otimes v, \theta\tilde{\varphi} \rangle.$$

Observe that $\langle u \otimes v, \theta\tilde{\varphi} \rangle$ is independent of the function θ chosen, and the mapping

$$C_0^\infty(\mathbf{R}^n) \ni \varphi \mapsto \langle u \otimes v, \tilde{\varphi} \rangle$$

is continuous. This discussion justifies the following definition:

The *convolution* $u * v$ of two distributions u and v in $\mathscr{D}'(\mathbf{R}^n)$, one of which has compact support, is a distribution on $\mathbf{R}^n$ defined by

$$\langle u * v, \varphi \rangle = \langle u \otimes v, \tilde{\varphi} \rangle, \qquad \varphi \in C_0^\infty(\mathbf{R}^n).$$

We state some basic facts concerning the convolution product:

1) $u * v = v * u$.

2) $\operatorname{supp}(u * v) \subset \operatorname{supp} u + \operatorname{supp} v = \{x + y;\ x \in \operatorname{supp} u,\ y \in \operatorname{supp} v\}$.

3) $D^\alpha(u * v) = (D^\alpha u) * v = u * (D^\alpha v)$.

4) If either $u \in \mathscr{D}'(\mathbf{R}^n)$, $v \in C_0^\infty(\mathbf{R}^n)$ or $u \in \mathscr{E}'(\mathbf{R}^n)$, $v \in C^\infty(\mathbf{R}^n)$, then we have $u * v \in C^\infty(\mathbf{R}^n)$ and

$$(u * v)(x) = \langle u_y, v(x - y) \rangle,$$

where u_y means that the distribution u operates on $v(x - y)$ as a function of y with x fixed.

5) Let ρ be a non-negative C^∞ function on $\mathbf{R}^n$ such that $\operatorname{supp} \rho = \{x \in \mathbf{R}^n; |x| \leq 1\}$ and $\int_{\mathbf{R}^n} \rho(x)\, dx = 1$, and let $\rho_\varepsilon(x) = \varepsilon^{-n} \rho(x/\varepsilon)$, $\varepsilon > 0$. If $u \in \mathscr{D}'(\mathbf{R}^n)$ (resp. $u \in \mathscr{E}'(\mathbf{R}^n)$), then the convolutions $u * \rho_\varepsilon$ are in $C^\infty(\mathbf{R}^n)$ (resp. $C_0^\infty(\mathbf{R}^n)$) and $u * \rho_\varepsilon \to u$ in $\mathscr{D}'(\mathbf{R}^n)$ (resp. $\mathscr{E}'(\mathbf{R}^n)$) as $\varepsilon \downarrow 0$. The functions $u * \rho_\varepsilon$ are called *regularizations* of u.

Jump Formula

If $x = (x_1, \ldots, x_n)$ is a point of $\mathbf{R}^n$, we write

$$x = (x', x_n), \qquad x' = (x_1, \ldots, x_{n-1}).$$

If $u \in C^\infty(\overline{\mathbf{R}^n_+})$, we define its extension u^0 to $\mathbf{R}^n$ by

$$u^0(x', x_n) = \begin{cases} u(x', x_n) & \text{if } x_n \geq 0, \\ 0 & \text{if } x_n < 0. \end{cases}$$

Then u^0 is a distribution on $\mathbf{R}^n$ and its j-th derivative $D_n^j(u^0)$ with respect to x_n is expressed as

$$D_n^j(u^0) = (D_n^j u)^0 + \frac{1}{i} \sum_{k=0}^{j-1} \gamma_{j-k-1} u \otimes D_n^k \delta(x_n). \tag{2}$$

Here $\gamma_k u$ is a C^∞ function on $\mathbf{R}^{n-1}_{x'}$ defined by

$$(\gamma_k u)(x') = \lim_{x_n \downarrow 0} D_n^k u(x', x_n),$$

and δ is the Dirac measure at the origin on $\mathbf{R}_{x_n}$.

More generally, if

$$P(x, D_x) = \sum_{j=0}^{m} P_j(x, D_{x'})\, D_n^j$$

is a differential operator of order m with C^∞ coefficients on $\mathbf{R}^n$, then we have

$$P(u^0) = (Pu)^0 + \frac{1}{i} \sum_{\ell+k+1 \leq m} P_{\ell+k+1}(x, D_{x'})\gamma_\ell u \otimes D_n^k \delta(x_n). \tag{3}$$

Here $P_j(x, D_{x'})$ is a differential operator of order $m - j$ with respect to x'. Formula (3) is referred to as the *jump formula*.

Regular Distributions with Respect to One Variable

If $x = (x_1, \ldots, x_n)$ is the variable in $\mathbf{R}^n$, we write

$$x = (x', x_n), \qquad x' = (x_1, \ldots, x_{n-1}),$$

so x' is the variable in $\mathbf{R}^{n-1}$.

A function $U(x_n)$, defined on $\mathbf{R}$ with values in $\mathscr{D}'(\mathbf{R}^{n-1}_{x'})$, is said to be *continuous* if, for every $\phi \in C_0^\infty(\mathbf{R}^{n-1}_{x'})$, the function $\langle U(x_n), \phi\rangle$ is continuous on $\mathbf{R}$.

We let

$$C(\mathbf{R}; \mathscr{D}'(\mathbf{R}^{n-1}_{x'})) = \text{the space of } \mathscr{D}'(\mathbf{R}^{n-1})\text{-valued continuous functions on } \mathbf{R}.$$

If $U \in C(\mathbf{R}; \mathscr{D}'(\mathbf{R}^{n-1}))$, we can associate injectively a distribution $u \in \mathscr{D}'(\mathbf{R}^n)$ by the formula

$$\langle u, \varphi\rangle = \int_{\mathbf{R}} \langle U(x_n), \varphi(\cdot, x_n)\rangle \, dx_n, \qquad \varphi \in C_0^\infty(\mathbf{R}^n).$$

Such a distribution u is said to be *continuous with respect to* x_n with values in $\mathscr{D}'(\mathbf{R}^{n-1}_{x'})$. We let

$$\gamma_0 u = U(0) \in \mathscr{D}'(\mathbf{R}^{n-1}_{x'}).$$

The distribution $\gamma_0 u$ is called the *sectional trace of order zero* of u on the hyperplane $\{x_n = 0\}$.

Let k be a positive integer. A function $U(x_n)$, defined on $\mathbf{R}$ with values in $\mathscr{D}'(\mathbf{R}^{n-1}_{x'})$, is said to be *of class* C^k if, for every $\phi \in C_0^\infty(\mathbf{R}^{n-1}_{x'})$, the function $\langle U(x_n), \phi\rangle$ is of class C^k on $\mathbf{R}$.

We let

$$C^k(\mathbf{R}; \mathscr{D}'(\mathbf{R}^{n-1})) = \text{the space of } \mathscr{D}'(\mathbf{R}^{n-1})\text{-valued } C^k \text{ functions on } \mathbf{R}.$$

If $U \in C^k(\mathbf{R}; \mathscr{D}'(\mathbf{R}^{n-1}))$, we have for $0 \leq j \leq k$:

$$\langle \partial_n^j u, \varphi\rangle = \int_{\mathbf{R}} \langle U^{(j)}(x_n), \varphi(\cdot, x_n)\rangle \, dx_n, \qquad \varphi \in C_0^\infty(\mathbf{R}^n).$$

This shows that the distribution $\partial_n^j u$ on $\mathbf{R}^n$ is the distribution associated with $U^{(j)} \in C(\mathbf{R}; \mathscr{D}'(\mathbf{R}^{n-1}))$. We say that u is *of class* C^k *with respect to* x_n, with values in $\mathscr{D}'(\mathbf{R}^{n-1}_{x'})$. We define the *sectional trace* $\gamma_j u$ *of order* j of u on the hyperplane $\{x_n = 0\}$ by

$$\gamma_j u = D_n^j U(0) \in \mathscr{D}'(\mathbf{R}^{n-1}), \qquad 0 \leq j \leq k.$$

We make no distinction between U and u for notational convenience.

It is obvious what we mean by $C^m([0, \infty); \mathscr{D}'(\mathbf{R}^{n-1}))$, $0 \leq m \leq \infty$. If $u \in C([0, \infty); \mathscr{D}'(\mathbf{R}^{n-1}))$, we define a distribution $u^0 \in \mathscr{D}'(\mathbf{R}^n)$ by the formula

$$\langle u^0, \varphi \rangle = \int_0^\infty \langle u(x_n), \varphi(\cdot, x_n) \rangle \, dx_n, \qquad \varphi \in C_0^\infty(\mathbf{R}^n).$$

The distribution u^0 is an extension of u to $\mathbf{R}^n$ which is equal to zero for $x_n < 0$.

If $u \in C^m([0, \infty); \mathscr{D}'(\mathbf{R}^{n-1}))$, we define its sectional traces $\gamma_j u$ $(0 \leq j \leq m)$ on the hyperplane $\{x_n = 0\}$ by

$$\gamma_j u = \lim_{x_n \downarrow 0} D_n^j u(\cdot, x_n) \qquad \text{in } \mathscr{D}'(\mathbf{R}^{n-1}).$$

Then it is easy to verify that formula (2) and hence the jump formula (3) can be extended to $C^m([0, \infty); \mathscr{D}'(\mathbf{R}^{n-1}))$.

The Fourier Transform

If $f \in L^1(\mathbf{R}^n)$, we define its *Fourier transform* $\hat{f}$ by the formula

$$\hat{f}(\xi) = \int_{\mathbf{R}^n} e^{-ix\cdot\xi} f(x) \, dx, \qquad \xi = (\xi_1, \ldots, \xi_n), \tag{4}$$

where $x \cdot \xi = x_1\xi_1 + \cdots + x_n\xi_n$. It follows from an application of the dominated convergence theorem (Theorem 1.19.3) that the function $\hat{f}$ is continuous on $\mathbf{R}^n$, and further we have

$$\|\hat{f}\|_\infty = \sup_{\xi \in \mathbf{R}^n} |\hat{f}(\xi)| \leq \|f\|_1.$$

We also denote $\hat{f}$ by $\mathscr{F}f$.

Similarly, if $g \in L^1(\mathbf{R}^n)$, we define

$$\check{g}(x) = \int_{\mathbf{R}^n} e^{ix\cdot\xi} g(\xi) \, đ\xi, \; đ\xi = \frac{1}{(2\pi)^n} \, d\xi.$$

The function $\check{g}$ is called the *inverse Fourier transform* of g. We also denote $\check{g}$ by $\mathscr{F}^* g$.

Now we introduce a subspace of $L^1(\mathbf{R}^n)$ which is invariant under the Fourier transform.

We let

$\mathscr{S}(\mathbf{R}^n)$ = the space of C^∞ functions φ on $\mathbf{R}^n$ such that, for any non-negative integer j, the quantity

$$p_j(\varphi) = \sup_{\substack{x \in \mathbf{R}^n \\ |\alpha| \le j}} \{(1 + |x|^2)^{j/2} |\partial^\alpha \varphi(x)|\}$$

is finite.

The space $\mathscr{S}(\mathbf{R}^n)$ is called the space of C^∞ functions on $\mathbf{R}^n$ *rapidly decreasing at infinity*. We equip the space $\mathscr{S}(\mathbf{R}^n)$ with the topology defined by the countable family $\{p_j\}$ of seminorms. It is easy to verify that $\mathscr{S}(\mathbf{R}^n)$ is complete; so it is a Fréchet space.

The next theorem summarizes the basic properties of the Fourier transform.

4.4.5 Theorem

(i) *The transforms* $\mathscr{F}$ *and* $\mathscr{F}^*$ *map* $\mathscr{S}(\mathbf{R}^n)$ *continuously into itself. Further we have*

$$\begin{cases} \widehat{D^\alpha \varphi}(\xi) = \xi^\alpha \hat{\varphi}(\xi), \\ D^\beta \hat{\varphi}(\xi) = \widehat{(-x)^\beta \varphi}(\xi), \end{cases} \qquad (\varphi \in \mathscr{S}(\mathbf{R}^n)),$$

for all multi-indices α *and* β.

(ii) *The transforms* $\mathscr{F}$ *and* $\mathscr{F}^*$ *are isomorphisms of* $\mathscr{S}(\mathbf{R}^n)$ *onto itself; more precisely* $\mathscr{F}\mathscr{F}^* = \mathscr{F}^*\mathscr{F} = I$ *on* $\mathscr{S}(\mathbf{R}^n)$. *In particular we have*

$$\varphi(x) = \int_{\mathbf{R}^n} e^{ix\cdot\xi} \hat{\varphi}(\xi)\, đ\xi, \qquad \varphi \in \mathscr{S}(\mathbf{R}^n). \tag{5}$$

(iii) *If* $\varphi, \psi \in \mathscr{S}(\mathbf{R}^n)$, *we have*

$$\int_{\mathbf{R}^n} \varphi(x)\hat{\psi}(x)\, dx = \int_{\mathbf{R}^n} \hat{\varphi}(\xi)\psi(\xi)\, d\xi, \tag{6}$$

$$\int_{\mathbf{R}^n} \varphi(x)\psi(x)\, dx = \int_{\mathbf{R}^n} \hat{\varphi}(\xi)\hat{\psi}(\xi)\, đ\xi. \tag{7}$$

Formula (5) is called the *Fourier inversion formula* and formula (7) is called the *Parseval formula*.

Tempered Distributions

For the spaces $C_0^\infty(\mathbf{R}^n)$, $\mathscr{S}(\mathbf{R}^n)$ and $C^\infty(\mathbf{R}^n)$, we have the following inclusions:

1) The injection of $C_0^\infty(\mathbf{R}^n)$ into $\mathscr{S}(\mathbf{R}^n)$ is continuous and the space $C_0^\infty(\mathbf{R}^n)$ is dense in $\mathscr{S}(\mathbf{R}^n)$.

2) The injection of $\mathscr{S}(\mathbf{R}^n)$ into $C^\infty(\mathbf{R}^n)$ is continuous and the space $\mathscr{S}(\mathbf{R}^n)$ is dense in $C^\infty(\mathbf{R}^n)$.

Hence the dual space $\mathscr{S}'(\mathbf{R}^n)$ of $\mathscr{S}(\mathbf{R}^n)$ can be identified with a linear subspace of $\mathscr{D}'(\mathbf{R}^n)$ containing $\mathscr{E}'(\mathbf{R}^n)$, by the identification of a continuous linear functional on $\mathscr{S}(\mathbf{R}^n)$ with its restriction to $C_0^\infty(\mathbf{R}^n)$. That is, we have

$$\mathscr{E}'(\mathbf{R}^n) \subset \mathscr{S}'(\mathbf{R}^n) \subset \mathscr{D}'(\mathbf{R}^n).$$

The elements of $\mathscr{S}'(\mathbf{R}^n)$ are called *tempered distributions* on $\mathbf{R}^n$. In other words, the tempered distributions are precisely those distributions on $\mathbf{R}^n$ that have continuous extensions to $\mathscr{S}(\mathbf{R}^n)$.

Now we give examples of tempered distributions. Roughly speaking, the tempered distributions are those which have at most polynomial growth at infinity, since the functions in $\mathscr{S}(\mathbf{R}^n)$ die out faster than any power of x at infinity.

1. The functions in $L^p(\mathbf{R}^n)$ $(1 \leq p \leq \infty)$ are tempered distributions.
2. A locally integrable function on $\mathbf{R}^n$ is a tempered distribution if it has at most polynomial growth at infinity.
3. If $u \in \mathscr{S}'(\mathbf{R}^n)$ and f is a C^∞ function on $\mathbf{R}^n$ all of whose derivatives have at most polynomial growth at infinity, then the product fu is a tempered distribution.
4. Any derivative of a tempered distribution is also a tempered distribution.

The importance of tempered distributions lies in the fact that they have Fourier transforms.

If $u \in \mathscr{S}'(\mathbf{R}^n)$, we define its *Fourier transform* $\mathscr{F}u$ by the formula

$$\langle \mathscr{F}u, \varphi \rangle = \langle u, \mathscr{F}\varphi \rangle, \qquad \varphi \in \mathscr{S}(\mathbf{R}^n). \tag{8}$$

Then we have $\mathscr{F}u \in \mathscr{S}'(\mathbf{R}^n)$, since the Fourier transform $\mathscr{F} : \mathscr{S}(\mathbf{R}^n) \to \mathscr{S}(\mathbf{R}^n)$ is an isomorphism. Further, in view of formula (6), it follows that the above definition (8) agrees with definition (4) if $u \in \mathscr{S}(\mathbf{R}^n)$. We also denote $\mathscr{F}u$ by $\hat{u}$.

Similarly, if $v \in \mathscr{S}'(\mathbf{R}^n)$, we define its *inverse Fourier transform* $\mathscr{F}^*v$ by the formula

$$\langle \mathscr{F}^*v, \psi \rangle = \langle v, \mathscr{F}^*\psi \rangle, \qquad \psi \in \mathscr{S}(\mathbf{R}^n).$$

The next theorem, which is a consequence of Theorem 4.4.5, summarizes the basic properties of Fourier transforms in the space $\mathscr{S}'(\mathbf{R}^n)$.

4.4.6 Theorem

(i) *The transforms* $\mathscr{F}$ *and* $\mathscr{F}^*$ *map* $\mathscr{S}'(\mathbf{R}^n)$ *continuously into itself. Further we have*

$$\begin{cases} \mathscr{F}(D^\alpha u) = \xi^\alpha \mathscr{F} u, \\ D^\beta(\mathscr{F} u) = \mathscr{F}((-x)^\beta u), \end{cases} \qquad (u \in \mathscr{S}'(\mathbf{R}^n)).$$

(ii) *The transforms* $\mathscr{F}$ *and* $\mathscr{F}^*$ *are isomorphisms of* $\mathscr{S}'(\mathbf{R}^n)$ *onto itself; more precisely,* $\mathscr{F}\mathscr{F}^* = \mathscr{F}^*\mathscr{F} = I$ *on* $\mathscr{S}'(\mathbf{R}^n)$.

(iii) *The transforms* $\mathscr{F}$ *and* $\mathscr{F}^*$ *are norm-preserving operators on* $L^2(\mathbf{R}^n)$ *and* $\mathscr{F}\mathscr{F}^* = \mathscr{F}^*\mathscr{F} = I$ *on* $L^2(\mathbf{R}^n)$.

Assertion (iii) is referred to as the *Plancherel theorem.*

As for distributions with compact support, we have:

4.4.7 Theorem. *If* $u \in \mathscr{E}'(\mathbf{R}^n)$, *then its Fourier transform* $\mathscr{F}u$ *is a* C^∞ *function on* $\mathbf{R}^n$ *given by*

$$\mathscr{F}u(\xi) = \langle u, e^{-ix\cdot\xi}\rangle, \qquad \xi \in \mathbf{R}^n.$$

Moreover, the function $\mathscr{F}u$ *is slowly increasing, that is, there exist constants* $C > 0$ *and* μ *such that*

$$|\mathscr{F}u(\xi)| \leq C(1 + |\xi|)^\mu, \qquad \xi \in \mathbf{R}^n.$$

4.5. Operators and Kernels

Let Ω_1 and Ω_2 be open subsets of $\mathbf{R}^{n_1}$ and $\mathbf{R}^{n_2}$, respectively. If $K \in \mathscr{D}'(\Omega_1 \times \Omega_2)$, we can define a continuous linear operator $A \in L(C_0^\infty(\Omega_2), \mathscr{D}'(\Omega_1))$ by the formula

$$\langle A\psi, \varphi\rangle = \langle K, \varphi \otimes \psi\rangle, \qquad \varphi \in C_0^\infty(\Omega_1), \qquad \psi \in C_0^\infty(\Omega_2).$$

We then write $A = \mathrm{Op}(K)$. Since the space $C_0^\infty(\Omega_1) \otimes C_0^\infty(\Omega_2)$ is sequentially dense in $C_0^\infty(\Omega_1 \times \Omega_2)$, it follows that the mapping

$$\mathscr{D}'(\Omega_1 \times \Omega_2) \ni K \mapsto \mathrm{Op}(K) \in L(C_0^\infty(\Omega_2), \mathscr{D}'(\Omega_1))$$

is injective. The next theorem asserts that it is also surjective.

4.5.1 Theorem (the Schwartz kernel theorem). *If A is a continuous linear operator on $C_0^\infty(\Omega_2)$ into $\mathscr{D}'(\Omega_1)$, then there exists a unique distribution $K_A \in \mathscr{D}'(\Omega_1 \times \Omega_2)$ such that $A = \mathrm{Op}(K_A)$.*

The distribution K_A is called the *kernel* of A. Formally we have

$$A\psi(x_1) = \int_{\Omega_2} K_A(x_1, x_2)\psi(x_2)\,dx_2, \qquad \psi \in C_0^\infty(\Omega_2).$$

If $A: C_0^\infty(\Omega_2) \to \mathscr{D}'(\Omega_1)$ is a continuous linear operator, we define its *transpose* A' by the formula

$$\langle A'\varphi, \psi\rangle = \langle \varphi, A\psi\rangle, \qquad \varphi \in C_0^\infty(\Omega_1), \qquad \psi \in C_0^\infty(\Omega_2).$$

Then the transpose A' is a continuous linear operator on $C_0^\infty(\Omega_1)$ into $\mathscr{D}'(\Omega_2)$. The kernel of A' is obtained from the kernel K_A of A by interchanging the roles of x_1 and x_2; formally this means that

$$A'\varphi(x_2) = \int_{\Omega_1} K_A(x_1, x_2)\varphi(x_1)\,dx_1, \qquad \varphi \in C_0^\infty(\Omega_1).$$

Clearly we have $(A')' = A$.

Since the spaces $C^\infty(\Omega_1)$ and $C_0^\infty(\Omega_2)$ are both reflexive, it follows that a linear operator A maps $C_0^\infty(\Omega_2)$ continuously into $C^\infty(\Omega_1)$ if and only if its transpose A' extends to a continuous linear operator on $\mathscr{E}'_b(\Omega_1)$ into $\mathscr{D}'_b(\Omega_2)$.

Similarly, if $A: C_0^\infty(\Omega_2) \to \mathscr{D}'(\Omega_1)$ is a continuous linear operator, we define its *adjoint* A^* by the formula

$$\langle A^*\varphi, \bar{\psi}\rangle = \langle \varphi, \overline{A\psi}\rangle, \qquad \varphi \in C_0^\infty(\Omega_1), \qquad \psi \in C_0^\infty(\Omega_2).$$

Then the adjoint A^* is a continuous linear operator on $C_0^\infty(\Omega_1)$ into $\mathscr{D}'(\Omega_2)$. The kernel of A^* is obtained from the kernel $\overline{K_A}$ by interchanging the roles of x_1 and x_2; formally this means that

$$A^*\varphi(x_2) = \int_{\Omega_1} \overline{K_A(x_1, x_2)}\varphi(x_1)\,dx_1, \qquad \varphi \in C_0^\infty(\Omega_1).$$

We also have $(A^*)^* = A$.

For example, if $\Omega_1 = \Omega_2$ is an open subset Ω of $\mathbf{R}^n$, and if

$$A = \sum_{|\alpha| \leq m} a_\alpha(x) D^\alpha$$

is a differential operator of order m with C^∞ coefficients on Ω, then we have

$$\begin{cases} A' = \sum_{|\alpha| \leq m} (-1)^{|\alpha|} D^\alpha(a_\alpha(x) \cdot), \\ A^* = \sum_{|\alpha| \leq m} (-1)^{|\alpha|} D^\alpha(\overline{a_\alpha(x)} \cdot). \end{cases}$$

This shows that A' and A^* are also differential operators of order m with C^∞ coefficients on Ω.

A continuous linear operator $A: C_0^\infty(\Omega_2) \to \mathscr{D}'(\Omega_1)$ is called a *regularizer* if it extends to a continuous linear operator on $\mathscr{E}'_b(\Omega_2)$ into $C^\infty(\Omega_1)$.

The next theorem gives a characterization of regularizers in terms of kernels.

4.5.2 Theorem. *A continuous linear operator $A: C_0^\infty(\Omega_2) \to \mathscr{D}'(\Omega_1)$ is a regularizer if and only if its kernel K_A is in $C^\infty(\Omega_1 \times \Omega_2)$.*

A continuous linear operator $A: C_0^\infty(\Omega_2) \to \mathscr{D}'(\Omega_1)$ is said to be *properly supported* if the restrictions to supp K_A of the projections $p_i: \Omega_1 \times \Omega_2 \to \Omega_i$ $(i = 1, 2)$ are proper mappings. It is easy to see that A is properly supported if and only if the following two conditions are satisfied:

(a) For any compact subset K_2 of Ω_2, there exists a compact subset K_1 of Ω_1 such that

$$\operatorname{supp} v \subset K_2 \quad \Rightarrow \quad \operatorname{supp} Av \subset K_1.$$

(b) For any compact subset K_1 of Ω_1, there exists a compact subset K_2 of Ω_2 such that

$$\operatorname{supp} v \cap K_2 = \varnothing \quad \Rightarrow \quad \operatorname{supp} Av \cap K_1 = \varnothing.$$

If A is properly supported, then it maps $C_0^\infty(\Omega_2)$ continuously into $\mathscr{E}'(\Omega_1)$, and further it extends uniquely to a continuous linear operator on $C^\infty(\Omega_2)$ into $\mathscr{D}'(\Omega_1)$.

We remark that A is properly supported if and only if A' (or A^*) is properly supported.

4.6. Distributions on a Manifold

Let M be an n-dimensional C^∞ manifold (without boundary) which satisfies the second axiom of countability. By virtue of Theorem 2.1.1, it follows that M is paracompact.

We let

$$C^\infty(M) = \text{the space of } C^\infty \text{ functions on } M.$$

We equip the space $C^\infty(M)$ with the topology defined by the family of seminorms

$$\varphi \mapsto p(\varphi \circ \chi^{-1}), \qquad (\varphi \in C^\infty(M)),$$

where (U, χ) ranges over all admissible charts on M and p ranges over all seminorms on $C^\infty(\chi(U))$ such as (4.2.2). By using a partition of unity, one can verify that the topology on $C^\infty(M)$ is defined by the family of seminorms associated with an atlas on M alone. But, since M satisfies the second axiom of countability, there exists an atlas on M consisting of countably many charts. This shows that $C^\infty(M)$ is metrizable. Further it is easy to see that $C^\infty(M)$ is complete; hence it is a Fréchet space.

If K is a compact subset of M, we let

$$C_K^\infty(M) = \text{the space of } C^\infty \text{ functions on } M \text{ with support in } K.$$

The space $C_K^\infty(M)$ is a closed subspace of $C^\infty(M)$. Further we let

$$C_0^\infty(M) = \bigcup_{K \subset M} C_K^\infty(M),$$

where K ranges over all compact subsets of M. We equip the space $C_0^\infty(M)$ with the inductive limit topology of the spaces $C_K^\infty(M)$.

We let

$$C^\infty(|M|) = \text{the space of } C^\infty \text{ densities on } M,$$

$$C_0^\infty(|M|) = \text{the space of } C^\infty \text{ densities on } M \text{ with compact support.}$$

Since M is paracompact, it follows from Theorems 2.8.2 and 2.11.1 that M admits a strictly positive C^∞ density μ. Hence we can identify $C^\infty(|M|)$ with $C^\infty(M)$ as linear topological spaces by the isomorphism

$$C^\infty(M) \to C^\infty(|M|),$$

$$\varphi \mapsto \varphi \cdot \mu.$$

Similarly, the space $C_0^\infty(|M|)$ can be identified with $C_0^\infty(M)$.

A *distribution* on M is a continuous linear functional on $C_0^\infty(|M|)$. The space of distributions on M is denoted by $\mathscr{D}'(M)$. Thus $\mathscr{D}'(M)$ is the dual space of $C_0^\infty(|M|)$. If $\varphi \in C_0^\infty(M)$ and $u \in \mathscr{D}'(M)$, we denote the action of u on $\varphi \cdot \mu$ by $\langle u, \varphi \cdot \mu \rangle$ or sometimes by $\langle \varphi \cdot \mu, u \rangle$.

A function u defined on M is said to be in $L^1_{\text{loc}}(M)$ if, for any admissible chart (U, χ) on M, the local representative $u \circ \chi^{-1}$ of u is in $L^1_{\text{loc}}(\chi(U))$. The elements of $L^1_{\text{loc}}(M)$ are called *locally integrable functions* on M. Every element u of $L^1_{\text{loc}}(M)$ defines a distribution on M by the formula

$$\langle u, \varphi \cdot \mu \rangle = \int_M u\varphi \cdot \mu, \qquad \varphi \in C_0^\infty(M).$$

We list some basic properties of distributions on a manifold:

1. If V is an open subset of M, then a distribution $u \in \mathscr{D}'(M)$ defines a distribution $u|_V \in \mathscr{D}'(V)$ by restriction to $C_0^\infty(|V|)$.
2. The space $\mathscr{D}'(M)$ has the sheaf property.
3. The space of distributions with compact support can be identified with the dual space $\mathscr{E}'(M)$ of $C^\infty(|M|)$.

We have the same topological properties of $\mathscr{D}'(M)$ and $\mathscr{E}'(M)$ as those of $\mathscr{D}'(\Omega)$ and $\mathscr{E}'(\Omega)$ stated in Section 4.4.

4.7. Differential Operators on a Manifold

Let M be an n-dimensional C^∞ manifold (without boundary). If P is a linear mapping of $C^\infty(M)$ into itself, and if (U, χ) is a chart on M, we let

$$P_\chi = (\chi^{-1})^* \circ (P|_U) \circ \chi^*,$$

where $P|_U$ is the restriction of P to U. Then P_χ is a linear mapping of $C^\infty(\chi(U))$ into itself, as represented in the following diagram:

$$\begin{array}{ccc} C^\infty(U) & \xrightarrow{P|_U} & C^\infty(U) \\ \chi^* \uparrow & & \downarrow (\chi^{-1})^* \\ C^\infty(\chi(U)) & \xrightarrow[P_\chi]{} & C^\infty(\chi(U)) \end{array}$$

A continuous linear mapping $P: C^\infty(M) \to C^\infty(M)$ is called a *differential operator* of order m on M, if, for any chart (U, χ) on M, the mapping P_χ is a differential operator of order m on $\chi(U) \subset \mathbf{R}^n$.

4.7.1 Example. Let (M, g) be an n-dimensional, Riemannian C^∞ manifold. The *Laplace–Beltrami operator*, or simply the *Laplacian* Δ_M of M, is a second-order differential operator defined in local coordinates by the formula

$$\Delta_M = \sum_{k,\ell=1}^{n} (\det(g_{ij}))^{-1/2} \frac{\partial}{\partial x_k} \left\{ (\det(g_{ij}))^{1/2} g^{k\ell} \frac{\partial}{\partial x_\ell} \right\},$$

where

$$g_{ij} = g\left(\frac{\partial}{\partial x_i}, \frac{\partial}{\partial x_j}\right),$$

$$(g^{ij}) = \text{the inverse matrix of } (g_{ij}).$$

If $M = \mathbf{R}^n$ with standard Euclidean metric $(g_{ij}) = (\delta_{ij})$, the operator Δ_M becomes the usual Laplacian

$$\Delta = \frac{\partial^2}{\partial x_1^2} + \cdots + \frac{\partial^2}{\partial x_n^2}.$$

4.8. Operators and Kernels on a Manifold

Let M and N be C^∞ manifolds equipped with strictly positive densities μ and ν, respectively.

If $K \in \mathscr{D}'(M \times N)$, we can define a continuous linear operator $A: C_0^\infty(N) \to \mathscr{D}'(M)$ by the formula

$$\langle A\psi, \varphi \cdot \mu \rangle = \langle K, \varphi \cdot \mu \otimes \psi \cdot \nu \rangle, \qquad \varphi \in C_0^\infty(M), \qquad \psi \in C_0^\infty(N).$$

If $A: C_0^\infty(N) \to \mathscr{D}'(M)$ is a continuous linear operator, we define its *transpose* A' by the formula

$$\langle A'\varphi, \psi \cdot \nu \rangle = \langle \varphi \cdot \mu, A\psi \rangle, \qquad \varphi \in C_0^\infty(M), \qquad \psi \in C_0^\infty(N).$$

Then the transpose A' is a continuous linear operator on $C_0^\infty(M)$ into $\mathscr{D}'(N)$. Also we have $(A')' = A$.

Similarly, we define the *adjoint* A^* of A by the formula

$$\langle A^*\varphi, \bar{\psi} \cdot \nu \rangle = \langle \varphi \cdot \mu, \overline{A\psi} \rangle, \qquad \varphi \in C_0^\infty(M), \qquad \psi \in C_0^\infty(N).$$

Then the adjoint A^* is a continuous linear operator on $C_0^\infty(M)$ into $\mathscr{D}'(N)$, and we have $(A^*)^* = A$.

We remark that the results in Section 4.5 extend to this case.

4.9. Domains of Class C^r

An open subset of $\mathbf{R}^n$ is called a *domain* if it is also connected. Let $0 \le r \le \infty$. A domain Ω in $\mathbf{R}^n$ with boundary $\partial\Omega$ is said to be *of class* C^r or a C^r *domain* if, at each point x' of $\partial\Omega$, there exist a neighborhood U of x' in $\mathbf{R}^n$ and a bijection χ of U onto $B = \{x = (x_1, \ldots, x_n) \in \mathbf{R}^n; |x| < 1\}$ such that:

(a) $\chi(U \cap \Omega) = B \cap \{x_n > 0\}$,
$\chi(U \cap \partial\Omega) = B \cap \{x_n = 0\}$.
(b) $\chi \in C^r(U)$, $\chi^{-1} \in C^r(B)$.

In other words, a C^r domain is an n-dimensional C^r manifold with boundary.

Note that a domain Ω is of class C^r if each point of the boundary $\partial\Omega$ has a neighborhood in which $\partial\Omega$ is the graph of a C^r function of $n - 1$ of the variables $x_1, \ldots, x_n$.

4.10. The Seeley Extension Theorem

The next theorem shows that if a domain Ω is of class C^∞, then the functions in $C^\infty(\bar{\Omega})$ are the restrictions to Ω of functions in $C^\infty(\mathbf{R}^n)$.

4.10.1 Theorem (Seeley). *Let Ω be either the half space $\mathbf{R}^n_+$ or a C^∞ domain in $\mathbf{R}^n$ with bounded boundary. Then there exists a continuous linear extension operator E: $C^\infty(\bar{\Omega}) \to C^\infty(\mathbf{R}^n)$. Further, the restriction of E to $C_0^\infty(\bar{\Omega})$ is a continuous linear extension operator on $C_0^\infty(\bar{\Omega})$ into $C_0^\infty(\mathbf{R}^n)$.*

Proof

(i) First we let $\Omega = \mathbf{R}^n_+$. The proof is based on the following:

4.10.2 Lemma. *There exists a function w in $\mathscr{S}(\mathbf{R})$ such that*

$$\begin{cases} \operatorname{supp} w \subset [1, \infty), \\ \displaystyle\int_1^\infty t^n w(t)\, dt = (-1)^n, \qquad n = 0, 1, 2, \ldots. \end{cases}$$

Assuming this lemma, whose proof we leave to the reader, we shall prove the theorem.

If $\varphi \in C^\infty(\overline{\mathbf{R}^n_+})$, we define

$$E\varphi(x', x_n) = \begin{cases} \varphi(x', x_n) & \text{if } x_n \geq 0, \\ \displaystyle\int_1^\infty w(s)\theta(-x_n s)\varphi(x', -sx_n)\, ds & \text{if } x_n < 0, \end{cases}$$

where $x = (x', x_n)$, $x' = (x_1, \ldots, x_{n-1})$ and $\theta \in C_0^\infty(\mathbf{R})$ with supp $\theta \subset [-2, 2]$ and $\theta(t) = 1$ for $|t| \leq 1$. Then it is easy to verify the following:

1. $E\varphi \in C^\infty(\mathbf{R}^n)$;
2. $E\colon C^\infty(\overline{\mathbf{R}^n_+}) \to C^\infty(\mathbf{R}^n)$ is continuous;
3. If supp $\varphi \subset \{x \in \mathbf{R}^n; |x'| \leq r, 0 \leq x_n \leq a\}$ for some $r > 0$ and $a > 0$, then supp $E\varphi \subset \{x \in \mathbf{R}^n; |x'| \leq r, |x_n| \leq a\}$.

This proves the theorem for the half space $\mathbf{R}^n_+$.

(ii) Now suppose that Ω is a C^∞ domain in $\mathbf{R}^n$ with bounded boundary $\partial\Omega$. Then we can choose a finite covering $\{U_j\}_{j=1}^N$ of $\partial\Omega$ by open subsets of $\mathbf{R}^n$, and C^∞ diffeomorphisms χ_j of U_j onto $B = \{x \in \mathbf{R}^n; |x| < 1\}$ such that the open sets $V_j = \chi_j^{-1}(\{x \in \mathbf{R}^n; |x'| < \frac{1}{2}, |x_n| < \sqrt{3}/2\})$, $1 \leq j \leq N$, form an open covering of $\Omega_\delta = \{x \in \Omega; \operatorname{dist}(x, \partial\Omega) < \delta\}$ for some $\delta > 0$. Further we can choose an open set V_0 in Ω, bounded away from $\partial\Omega$, such that $\Omega \subset (\bigcup_{j=1}^N V_j) \cup V_0$. Let $\{\omega_j\}_{j=0}^N$ be a partition of unity subordinate to the covering $\{V_j\}_{j=0}^N$. If $\varphi \in C^\infty(\bar{\Omega})$, we define

$$E\varphi = \omega_0 \varphi + \sum_{j=1}^N \chi_j^*(E((\chi_j^{-1})^*(\omega_j \varphi))).$$

Then it is easy to verify that this operator E enjoys the desired properties.

Notes

Schwartz [1] and Gelfand-Shilov [1] are the classics for distribution theory; see also Barros-Neto [1] and Treves [1]. Our treatment here follows the expositions of Chazarain–Piriou [1] and Hörmander [1].

Section 4.3: A characterization of differential operators, Theorem 4.3.1, is due to Peetre [1].

Section 4.6: Distributions on a manifold were first studied by de Rham [1].

Section 4.10: The extension theorem, Theorem 4.10.1, is due to Seeley [1], although our proof differs somewhat from that of Seeley.

5 Sobolev Spaces

One of the most useful ways of measuring differentiability properties of functions on $\mathbf{R}^n$ is in terms of L^2-norms, and is provided by the Sobolev spaces on $\mathbf{R}^n$. The great advantage of this approach lies in the fact that the Fourier transform works very well in the Hilbert space $L^2(\mathbf{R}^n)$. The purpose of this chapter is to summarize the basic definitions and results about Sobolev spaces which will be needed for the study of boundary value problems in Chapter 8.

5.1. The Spaces $H^s(\mathbf{R}^n)$

If $s \in \mathbf{R}$, we let

$H^s(\mathbf{R}^n)$ = the space of distributions $u \in \mathscr{S}'(\mathbf{R}^n)$ such that $\hat{u} = \mathscr{F}u$ is a locally integrable function and $(1 + |\xi|^2)^{s/2}\hat{u} \in L^2(\mathbf{R}^n)$.

We equip the space $H^s(\mathbf{R}^n)$ with the inner product

$$(u, v)_s = \int_{\mathbf{R}^n} (1 + |\xi|^2)^s \hat{u}(\xi)\overline{\hat{v}(\xi)}\, đ\xi, \qquad đ\xi = \frac{1}{(2\pi)^n}\, d\xi,$$

and with the associated norm

$$\|u\|_s = \left(\int_{\mathbf{R}^n} (1 + |\xi|^2)^s |\hat{u}(\xi)|^2 \, d\!\!{}^-\xi \right)^{1/2}.$$

The space $H^s(\mathbf{R}^n)$ is called the *Sobolev space* of order s. Roughly speaking, the order s "counts" the number of L^2-derivatives of elements in $H^s(\mathbf{R}^n)$ (cf. Theorems 5.1.1 and 5.1.2 below).

We define a linear map

$$\Lambda^s \colon \mathscr{S}'(\mathbf{R}^n) \to \mathscr{S}'(\mathbf{R}^n)$$

by the formula

$$\Lambda^s u = \mathscr{F}^*((1 + |\xi|^2)^{s/2} \mathscr{F} u), \qquad u \in \mathscr{S}'(\mathbf{R}^n).$$

This can be visualized as follows:

$$\mathscr{S}'(\mathbf{R}^n) \xrightarrow{\mathscr{F}} \mathscr{S}'(\mathbf{R}^n) \xrightarrow{(1+|\xi|^2)^{s/2}} \mathscr{S}'(\mathbf{R}^n) \xrightarrow{\mathscr{F}^*} \mathscr{S}'(\mathbf{R}^n).$$

Thus the map Λ^s is an isomorphism of $\mathscr{S}'(\mathbf{R}^n)$ onto itself, and its inverse is the map Λ^{-s}. Further it follows from an application of the Plancherel theorem (Theorem 4.4.6) that:

(a) $u \in H^s(\mathbf{R}^n)$ if and only if $\Lambda^s u \in L^2(\mathbf{R}^n)$;
(b) $(u, v)_s = \int_{\mathbf{R}^n} \Lambda^s u(x) \overline{\Lambda^s v(x)} \, dx$.

This shows that Λ^s is an isometric isomorphism of $H^s(\mathbf{R}^n)$ onto $L^2(\mathbf{R}^n)$. Hence the Sobolev space $H^s(\mathbf{R}^n)$ is a Hilbert space. In particular we have

$$H^0(\mathbf{R}^n) = L^2(\mathbf{R}^n).$$

We list some basic topological properties of $H^s(\mathbf{R}^n)$:

1) If $s > t$, then we have the inclusions

$$\mathscr{S}(\mathbf{R}^n) \subset H^s(\mathbf{R}^n) \subset H^t(\mathbf{R}^n) \subset \mathscr{S}'(\mathbf{R}^n)$$

with continuous injections.

We let

$$H^\infty(\mathbf{R}^n) = \bigcap_{s \in \mathbf{R}} H^s(\mathbf{R}^n),$$
$$H^{-\infty}(\mathbf{R}^n) = \bigcup_{s \in \mathbf{R}} H^s(\mathbf{R}^n).$$

Then we have the inclusions

$$\mathscr{S}(\mathbf{R}^n) \subset H^{\infty}(\mathbf{R}^n),$$
$$\mathscr{E}'(\mathbf{R}^n) \subset H^{-\infty}(\mathbf{R}^n).$$

The second inclusion follows from Theorem 4.4.7.

2) The space $\mathscr{S}(\mathbf{R}^n)$ is dense in $H^s(\mathbf{R}^n)$ for each $s \in \mathbf{R}$.

In fact, since $\mathscr{S}(\mathbf{R}^n)$ is dense in $L^2(\mathbf{R}^n)$, it follows that $\mathscr{S}(\mathbf{R}^n) = \Lambda^{-s}(\mathscr{S}(\mathbf{R}^n))$ is dense in $H^s(\mathbf{R}^n) = \Lambda^{-s}(L^2(\mathbf{R}^n))$.

The next two theorems give a direct description of $H^s(\mathbf{R}^n)$ when $s > 0$.

5.1.1 Theorem. *Let m be a positive integer. Then the Sobolev space $H^m(\mathbf{R}^n)$ is the space of functions $u \in L^2(\mathbf{R}^n)$ such that $D^\alpha u \in L^2(\mathbf{R}^n)$ for $|\alpha| \le m$. Furthermore, the norm $\|u\|_m$ is equivalent to the norm*

$$\left(\sum_{|\alpha| \le m} \int_{\mathbf{R}^n} |D^\alpha u(x)|^2 \, dx \right)^{1/2}.$$

5.1.2 Theorem. *Let $s = m + \sigma$, where m is a positive integer and $0 < \sigma < 1$. Then the Sobolev space $H^s(\mathbf{R}^n)$ is the space of functions $u \in H^m(\mathbf{R}^n)$ such that*

$$\iint_{\mathbf{R}^n \times \mathbf{R}^n} \frac{|D^\alpha u(x) - D^\alpha u(y)|^2}{|x - y|^{n+2\sigma}} \, dx \, dy < +\infty$$

for $|\alpha| = m$. Furthermore, the norm $\|u\|_s$ is equivalent to the norm

$$\left(\|u\|_m^2 + \sum_{|\alpha| = m} \iint_{\mathbf{R}^n \times \mathbf{R}^n} \frac{|D^\alpha u(x) - D^\alpha u(y)|^2}{|x - y|^{n+2\sigma}} \, dx \, dy \right)^{1/2}.$$

The next theorem states that the elements of $H^s(\mathbf{R}^n)$ are smooth in the classical sense for sufficiently large $s > 0$.

5.1.3 Theorem (Sobolev). *If $s > n/2 + k$, where k is a non-negative integer, then we have the inclusion*

$$H^s(\mathbf{R}^n) \subset C^k(\mathbf{R}^n)$$

with continuous injection.

This theorem is one of many *Sobolev imbedding theorems*.

Since the Sobolev space $H^s(\mathbf{R}^n)$ is a Hilbert space, it is its own dual space. But it is more useful to consider the following characterization of the dual space of $H^s(\mathbf{R}^n)$:

5.1.4 Theorem. *The bilinear form* $\langle\ ,\ \rangle$ *on the product space* $\mathscr{S}(\mathbf{R}^n) \times \mathscr{S}(\mathbf{R}^n)$ *defined by*

$$\{u, v\} \mapsto \langle u, v\rangle = \int_{\mathbf{R}^n} u(x)v(x)\, dx$$

extends uniquely to a continuous bilinear form $\langle\ ,\ \rangle$ *on the product space* $H^s(\mathbf{R}^n) \times H^{-s}(\mathbf{R}^n)$ *for each* $s \in \mathbf{R}$, *given by*

$$\{u, v\} \mapsto \langle u, v\rangle = \int_{\mathbf{R}^n} \hat{u}(\xi)\hat{v}(-\xi)\, đ\xi.$$

This bilinear form on the space $H^s(\mathbf{R}^n) \times H^{-s}(\mathbf{R}^n)$ *permits us to identify the strong dual space of* $H^s(\mathbf{R}^n)$ *with* $H^{-s}(\mathbf{R}^n)$.

If F is a closed subset of $\mathbf{R}^n$, we let

$H^s_F(\mathbf{R}^n) =$ the subspace of $H^s(\mathbf{R}^n)$ consisting of the elements with support in F.

Since the injection $H^s(\mathbf{R}^n) \to \mathscr{D}'(\mathbf{R}^n)$ is continuous, it follows that $H^s_F(\mathbf{R}^n)$ is a closed subspace of $H^s(\mathbf{R}^n)$; hence it is a Hilbert space.

The next theorem is a Sobolev space version of the Ascoli-Arzelà theorem.

5.1.5 Theorem (Rellich). *Let* K *be a compact subset of* $\mathbf{R}^n$. *If* $s > t$, *then the injection* $H^s_K(\mathbf{R}^n) \to H^t_K(\mathbf{R}^n)$ *is compact.*

If $\|\cdot\|_{s_1}$ and $\|\cdot\|_{s_2}$ are two Sobolov norms, then the intermediate norms between them are estimated as follows:

5.1.6 Theorem. *Let* s_1, s, s_2 *be real numbers such that* $s_1 < s < s_2$. *For every* $\varepsilon > 0$, *there exists a constant* $C_\varepsilon > 0$ *such that*

$$\|u\|_s^2 \le \varepsilon\|u\|_{s_2}^2 + C_\varepsilon\|u\|_{s_1}^2, \qquad u \in H^{s_2}(\mathbf{R}^n).$$

This inequality is called the *interpolation inequality.*

5.2. The Spaces $H^s_{\mathrm{loc}}(\Omega)$

Now we study distributions which behave locally just like the distributions in $H^s(\mathbf{R}^n)$. In doing so, the next theorem plays a fundamental role.

5.2.1 Theorem. *The multiplication*

$$\{\varphi, u\} \mapsto \varphi u$$

is a continuous, bilinear mapping of $\mathscr{S}(\mathbf{R}^n) \times \mathscr{S}(\mathbf{R}^n)$ *into* $H^s(\mathbf{R}^n)$ *for each* $s \in \mathbf{R}$; *more precisely we have*

$$\|\varphi u\|_s \leq 2^{|s|/2} \|u\|_s \left(\int_{\mathbf{R}^n} (1 + |\xi|^2)^{|s|/2} |\hat{\varphi}(\xi)| \, d\xi \right).$$

If Ω is an open subset of $\mathbf{R}^n$, we let

$H^s_{\text{loc}}(\Omega)$ = the space of distributions $u \in \mathscr{D}'(\Omega)$ such that $\varphi u \in H^s(\mathbf{R}^n)$ for all $\varphi \in C_0^\infty(\Omega)$.

We equip the space $H^s_{\text{loc}}(\Omega)$ with the topology defined by the family of seminorms

$$u \mapsto \|\varphi u\|_s,$$

where φ ranges over the space $C_0^\infty(\Omega)$. Let $\{K_j\}$ be an exhaustive sequence of compact subsets of Ω. If we take a sequence $\{\varphi_j\}$ in $C_0^\infty(\Omega)$ such that $\varphi_j = 1$ on K_j, then the topology on $H^s_{\text{loc}}(\Omega)$ is defined by the countably many seminorms $u \mapsto \|\varphi_\tau u\|_s$ alone. In fact, for every $\varphi \in C_0^\infty(\Omega)$ one can take j so large that $\varphi_j \varphi = \varphi$. Then it follows from an application of Theorem 5.2.1 that

$$\|\varphi u\|_s = \|\varphi \varphi_j u\|_s \leq C_s \|\varphi_j u\|_s,$$

where $C_s > 0$ is a constant independent of φ_j. This shows that $H^s_{\text{loc}}(\Omega)$ is metrizable. Further, by virtue of the completeness of the spaces $H^s(\mathbf{R}^n)$ and $\mathscr{D}'(\mathbf{R}^n)$, one can easily check that $H^s_{\text{loc}}(\Omega)$ is complete. Hence the space $H^s_{\text{loc}}(\Omega)$ is a Fréchet space.

Here are some basic topological properties of $H^s_{\text{loc}}(\Omega)$:

1) We have the inclusions

$$C^\infty(\Omega) \subset H^s_{\text{loc}}(\Omega) \subset \mathscr{D}'(\Omega)$$

with continuous injections. Further, the space $C_0^\infty(\Omega)$ is dense in $H^s_{\text{loc}}(\Omega)$ for each $s \in \mathbf{R}$.

2) (Sobolev) If $s > n/2 + k$, where k is a non-negative integer, then we have the inclusion

$$H^s_{\text{loc}}(\Omega) \subset C^k(\Omega)$$

with continuous injection. Further we have

$$C^\infty(\Omega) = \bigcap_{s \in R} H^s_{\mathrm{loc}}(\Omega).$$

We let

$H^s_{\mathrm{comp}}(\Omega)$ = the union of the spaces $H^s_K(\mathbf{R}^n)$ where K ranges over all compact subsets of Ω.

We equip the space $H^s_{\mathrm{comp}}(\Omega)$ with the inductive limit topology of the spaces $H^s_K(\mathbf{R}^n)$.

We define a bilinear form $\langle\ ,\ \rangle$ on the product space $H^s_{\mathrm{loc}}(\Omega) \times H^{-s}_{\mathrm{comp}}(\Omega)$ by the formula

$$\{u, v) \mapsto \langle u, v\rangle = \langle \varphi u, v\rangle, \tag{1}$$

where φ is a function in $C_0^\infty(\Omega)$ such that $\varphi = 1$ in a neighborhood of supp v, and $\langle\ ,\ \rangle$ on the right-hand side is the pairing of $H^s(\mathbf{R}^n)$ and $H^{-s}(\mathbf{R}^n)$. It is easy to verify that the quantity $\langle \varphi u, v\rangle$ does not depend on the function φ chosen.

Then we have:

5.2.2 Theorem. *The spaces $H^s_{\mathrm{loc}}(\Omega)$ and $H^{-s}_{\mathrm{comp}}(\Omega)$ are dual to each other with respect to the bilinear pairing of $H^s_{\mathrm{loc}}(\Omega)$ and $H^{-s}_{\mathrm{comp}}(\Omega)$ defined by formula* (1).

The characterization of $H^s(\mathbf{R}^n)$ in terms of L^2-norms in Theorems 5.1.1 and 5.1.2 allows us to prove the invariance of the space $H^s_{\mathrm{loc}}(\Omega)$ under C^∞ diffeomorphisms.

Let Ω_1, Ω_2 be two open subsets of $\mathbf{R}^n$ and $\chi: \Omega_1 \to \Omega_2$ a C^∞ diffeomorphism. If $v \in \mathscr{D}'(\Omega_2)$, we define a distribution $\chi^* v \in \mathscr{D}'(\Omega_1)$ by the formula (cf. formula (2.12.1))

$$\langle \chi^* v, \varphi\rangle = \langle v, \varphi \circ \chi^{-1} \cdot |\det(J(\chi^{-1}))|\rangle, \qquad \varphi \in C_0^\infty(\Omega_1),$$

where $J(\chi^{-1})$ is the Jacobian matrix of χ^{-1}. The distribution $\chi^* v$ is called the *inverse image* of v under χ.

Then we have:

5.2.3 Theorem. *Let $\chi: \Omega_1 \to \Omega_2$ be a C^∞ diffeomorphism. Then the mapping $v \mapsto \chi^* v$ is an isomorphism of $H^s_{\mathrm{loc}}(\Omega_2)$ onto $H^s_{\mathrm{loc}}(\Omega_1)$, and its inverse is the mapping $u \mapsto (\chi^{-1})^* u$.*

5.3. The Spaces $H^s(M)$

Theorem 5.2.3 allows us to define $H^s_{\text{loc}}(M)$, where M is a manifold, as follows:

Let M be an n-dimensional C^∞ manifold which satisfies the second axiom of countability. We let

$H^s_{\text{loc}}(M) =$ the space of distributions $u \in \mathscr{D}'(M)$ such that, for any admissible chart (U, χ) on M, the inverse image $(\chi^{-1})^*(u|_U)$ of $u|_U$ under χ^{-1} belongs to $H^s_{\text{loc}}(\chi(U))$.

We equip the space $H^s_{\text{loc}}(M)$ with the topology defined by the family of seminorms

$$u \mapsto \|\tilde{\varphi} \cdot (\chi^{-1})^*(u|_U)\|_s,$$

where (U, χ) ranges over all admissible charts on M and $\tilde{\varphi}$ ranges over the space $C_0^\infty(\chi(U))$. Then the space $H^2_{\text{loc}}(M)$ is a Fréchet space.

Now suppose that M is an n-dimensional *compact* C^∞ manifold. By the compactness of M, one can find an atlas $\{(U_j, \chi_j)\}_{j=1}^N$ consisting of finitely many charts on M. Let $\{\varphi_j\}_{j=1}^N$ be a partition of unity subordinate to the covering $\{U_j\}_{j=1}^N$. Then the topology on $H^s_{\text{loc}}(M)$ can be defined by the norm associated with the inner product

$$(u, v)_s = \sum_{j=1}^{N} ((\chi_j^{-1})^*(\varphi_j u), (\chi_j^{-1})^*(\varphi_j v))_s,$$

where $(\ ,\ )_s$ on the right-hand side is the inner product in $H^s(\mathbf{R}^n)$. Hence the space $H^s_{\text{loc}}(M)$ is a Hilbert space.

In the case when M is *compact*, we write

$$H^s(M) = H^s_{\text{loc}}(M).$$

Observe that all the results we stated about $H^s(\mathbf{R}^n)$ in Section 5.1 are also true for $H^s(M)$, since the spaces $H^s(M)$ are defined to be locally the spaces $H^s(\mathbf{R}^n)$.

We summarize some basic topological properties of $H^s(M)$:

1) If $s > t$, then we have the inclusions

$$C^\infty(M) \subset H^s(M) \subset H^t(M) \subset \mathscr{D}'(M)$$

with continuous injections. Further we have

$$\mathscr{D}'(M) = \bigcup_{s \in \mathbf{R}} H^s(M).$$

2) The space $C^\infty(M)$ is dense in $H^s(M)$ for each $s \in \mathbf{R}$.

3) (Sobolev) If $s > n/2 + k$, where k is a non-negative integer, then we have the inclusion

$$H^s(M) \subset C^k(M)$$

with continuous injection. Further we have

$$C^\infty(M) = \bigcap_{s \in \mathbf{R}} H^s(M).$$

4) Let μ be a strictly positive density on M. The bilinear form $\langle\ ,\ \rangle$ on the product space $C^\infty(M) \times C^\infty(M)$ defined by

$$\{u, v\} \mapsto \langle u, v\rangle = \int_M u(x)v(x)\, d\mu(x)$$

extends uniquely to a continuous bilinear form $\langle\ ,\ \rangle$ on the product space $H^s(M) \times H^{-s}(M)$ for each $s \in \mathbf{R}$. The spaces $H^s(M)$ and $H^{-s}(M)$ are dual to each other with respect to this bilinear pairing of $H^s(M)$ and $H^{-s}(M)$.

Similarly, the spaces $H^s(M)$ and $H^{-s}(M)$ are antidual to each other with respect to an extension of the sesquilinear form $(\ ,\ )$ on the product space $C^\infty(M) \times C^\infty(M)$ defined by

$$\{u, v\} \mapsto (u, v) = \int_M u(x)\overline{v(x)}\, d\mu(x).$$

We denote again by $(\ ,\ )$ this sesquilinear form on the product space $H^s(M) \times H^{-s}(M)$. We remark that

$$(u, v) = \langle u, \bar{v}\rangle, \qquad u \in H^s(M), \qquad v \in H^{-s}(M).$$

5) (Rellich) If $s > t$, then the injection

$$H^s(M) \to H^t(M)$$

is compact.

6) Let s_1, s, s_2 be real numbers such that $s_1 < s < s_2$. For every $\varepsilon > 0$, there exists a constant $C_\varepsilon > 0$ such that

$$\|u\|_s^2 \le \varepsilon \|u\|_{s_2}^2 + C_\varepsilon \|u\|_{s_1}^2, \qquad u \in H^{s_2}(M).$$

5.4. The Spaces $H^s(\overline{\mathbf{R}^n_+})$

Preparatory to studying Sobolev spaces on a C^∞ manifold with boundary, we consider Sobolev spaces on the half space $\overline{\mathbf{R}^n_+} = \{(x_1, \ldots, x_n) \in \mathbf{R}^n;\ x_n \geq 0\}$.

We define the *restriction* map

$$\rho : H^s(\mathbf{R}^n) \to \mathscr{D}'(\mathbf{R}^n_+)$$

by

$$\rho u = u|_{\mathbf{R}^n_+}, \qquad u \in H^s(\mathbf{R}^n).$$

Then the null space $\{u \in H^s(\mathbf{R}^n);\ \rho u = 0\}$ of ρ is the closed subspace $H^s_{\mathbf{R}^n \setminus \mathbf{R}^n_+}(\mathbf{R}^n)$ of $H^s(\mathbf{R}^n)$. Hence we have the assertion:

$$\left\{\begin{array}{l}\text{The factor space } H^s(\mathbf{R}^n)/H^s_{\mathbf{R}^n \setminus \mathbf{R}^n_+}(\mathbf{R}^n) \text{ is isomorphic}\\ \text{to the range } \{\rho u;\ u \in H^s(\mathbf{R}^n)\} \text{ of } \rho.\end{array}\right. \tag{1}$$

This leads us to the following definition of a Sobolev space on $\overline{\mathbf{R}^n_+}$:

$H^s(\overline{\mathbf{R}^n_+})$ = the space of distributions $u \in \mathscr{D}'(\mathbf{R}^n_+)$ such that there exists a distribution $U \in H^s(\mathbf{R}^n)$ with $\rho U = u$.

We equip the space $H^s(\overline{\mathbf{R}^n_+})$ with the norm

$$\|u\|_s = \inf \|U\|_s,$$

where the infimum is taken over all such U.

On the other hand, since we have the orthogonal decomposition

$$H^s(\mathbf{R}^n) = H^s_{\mathbf{R}^n \setminus \mathbf{R}^n_+}(\mathbf{R}^n) \oplus (H^s_{\mathbf{R}^n \setminus \mathbf{R}^n_+}(\mathbf{R}^n))^\perp,$$

it follows from an application of Theorem 3.2.18 that:

$$\left\{\begin{array}{l}\text{The factor space } H^s(\mathbf{R}^n)/H^s_{\mathbf{R}^n \setminus \mathbf{R}^n_+}(\mathbf{R}^n) \text{ is isomorphic}\\ \text{to the space } (H^s_{\mathbf{R}^n \setminus \mathbf{R}^n_+}(\mathbf{R}^n))^\perp.\end{array}\right. \tag{2}$$

Therefore, combining assertions (1) and (2), we obtain that the space $H^s(\overline{\mathbf{R}^n_+})$ is isomorphic to the space $(H^s_{\mathbf{R}^n \setminus \mathbf{R}^n_+}(\mathbf{R}^n))^\perp$. Hence the Sobolev space $H^s(\overline{\mathbf{R}^n_+})$ admits a Hilbert space structure. We remark that, for every $u \in H^s(\overline{\mathbf{R}^n_+})$, there exists a unique distribution $U \in (H^s_{\mathbf{R}^n \setminus \mathbf{R}^n_+}(\mathbf{R}^n))^\perp$ such that $\rho U = u$ and $\|U\|_s = \|u\|_s$.

The characterization of $H^s(\mathbf{R}^n)$ in terms of L^2-norms in Theorems 5.1.1 and 5.1.2 allows us to obtain the following:

5.4.1 Theorem. *If $s \geq 0$, then the Seeley extension operator*

$$E: C_0^\infty(\overline{\mathbf{R}_+^n}) \to C_0^\infty(\mathbf{R}^n)$$

extends uniquely to a continuous linear extension operator

$$E: H^s(\overline{\mathbf{R}_+^n}) \to H^s(\mathbf{R}^n).$$

The next theorem gives a direct description of $H^s(\overline{\mathbf{R}_+^n})$ when s is a non-negative integer.

5.4.2 Theorem. *If m is a non-negative integer, then the Sobolev space $H^m(\overline{\mathbf{R}_+^n})$ is the space of functions $u \in L^2(\mathbf{R}_+^n)$ such that $D^\alpha u \in L^2(\mathbf{R}_+^n)$ for $|\alpha| \leq m$. Furthermore, the norm $\|u\|_m$ is equivalent to the norm*

$$\left(\sum_{|\alpha| \leq m} \int_{\mathbf{R}_+^n} |D^\alpha u(x)|^2 \, dx \right)^{1/2}.$$

Here are some basic topological properties of $H^s(\overline{\mathbf{R}_+^n})$:

1) We have the inclusions

$$C_0^\infty(\overline{\mathbf{R}_+^n}) \subset H^s(\overline{\mathbf{R}_+^n}) \subset \mathscr{D}'(\mathbf{R}_+^n)$$

with continuous injections.

2) The space $C_0^\infty(\overline{\mathbf{R}_+^n})$ is dense in $H^s(\overline{\mathbf{R}_+^n})$ for each $s \in \mathbf{R}$.

3) (Sobolev) If $s > n/2 + k$, where k is a non-negative integer, then we have the inclusion

$$H^s(\overline{\mathbf{R}_+^n}) \subset C^k(\overline{\mathbf{R}_+^n})$$

with continuous injection.

We define a bilinear form $\langle \ , \ \rangle$ on the product space $H^s(\overline{\mathbf{R}_+^n}) \times H_{\overline{\mathbf{R}_+^n}}^{-s}(\mathbf{R}^n)$ by the formula

$$\langle u, v \rangle = \langle \tilde{u}, v \rangle, \tag{3}$$

where $\tilde{u}$ is an extension of u in $H^s(\mathbf{R}^n)$ and $\langle \ , \ \rangle$ on the right-hand side is the pairing of $H^s(\mathbf{R}^n)$ and $H^{-s}(\mathbf{R}^n)$. One can easily verify that the quantity $\langle \tilde{u}, v \rangle$ does not depend on the extension $\tilde{u}$ chosen.

Then we have:

5.4.3 Theorem. *The spaces $H^s(\overline{\mathbf{R}^n_+})$ and $H^{-s}_{\overline{\mathbf{R}^n_+}}(\mathbf{R}^n)$ are dual to each other with respect to the bilinear pairing of $H^s(\overline{\mathbf{R}^n_+})$ and $H^{-s}_{\overline{\mathbf{R}^n_+}}(\mathbf{R}^n)$ defined by formula* (3).

5.5. The Spaces $H^s(\bar{\Omega})$

Now let Ω be a bounded, C^∞ domain in $\mathbf{R}^n$. Its closure $\bar{\Omega}$ is an n-dimensional, compact C^∞ manifold with boundary. By Theorems 2.13.2 and 2.13.3, we may suppose that (cf. Figure 5-1):

(a) The domain Ω is a relatively compact open subset of an n-dimensional compact C^∞ manifold M without boundary, in which Ω has a C^∞ boundary $\partial\Omega$.

(b) In a neighborhood W of $\partial\Omega$ in M, a normal coordinate t is chosen so that the points of W are represented as (x', t), $x' \in \partial\Omega$, $-1 < t < 1$; $t > 0$ in Ω, $t < 0$ in $M \setminus \bar{\Omega}$ and $t = 0$ only on $\partial\Omega$.

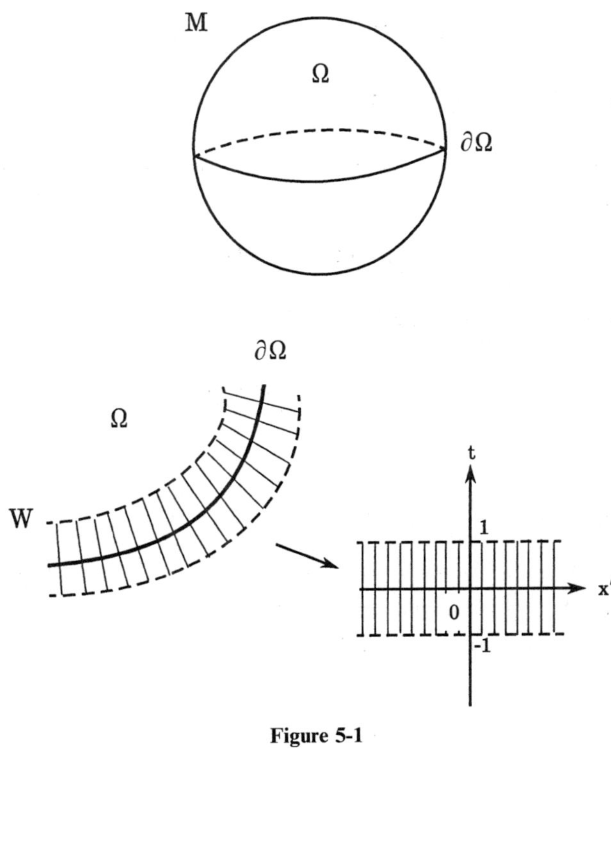

Figure 5-1

(c) The manifold M is equipped with a strictly positive density μ which, on W, is the product of a strictly positive density ω on $\partial\Omega$ and the Lebesgue measure dt on $(-1, 1)$.

The Sobolev spaces $H^s(\bar{\Omega})$ are defined similarly to the way in which the spaces $H^s(\overline{\mathbf{R}^n_+})$ were defined, replacing $\mathbf{R}^n$ by M. That is, we let

$H^s(\bar{\Omega})$ = the space of distributions $u \in \mathscr{D}'(\Omega)$ such that there exists a distribution $U \in H^s(M)$ with $\rho U = u$. Here ρ is the restriction map to Ω.

We equip the space $H^s(\bar{\Omega})$ with the norm

$$\|u\|_s = \inf \|U\|_s,$$

where the infimum is taken over all such U.

All the results we stated about the spaces $H^s(\overline{\mathbf{R}^n_+})$ in Section 5.4 are also true for the spaces $H^s(\bar{\Omega})$. We summarize some basic topological properties of $H^s(\bar{\Omega})$:

1) The Sobolev space $H^s(\bar{\Omega})$ is a Hilbert space.
2) We have the inclusions

$$C^\infty(\bar{\Omega}) \subset H^s(\bar{\Omega}) \subset \mathscr{D}'(\Omega)$$

with continuous injections. Further, the space $C^\infty(\bar{\Omega})$ is dense in $H^s(\bar{\Omega})$ for each $s \in \mathbf{R}$.

3) (Sobolev) If $s > n/2 + k$, where k is a non-negative integer, then we have the inclusion

$$H^s(\bar{\Omega}) \subset C^k(\bar{\Omega})$$

with continuous injection. Further we have

$$C^\infty(\bar{\Omega}) = \bigcap_{s \in \mathbf{R}} H^s(\bar{\Omega}).$$

4) We let

$H^s_{\bar{\Omega}}(M)$ = the subspace of $H^s(M)$ consisting of the elements with support in $\bar{\Omega}$.

Since the injection $H^s(M) \to \mathscr{D}'(\Omega)$ is continuous, it follows that $H^s_{\bar{\Omega}}(M)$ is a closed subspace of $H^s(M)$; hence it is a Hilbert space.

We define a bilinear form $\langle\ ,\ \rangle$ on the product space $H^s(\bar{\Omega}) \times H^{-s}_{\bar{\Omega}}(M)$ by the formula

$$\langle u, v\rangle = \langle \tilde{u}, v\rangle, \tag{1}$$

where $\tilde{u}$ is an extension of u in $H^s(M)$ and $\langle\ ,\ \rangle$ on the right-hand side is the pairing of $H^s(M)$ and $H^{-s}(M)$. It is easy to verify that the quantity $\langle\tilde{u}, v\rangle$ is independent of the extension $\tilde{u}$ chosen.

Then we have:

5.5.1 Theorem. *The spaces $H^s(\bar{\Omega})$ and $H_{\bar{\Omega}}^{-s}(M)$ are dual to each other with respect to the bilinear pairing of $H^s(\bar{\Omega})$ and $H_{\bar{\Omega}}^{-s}(M)$ defined by formula* (1).

5) By covering a neighborhood of $\partial\Omega$ with local charts and locally using the Seeley extension operator (cf. the proof of Theorem 4.10.1), we can obtain an extension operator

$$E: C^\infty(\bar{\Omega}) \to C^\infty(M).$$

If $s \geq 0$, then this operator E extends uniquely to a continuous linear extension operator

$$E: H^s(\bar{\Omega}) \to H^s(M).$$

The next proposition follows from the proof of Theorem 4.10.1.

5.5.2 Proposition. *Let $E': H^{-s}(M) \to H_{\bar{\Omega}}^{-s}(M)$ be the transpose of the extension operator $E: H^s(\bar{\Omega}) \to H^s(M)$ $(s \geq 0)$. If $u \in H^{-s}(M)$ is of class C^∞ up to $\partial\Omega$ in Ω and also in $M \setminus \bar{\Omega}$, then $E'u \in H_{\bar{\Omega}}^{-s}(M)$ is of class C^∞ up to $\partial\Omega$ in Ω.*

6) (Rellich) If $s \geq 0$ and $s > t$, then the injection $H^s(\bar{\Omega}) \to H^t(\bar{\Omega})$ is compact.

In fact, it suffices to note that the injection $H^s(\bar{\Omega}) \to H^t(\bar{\Omega})$ can be written as

$$H^s(\bar{\Omega}) \xrightarrow{E} H^s(M) \to H^t(M) \xrightarrow{\rho} H^t(\bar{\Omega}).$$

7) Let s_1, s, s_2 be real numbers such that $0 \leq s_1 < s < s_2$. For every $\varepsilon > 0$, there exists a constant $C_\varepsilon > 0$ such that

$$\|u\|_s^2 \leq \varepsilon\|u\|_{s_2}^2 + C_\varepsilon\|u\|_{s_1}^2, \qquad u \in H^{s_2}(\bar{\Omega}).$$

In fact, we have with a constant $C > 0$ independent of ε:

$$\begin{aligned}\|u\|_s^2 &\leq \|Eu\|_s^2 \\ &\leq \varepsilon\|Eu\|_{s_2}^2 + C'_\varepsilon\|Eu\|_{s_1}^2 \\ &\leq C(\varepsilon\|u\|_{s_2}^2 + C'_\varepsilon\|u\|_{s_1}^2).\end{aligned}$$

5.6. Trace Theorems

We study the restrictions to the hyperplane $\{x_n = 0\}$ of functions in $H^s(\overline{\mathbf{R}^n_+})$.

If $x = (x_1, \ldots, x_n)$ is a point of $\mathbf{R}^n$, we write

$$x = (x', x_n), \qquad x' = (x_1, \ldots, x_{n-1}).$$

If j is a non-negative integer, we define the *trace* map

$$\gamma_j\colon C_0^\infty(\overline{\mathbf{R}^n_+}) \to C_0^\infty(\mathbf{R}^{n-1})$$

by the formula

$$\gamma_j u(x') = \lim_{x_n \downarrow 0} D_n^j u(x', x_n), \qquad u \in C_0^\infty(\overline{\mathbf{R}^n_+}).$$

Then we have:

5.6.1 Theorem. *If $0 \le j < s - \frac{1}{2}$, then the trace map $\gamma_j\colon C_0^\infty(\overline{\mathbf{R}^n_+}) \to C_0^\infty(\mathbf{R}^{n-1})$ extends uniquely to a continuous linear map $\gamma_j\colon H^s(\overline{\mathbf{R}^n_+}) \to H^{s-j-1/2}(\mathbf{R}^{n-1})$. Furthermore, if $u \in H^s(\overline{\mathbf{R}^n_+})$, then the mapping $x_n \mapsto D_n^j u(\cdot, x_n)$ is a continuous function on $[0, \infty)$ with values in $H^{s-j-1/2}(\mathbf{R}^{n-1})$.*

The next theorem shows that the result of Theorem 5.6.1 is sharp.

5.6.2 Theorem. *If $0 \le j < s - \frac{1}{2}$, then the trace map*

$$H^s(\overline{\mathbf{R}^n_+}) \to \prod_{0 \le j < s - 1/2} H^{s-j-1/2}(\mathbf{R}^{n-1})$$

$$u \mapsto (\gamma_j u)_{0 \le j < s - 1/2}$$

is surjective.

Now let Ω be a bounded, C^∞ domain in $\mathbf{R}^n$. Its closure $\bar{\Omega}$ is an n-dimensional compact C^∞ manifold with boundary $\partial\Omega$. We may suppose that (cf. Figure 5-1):

(a) The domain Ω is a relatively compact open subset of an n-dimensional, compact C^∞ manifold M without boundary.

(b) In a neighborhood W of $\partial\Omega$ in M, a normal coordinate t is chosen so that the points of W are represented as (x', t), $x' \in \partial\Omega$, $-1 < t < 1$; $t > 0$ in Ω, $t < 0$ in $M \setminus \bar{\Omega}$ and $t = 0$ only on $\partial\Omega$.

(c) The manifold M is equipped with a strictly positive density μ which, on W, is the product of a strictly positive density ω on $\partial\Omega$ and the Lebesgue measure dt on $(-1, 1)$.

If j is a non-negative integer, we define the *trace* map

$$\gamma_j\colon C^\infty(\bar{\Omega}) \to C^\infty(\partial\Omega)$$

by the formula

$$\gamma_j u(x') = \lim_{t \downarrow 0} D_t^j u(x', t), \qquad u \in C^\infty(\bar{\Omega}).$$

Then Theorems 5.6.1 and 5.6.2 extend to this case:

5.6.3 Theorem. *If $0 \le j < s - \frac{1}{2}$, then the trace map $\gamma_j\colon C^\infty(\bar{\Omega}) \to C^\infty(\partial\Omega)$ extends uniquely to a continuous linear map $\gamma_j\colon H^s(\bar{\Omega}) \to H^{s-j-1/2}(\partial\Omega)$. Furthermore, if $u \in H^s(\bar{\Omega})$, then the mapping $t \mapsto D_t^j u(\cdot, t)$ is a continuous function on $[0, 1)$ with values in $H^{s-j-1/2}(\partial\Omega)$.*

5.6.4 Theorem. *If $0 \le j < s - \frac{1}{2}$, then the trace map*

$$H^s(\bar{\Omega}) \to \prod_{0 \le j < s - 1/2} H^{s-j-1/2}(\partial\Omega)$$

$$u \mapsto (\gamma_j u)_{0 \le j < s-1/2}$$

is surjective.

Let k be a non-negative integer. A distribution $u \in \mathscr{D}'(\Omega)$ is said to have *sectional traces* on $\partial\Omega$ up to order k if the mapping $t \mapsto u(\cdot, t)$ is a C^k function on $[0, 1)$ with values in $\mathscr{D}'(\partial\Omega)$. This is equivalent to saying that, for every $\phi \in C^\infty(\partial\Omega)$, the function $\langle u(\cdot, t), \phi \cdot \omega \rangle$ is of class C^k on $[0, 1)$. Then we define the *sectional trace* $\gamma_j u$ of order j $(0 \le j \le k)$ on $\partial\Omega$ by

$$\gamma_j u = \lim_{t \downarrow 0} D_t^j u(\cdot, t) \qquad \text{in } \mathscr{D}'(\partial\Omega).$$

Theorem 5.6.3 tells us that if $u \in H^s(\bar{\Omega})$ with $s > \frac{1}{2}$, then u has sectional traces $\gamma_j u$ on $\partial\Omega$ up to order $k < s - \frac{1}{2}$, and $\gamma_j u \in H^{s-j-1/2}(\partial\Omega)$.

The next result is useful for the interpretation and study of boundary conditions in terms of distributions, and will be important in Chapter 8.

5.6.5 Theorem. *Let Ω be a bounded domain in $\mathbf{R}^n$ with C^∞ boundary $\partial\Omega$, and let*

$$A = \sum_{i,j=1}^{n} a^{ij}(x) \frac{\partial^2}{\partial x_i \, \partial x_j} + \sum_{i=1}^{n} b^i(x) \frac{\partial}{\partial x_i} + c(x)$$

be a second-order differential operator with real C^∞ coefficients. Suppose that the boundary $\partial\Omega$ is non-characteristic *with respect to the operator A, that is,*

$$\sum_{i,j=1}^{n} a^{ij}(x')\nu_i\nu_j > 0 \qquad \text{on } \partial\Omega,$$

where $\nu = (\nu_1, \ldots, \nu_n)$ is the unit normal to $\partial\Omega$ at x'.

If $u \in H^\sigma(\bar{\Omega})$ and $Au \in H^s(\bar{\Omega})$ with $\sigma \le s + 2$, then u has sectional traces $\gamma_j u$ on $\partial\Omega$ up to order $k < s + \frac{3}{2}$. Moreover, we have $\gamma_j u \in H^{\sigma - j - 1/2}(\partial\Omega)$ and

$$|\gamma_j u|_{H^{\sigma-j-1/2}(\partial\Omega)} \le C_{\sigma,s}(\|Au\|_{H^s(\bar{\Omega})} + \|u\|_{H^\sigma(\bar{\Omega})}) \tag{1}$$

with some constant $C_{\sigma,s} > 0$.

5.6.6 Remark. Theorem 5.6.5 is an expression of the fact that if we know about the derivatives of the solution u of $Au = f$ in tangential directions, then we can derive information about the normal derivatives $\gamma_j u$ by means of the equation $Au = f$.

If $u \in \mathscr{D}'(\Omega)$ has a sectional trace of order zero on $\partial\Omega$, we can define its extension u^0 in $\mathscr{D}'(M)$ as follows: choose functions $\theta \in C_0^\infty(W)$ and $\psi \in C_0^\infty(\Omega)$ such that $\theta + \psi = 1$ on $\bar{\Omega}$, and define u^0 by the formula

$$\langle u^0, \varphi \cdot \mu \rangle = \int_0^1 \langle u(t), (\theta\varphi)(\cdot, t) \cdot \omega \rangle \, dt + \langle u, \psi\varphi \cdot \mu \rangle, \qquad \varphi \in C^\infty(M).$$

The distribution u^0 is an extension of u to M which is equal to zero in $M \setminus \bar{\Omega}$.

If $v \in \mathscr{D}'(\partial\Omega)$, we define a *multiple layer* $v \otimes D_t^j \delta$ $(j = 0, 1, \ldots)$ by the formula

$$\langle v \otimes D_t^j \delta, \varphi \cdot \mu \rangle = (-1)^j < v, D_t^j \varphi(\cdot, 0) \cdot \omega \rangle, \qquad \varphi \in C^\infty(M).$$

It is clear that $v \otimes D_t^j \delta$ is a distribution on M with support in $\partial\Omega$.

Let P be a differential operator of order m with C^∞ coefficients on M. In a neighborhood of $\partial\Omega$, we can write $P = P(x, D_x)$ uniquely in the form

$$P(x, D_x) = \sum_{j=0}^{m} P_j(x, D_{x'}) \, D_t^j, \qquad x = (x', t),$$

where $P_j(x, D_{x'})$ is a differential operator of order $m - j$ acting along the surfaces parallel to the boundary $\partial\Omega$.

Now it is easy to see that formulas (4.4.2) and (4.4.3) extend to this case:

1) If $u \in \mathscr{D}'(\Omega)$ has sectional traces on $\partial\Omega$ up to order j, then we have

$$D_t^j(u^0) = (D_t^j u)^0 + \frac{1}{i}\sum_{k=0}^{j-1} \gamma_{j-k-1} u \otimes D_t^k \delta. \tag{2}$$

2) If $u \in \mathscr{D}'(\Omega)$ has sectional traces on $\partial\Omega$ up to order m, then we have

$$P(u^0) = (Pu)^0 + \frac{1}{i}\sum_{\ell+k+1 \le m} P_{\ell+k+1}(x, D_{x'})\gamma_\ell u \otimes D_t^k \delta. \tag{3}$$

5.7. Sobolev Spaces and Regularizations

We introduce a two-parameter family of norms on the Sobolev spaces $H^s(\mathbf{R}^n)$.

If $m > 0$ and $0 < \rho < 1$, we let

$$\|u\|_{(s,m,\rho)}^2 = \int_{\mathbf{R}^n} (1 + |\xi|^2)^s (1 + |\rho\xi|^2)^{-m} |\hat{u}(\xi)|^2 \, đ\xi. \tag{1}$$

We list two results which follow at once:

1) For all $u \in H^{s-m}(\mathbf{R}^n)$, we have

$$\rho^m \|u\|_{(s,m,\rho)} \le \|u\|_{s-m} \le \|u\|_{(s,m,\rho)},$$

that is, the norm $\|u\|_{(s,m,\rho)}$ is equivalent to the norm $\|u\|_{s-m}$.

2) If $u \in H^s(\mathbf{R}^n)$, then we have

$$\|u\|_{(s,m,\rho)} \uparrow \|u\|_s \qquad \text{as } \rho \downarrow 0,$$

so that

$$\|u\|_s = \sup_{0<\rho<1} \|u\|_{(s,m,\rho)}.$$

This is an immediate consequence of the dominated convergence theorem (Theorem 1.19.3).

The next lemma explains a motivation of introduction of the norm $\|\cdot\|_{(s,m,\rho)}$.

5.7.1 Lemma. *If $u \in H^{s-m}(\mathbf{R}^n)$, and there exists a constant $C > 0$ independent of ρ such that*

$$\|u\|_{(s,m,\rho)} \le C,$$

then u is in $H^s(\mathbf{R}^n)$ and $\|u\|_s \le C$.

This lemma follows immediately from the monotone convergence theorem (Theorem 1.19.1).

The next theorem gives an equivalent expression for the norm $\|\cdot\|_{(s,m,\rho)}$.

5.7.2 Theorem. *For all $u \in H^{s-m}(\mathbf{R}^n)$, we have*

$$\begin{aligned} \|u\|_{(s,m,\rho)} &\le \inf_{\substack{u'+u''=u \\ u',u'' \in H^{-\infty}(\mathbf{R}^n)}} \left\{ \frac{1}{\rho^m} \|u'\|_{s-m} + \|u''\|_s \right\} \\ &\le 2^{(m+1)/2} \|u\|_{(s,m,\rho)}. \end{aligned}$$

Now let χ be a function in $C_0^\infty(\mathbf{R}^n)$ satisfying the following conditions:

(a) For a non-negative integer k, we have

$$\hat{\chi}(\xi) = O(|\xi|^k) \qquad \text{as } \xi \to 0. \tag{2}$$

(b) If $\hat{\chi}(t\xi) = 0$ for all $t \in \mathbf{R}$, then $\xi = 0$.

For example, we may take

$$\chi = \Delta^k \theta,$$

where θ is a function in $C_0^\infty(\mathbf{R}^n)$ such that $\hat{\theta}(0) = \int_{\mathbf{R}^n} \theta(x)\,dx \ne 0$. Further we let

$$\chi_\varepsilon(x) = \frac{1}{\varepsilon^n} \chi\left(\frac{x}{\varepsilon}\right), \qquad \varepsilon > 0.$$

The next theorem gives another equivalent expression for the norm $\|u\|_{(s,m,\rho)}$ in terms of the regularizations $u * \chi_\varepsilon$ of u.

5.7.3 Theorem. *Suppose that the function χ satisfies condition (2) for $k > s$. Then, for any $s_1 \in \mathbf{R}$ and $t < s + s_1 - m$, there exist constants $C_{s,s_1,t} > 0$ and $C'_{s,s_1,t} > 0$ independent of ρ such that we have*

$$\|u\|^2_{(s+s_1,m,\rho)} \leq C_{s,s_1,t}\left(\int_0^1 \|u * \chi_\varepsilon\|^2_{s_1}\left(1 + \frac{\rho^2}{\varepsilon^2}\right)^{-m} \varepsilon^{-2s}\frac{d\varepsilon}{\varepsilon} + \|u\|_t^2\right)$$
$$\leq C'_{s,s_1,t}\|u\|^2_{(s+s_1,m,\rho)}$$

for all $u \in H^{s+s_1-m}(\mathbf{R}^n)$.

Now let M be an n-dimensional *compact* C^∞ manifold without boundary. If $m > 0$ and $0 < \rho < 1$, we define a norm $\|\cdot\|_{(s,m,\rho)}$ on the Sobolev space $H^{s-m}(M)$ by the formula

$$\|u\|_{(s,m,\rho)} = \inf_{\substack{u'+u''=u \\ u',u'' \in \mathcal{D}'(M)}} \left\{\frac{1}{\rho^m}\|u'\|_{s-m} + \|u''\|_s\right\}.$$

Then the above results are also true for the spaces $H^s(M)$. More precisely, we have the following (cf. Hörmander [1], [2]):

1) The norm $\|\cdot\|_{(s,m,\rho)}$ increases as $\rho \downarrow 0$, and we have

$$\|u\|_s = \sup_{0<\rho<1} \|u\|_{(s,m,\rho)}$$

if $u \in H^s(M)$.

2) The norm $\|\cdot\|_{(s,m,\rho)}$ has locally an equivalent expression such as formula (1), hence it is equivalent to the norm $\|\cdot\|_{s-m}$.

3) Lemma 5.7.1 extends to this case:

5.7.4 Lemma. *If $u \in H^{s-m}(M)$, and there exists a constant $C > 0$ independent of ρ such that*

$$\|u\|_{(s,m,\rho)} \leq C,$$

then u is in $H^s(M)$ and $\|u\|_s \leq C$.

4) Let (U, ψ) be a chart on M with $\psi(x) = (x_1, \ldots, x_n)$, and take a function $\chi \in C_0^\infty(U)$ which satisfies conditions (a) and (b).

The next theorem is a local version of Theorem 5.7.3.

5.7.5 Theorem. *Let K be a compact subset of U and $0 < r < \operatorname{dist}(K, \partial U)$. Choose χ as above with support in the ball $\{x \in U; |\psi(x)| < r\}$, and $k > s$. Then, for any $s_1 \in \mathbf{R}$ and $t < s + s_1 - m$, there exist constants $C_{s,s_1,t} > 0$ and $C'_{s,s_1,t} > 0$ independent of ρ such that we have*

$$\|u\|^2_{(s+s_1,m,\rho)} \leq C_{s,s_1,t}\left(\int_0^1 \|u * \chi_\varepsilon\|^2_{s_1}\left(1 + \frac{\rho^2}{\varepsilon^2}\right)^{-m} \varepsilon^{-2s} \frac{d\varepsilon}{\varepsilon} + \|u\|^2_t\right)$$

$$\leq C'_{s,s_1,t}\|u\|^2_{(s+s_1,m,\rho)}$$

for all $u \in H^{s+s_1-m}(M)$ with support in K.

Notes

Our treatment of Sobolev spaces is adapted from Chazarain-Piriou [1] and Hörmander [1]. For more leisurely treatments of Sobolev spaces, the reader is referred to Adams [1] and Lions-Magenes [1].

Section 5.6: The trace theorem, Theorem 5.6.5, is due to Hörmander [2].

Section 5.7: The two-parameter family $\|\cdot\|_{(s,1,\rho)}$ of norms was introduced by Hörmander [1], and was used to prove regularity theorems for linear partial differential equations. See also Hörmander [2], Fediĭ [1] and Oleĭnik-Radkevič [1].

6 The Calculus of Pseudo-Differential Operators

In recent years there has been a trend in the theory of partial differential equations towards constructive methods. The development of the theory of pseudo-differential operators has made possible such an approach to the study of (non-)elliptic differential operators. The class of pseudo-differential operators is essentially the smallest algebra of operators which contains all differential operators, all fundamental solutions of elliptic differential operators and all integral operators with C^∞ kernel.

In this chapter we define pseudo-differential operators and study their basic properties such as the behavior of transposes, adjoints and compositions of such operators, and the effect of a change of coordinates on such operators. Furthermore we discuss in detail, via functional analysis, the behavior of elliptic pseudo-differential operators on Sobolev spaces, and formulate classical surface and volume potentials in terms of pseudo-differential operators. This calculus of pseudo-differential operators will be applied to elliptic boundary value problems in Chapter 8. Finally we give Gårding's inequality and related inequalities, and describe two classes of hypoelliptic pseudo-differential operators which arise in the construction of Feller semigroups in Chapter 10.

6.1. Symbol Classes

Let Ω be an open subset of $\mathbf{R}^n$. If

$$P(x, D) = \sum_{|\alpha| \le m} a_\alpha(x) D^\alpha$$

is a differential operator of order m with C^∞ coefficients on Ω, then we have, by Theorem 4.4.5,

$$P(x, D)u(x) = \int_{\mathbf{R}^n} e^{ix\cdot\xi} p(x, \xi)\hat{u}(\xi)\, đ\xi, \qquad u \in C_0^\infty(\Omega), \tag{1}$$

where

$$p(x, \xi) = \sum_{|\alpha| \le m} a_\alpha(x)\xi^\alpha.$$

We shall use the Fourier integral representation (1) to define pseudo-differential operators, taking $p(x, \xi)$ to belong to a wider class of functions than polynomials.

If $m \in \mathbf{R}$, we let

$S^m(\Omega \times \mathbf{R}^N) =$ the set of all functions $a \in C^\infty(\Omega \times \mathbf{R}^N)$ with the property that for any compact $K \subset \Omega$ and multi-indices α, β there exists a constant $C_{K,\alpha,\beta} > 0$ such that

$$|\partial_x^\beta \partial_\theta^\alpha a(x, \theta)| \le C_{K,\alpha,\beta}(1 + |\theta|)^{m-|\alpha|}, \qquad x \in K, \theta \in \mathbf{R}^N.$$

The elements of $S^m(\Omega \times \mathbf{R}^N)$ are called *symbols* of order m. We drop the $\Omega \times \mathbf{R}^N$ and use S^m when the context is clear.

6.1.1 Examples

1. A polynomial $p(x, \xi) = \sum_{|\alpha| \le m} a_\alpha(x)\xi^\alpha$ of order m with coefficients in $C^\infty(\Omega)$ is in $S^m(\Omega \times \mathbf{R}^n)$.
2. If $m \in \mathbf{R}$, the function

$$\Omega \times \mathbf{R}^n \ni (x, \xi) \mapsto (1 + |\xi|^2)^{m/2}$$

is in $S^m(\Omega \times \mathbf{R}^n)$.

3. A function $a \in C^\infty(\Omega \times (\mathbf{R}^N \setminus \{0\}))$ is said to be *positively homogeneous of degree m* in the variable θ if it satisfies

$$a(x, t\theta) = t^m a(x, \theta), \qquad t > 0, \qquad \theta \in \mathbf{R}^N \setminus \{0\}.$$

If $a(x, \theta)$ is positively homogeneous of degree m in θ, and if $\varphi(\theta)$ is a C^∞ function such that $\varphi(\theta) = 0$ for $|\theta| \le 1/2$ and $\varphi(\theta) = 1$ for $|\theta| \ge 1$, then the function $\varphi(\theta)a(x, \theta)$ is in $S^m(\Omega \times \mathbf{R}^N)$.

If K is a compact subset of Ω and j is a non-negative integer, we define a seminorm $C_{K,j,m}$ on $S^m(\Omega \times \mathbf{R}^N)$ by

$$S^m(\Omega \times \mathbf{R}^N) \ni a \mapsto C_{K,j,m}(a) = \sup_{\substack{x \in K \\ \theta \in \mathbf{R}^N \\ |\alpha|+|\beta| \le j}} \frac{|\partial_x^\beta \partial_\theta^\alpha a(x, \theta)|}{(1 + |\theta|)^{m-|\alpha|}}.$$

We equip the space $S^m(\Omega \times \mathbf{R}^N)$ with the topology defined by the family $\{C_{K,j,m}\}$ of seminorms where K ranges over all compact subsets of Ω and $j = 0, 1, \ldots$. It is easy to verify that $S^m(\Omega \times \mathbf{R}^N)$ is a Fréchet space.

We set

$$S^\infty(\Omega \times \mathbf{R}^N) = \bigcup_{m \in \mathbf{R}} S^m(\Omega \times \mathbf{R}^N),$$

$$S^{-\infty}(\Omega \times \mathbf{R}^N) = \bigcap_{m \in \mathbf{R}} S^m(\Omega \times \mathbf{R}^N).$$

We remark that a function $a \in C^\infty(\Omega \times \mathbf{R}^N)$ which vanishes for sufficiently large $|\theta|$ is in $S^{-\infty}(\Omega \times \mathbf{R}^N)$.

We list some facts which follow at once:

1. $m \le m' \Rightarrow S^{-\infty} \subset S^m \subset S^{m'} \subset S^\infty$.
2. $a \in S^m \Rightarrow \partial_x^\beta \partial_\theta^\alpha a \in S^{m-|\alpha|}$.
3. $a \in S^m,\ b \in S^{m'} \Rightarrow ab \in S^{m+m'}$.

In particular, it follows that S^∞ is a commutative algebra and that $S^{-\infty}$ is an ideal of S^∞.

The next theorem gives a meaning to a formal sum of symbols of decreasing order.

6.1.2 Theorem. *Let* $a_j \in S^{m_j}(\Omega \times \mathbf{R}^N)$, $m_j \downarrow -\infty$, $j = 0, 1, \ldots$. *Then there exists a symbol* $a \in S^{m_0}(\Omega \times \mathbf{R}^N)$, *unique modulo* $S^{-\infty}(\Omega \times \mathbf{R}^N)$, *such that for all* $k > 0$ *we have*

$$a - \sum_{j=0}^{k-1} a_j \in S^{m_k}(\Omega \times \mathbf{R}^N). \tag{2}$$

If formula (2) holds, we write

$$a \sim \sum_{j=0}^{\infty} a_j.$$

The formal sum $\sum_j a_j$ is called an *asymptotic expansion* of a.

A symbol $a(x, \theta) \in S^m(\Omega \times \mathbf{R}^N)$ is said to be *classical* if there exist C^∞ functions $a_j(x, \theta)$, positively homogeneous of degree $m - j$ in θ for $|\theta| \geq 1$, such that

$$a \sim \sum_{j=0}^{\infty} a_j.$$

We remark that the homogeneous functions a_j are uniquely determined (for $|\theta| \geq 1$) for a. The homogeneous function a_0 of degree m is called the *principal part* of a.

We let

$$S_{\text{cl}}^m(\Omega \times \mathbf{R}^N) = \text{the set of all classical symbols of order } m.$$

For example, the symbols in Examples 6.1.1 are all classical, and they have respectively as principal part the following functions:

1. $p_m(x, \xi) = \sum_{|\alpha|=m} a_\alpha(x)\xi^\alpha$.
2. $|\xi|^m$.
3. $a(x, \theta)$.

A symbol $a \in S^m(\Omega \times \mathbf{R}^N)$ is said to be *elliptic* of order m if there exists a symbol $b \in S^{-m}(\Omega \times \mathbf{R}^N)$ such that

$$ab \equiv 1 \qquad \text{mod } S^{-1}(\Omega \times \mathbf{R}^N).$$

We give a useful criterion for ellipticity:

6.1.3 Theorem. *A symbol $a(x, \theta)$ in $S^m(\Omega \times \mathbf{R}^N)$ is elliptic if and only if, for any compact $K \subset \Omega$, there exists a constant $c_K > 0$ such that*

$$|a(x, \theta)| \geq c_K(1 + |\theta|)^m, \qquad x \in K, \qquad |\theta| \geq \frac{1}{c_K}.$$

There is a simple criterion in the case of classical symbols.

6.1.4 Corollary. *Let $a(x, \theta) \in S_{\text{cl}}^m(\Omega \times \mathbf{R}^N)$ with principal part $a_0(x, \theta)$. Then $a(x, \theta)$ is elliptic if and only if we have*

$$a_0(x, \theta) \neq 0, \qquad x \in \Omega, \qquad |\theta| = 1.$$

For example, a polynomial $p(x, \xi) = \sum_{|\alpha| \leq m} a_\alpha(x)\xi^\alpha$ of order m is elliptic if and only if $p_m(x, \xi) = \sum_{|\alpha|=m} a_\alpha(x)\xi^\alpha \neq 0$ for all $(x, \xi) \in \Omega \times (\mathbf{R}^n \setminus \{0\})$.

6.2. Phase Functions

Let Ω be an open subset of $\mathbf{R}^n$. A function $\varphi(x, \theta)$ in $C^\infty(\Omega \times (\mathbf{R}^N \setminus \{0\}))$ is called a *phase function* on the space $\Omega \times (\mathbf{R}^N \setminus \{0\})$ if it satisfies the following three conditions:

(i) φ is real-valued.
(ii) φ is positively homogeneous of degree one in the variable θ.
(iii) The differential $d\varphi$ does not vanish on $\Omega \times (\mathbf{R}^N \setminus \{0\})$.

6.2.1 Example. Let U be an open subset of $\mathbf{R}^p$ and $\Omega = U \times U$. The function

$$\varphi(x, y, \xi) = (x - y)\cdot\xi$$

is a phase function on the space $\Omega \times (\mathbf{R}^N \setminus \{0\})$ $(n = 2p, N = p)$.

The next lemma will play a fundamental role in defining oscillatory integrals in Section 6.3.

6.2.2 Lemma. *If φ is a phase function on $\Omega \times (\mathbf{R}^N \setminus \{0\})$, then there exists a first-order differential operator*

$$L = \sum_{j=1}^{N} a_j(x, \theta) \frac{\partial}{\partial \theta_j} + \sum_{k=1}^{n} b_k(x, \theta) \frac{\partial}{\partial x_k} + c(x, \theta)$$

such that

$$L(e^{i\varphi}) = e^{i\varphi},$$

and its coefficients enjoy the following properties:

$$a_j \in S^0; \qquad b_k, c \in S^{-1}.$$

Furthermore, the transpose L' of L has coefficients a'_j, b'_k, c' in the same symbol classes as a_j, b_k, c.

6.3. Oscillatory Integrals

Let Ω be an open subset of $\mathbf{R}^n$ and $\varphi(x, \theta)$ a phase function on $\Omega \times (\mathbf{R}^N \setminus \{0\})$.

For any symbol $a(x, \theta) \in S^\infty(\Omega \times \mathbf{R}^N) = \bigcup_{m \in \mathbf{R}} S^m(\Omega \times \mathbf{R}^N)$, we wish to give a meaning to the integral

$$I_\varphi(au) = \iint e^{i\varphi(x, \theta)} a(x, \theta) u(x)\, dx\, d\theta, \qquad u \in C_0^\infty(\Omega). \tag{1}$$

It is clear that if $a \in S^m(\Omega \times \mathbf{R}^N)$ with $m < -N$, this integral is absolutely covergent.

We consider the general case. By Lemma 6.2.2, we can replace $e^{i\varphi}$ in formula (1) by $L(e^{i\varphi})$. Then a *formal* integration by parts gives us that

$$I_\varphi(au) = \iint e^{i\varphi(x,\theta)} L'(a(x, \theta)u(x))\, dx\, d\theta.$$

But the properties of the coefficients of L' imply that L' maps S^r continuously into S^{r-1} for all $r \in \mathbf{R}$. Continuing this process, one can reduce the growth of the integrand at infinity until it becomes integrable. In this way we can give a meaning to the integral (1) for all $a \in S^\infty(\Omega \times \mathbf{R}^N)$.

More precisely, we have:

6.3.1 Theorem

(i) *The linear functional*

$$S^{-\infty}(\Omega \times \mathbf{R}^N) \ni a \mapsto I_\varphi(au) \in \mathbf{C}$$

extends uniquely to a linear functional ℓ on $S^\infty(\Omega \times \mathbf{R}^N)$ whose restriction to each $S^m(\Omega \times \mathbf{R}^N)$ is continuous. Furthermore, the restriction of ℓ to $S^m(\Omega \times \mathbf{R}^N)$ is expressed as

$$\ell(a) = \iint e^{i\varphi(x,\theta)} (L')^k(a(x, \theta)u(x))\, dx\, d\theta,$$

where $k > m + N$.

(ii) *For any fixed $a \in S^m(\Omega \times \mathbf{R}^N)$, the mapping*

$$C_0^\infty(\Omega) \ni u \mapsto I_\varphi(au) = \ell(a) \in \mathbf{C} \tag{2}$$

is a distribution of order $\leq k$ for $k > m + N$.

We call the linear functional ℓ on S^∞ an *oscillatory integral*, but use the standard notation as in formula (1). The distribution (2) is called the *Fourier integral distribution* associated with the phase function φ and the amplitude a, and is denoted by

$$\int e^{i\varphi(x,\theta)} a(x, \theta)\, d\theta.$$

Oscillatory integrals depending on a parameter behave like ordinary integrals. In fact, we have:

6.3.2 Theorem. *Let Y be an open subset of $\mathbf{R}^p$ and let $\varphi(x, y, \theta)$ be a phase function on $\Omega \times Y \times (\mathbf{R}^N \setminus \{0\})$ such that*

$$d_{x,\theta}\, \varphi(x, y, \theta) \neq 0 \qquad \text{on } \Omega \times Y \times (\mathbf{R}^N \setminus \{0\}).$$

If $a \in S^m(\Omega \times Y \times \mathbf{R}^N)$, we let

$$F(y) = \iint e^{i\varphi(x, y, \theta)} a(x, y, \theta) u(x, y)\, dx\, d\theta, \qquad u \in C_0^\infty(\Omega \times Y).$$

Then we have the following:

(i) *The distribution F is in $C_0^\infty(Y)$ and one may differentiate under the integral sign.*

(ii) $\displaystyle\int F(y)\, dy = \iiint e^{i\varphi(x, y, \theta)} a(x, y, \theta) u(x, y)\, dx\, dy\, d\theta.$

If u is a distribution on Ω, the *singular support* of u is the smallest closed subset of Ω outside of which u is C^∞. The singular support of u is denoted by sing supp u.

The next theorem estimates the singular support of a Fourier integral distribution.

6.3.3 Theorem. *If φ is a phase function on $\Omega \times (\mathbf{R}^N \setminus \{0\})$ and if $a \in S^\infty(\Omega \times \mathbf{R}^N)$, then the distribution*

$$A = \int e^{i\varphi(x, \theta)} a(x, \theta)\, d\theta \in \mathscr{D}'(\Omega)$$

satisfies

$$\text{sing supp } A \subset \{x \in \Omega;\ d_\theta\, \varphi(x, \theta) = 0 \textit{ for some } \theta \in \mathbf{R}^N \setminus \{0\}\}.$$

6.4. Fourier Integral Operators

Let U and V be open subsets of $\mathbf{R}^p$ and $\mathbf{R}^q$, respectively. If $\varphi(x, y, \theta)$ is a phase function on $U \times V \times (\mathbf{R}^N \setminus \{0\})$ and if $a(x, y, \theta) \in S^\infty(U \times V \times \mathbf{R}^N)$, then there is associated a distribution $K \in \mathscr{D}'(U \times V)$ defined by

$$K = \int e^{i\varphi(x, y, \theta)} a(x, y, \theta)\, d\theta.$$

Applying Theorem 6.3.3 to our situation, we obtain that

$$\text{sing supp } K \subset \{(x, y) \in U \times V;\ d_\theta\, \varphi(x, y, \theta) = 0 \text{ for some } \theta \in \mathbf{R}^N \setminus \{0\}\}.$$

The distribution K defines a continuous linear operator

$$A\colon C_0^\infty(V) \to \mathscr{D}'(U)$$

by the formula

$$\langle Av, u\rangle = \langle K, u \otimes v\rangle, \qquad u \in C_0^\infty(U), \qquad v \in C_0^\infty(V).$$

The operator A is called the *Fourier integral operator* associated with the phase function φ and the amplitude a, and is denoted by

$$Av(x) = \iint e^{i\varphi(x, y, \theta)} a(x, y, \theta) v(y)\, dy\, d\theta.$$

The next theorem summarizes some basic properties of the operator A.

6.4.1 Theorem

(i) *If $d_{y,\theta}\, \varphi(x, y, \theta) \neq 0$ on $U \times V \times (\mathbf{R}^N \setminus \{0\})$, then the operator A maps $C_0^\infty(V)$ continuously into $C^\infty(U)$.*

(ii) *If $d_{x,\theta}\, \varphi(x, y, \theta) \neq 0$ on $U \times V \times (\mathbf{R}^N \setminus \{0\})$, then the operator A extends to a continuous linear operator on $\mathscr{E}'(V)$ into $\mathscr{D}'(U)$.*

(iii) *If $d_{y,\theta}\, \varphi(x, y, \theta) \neq 0$ and $d_{x,\theta}\, \varphi(x, y, \theta) \neq 0$ on $U \times V \times (\mathbf{R}^N \setminus \{0\})$, then we have, for all $v \in \mathscr{E}'(V)$,*

$$\text{sing supp } Av \subset \{x \in U;\ d_\theta\, \varphi(x, y, \theta) = 0 \textit{ for some } y \in \text{sing supp } v \textit{ and } \theta \in \mathbf{R}^N \setminus \{0\}\}.$$

6.5. Pseudo-Differential Operators

Definition and Basic Properties

Let Ω be an open subset of $\mathbf{R}^n$ and $m \in \mathbf{R}$. A *pseudo-differential operator* of order m on Ω is a Fourier integral operator of the form

$$Au(x) = \iint e^{i(x-y)\cdot\xi} a(x, y, \xi) u(y)\, dy\, đ\xi, \qquad u \in C_0^\infty(\Omega), \tag{1}$$

with some $a \in S^m(\Omega \times \Omega \times \mathbf{R}^n)$. In other words, a pseudo-differential operator of order m is a Fourier integral operator associated with the phase function $\varphi(x, y, \xi) = (x - y)\cdot\xi$ and some amplitude $a \in S^m(\Omega \times \Omega \times \mathbf{R}^n)$.

We let

$$L^m(\Omega) = \text{the set of all pseudo-differential operators of order } m \text{ on } \Omega.$$

Applying Theorems 6.4.1 and 6.3.3 to our situation, we obtain the following results:

1) A pseudo-differential operator A maps $C_0^\infty(\Omega)$ continuously into $C^\infty(\Omega)$, and it extends to a continuous linear operator $A: \mathscr{E}'(\Omega) \to \mathscr{D}'(\Omega)$.

2) The kernel K_A of a pseudo-differential operator A satisfies

$$\operatorname{sing\,supp} K_A \subset \{(x, x); x \in \Omega\},$$

that is, the kernel K_A is C^∞ off the diagonal $\{(x, x); x \in \Omega\}$ in $\Omega \times \Omega$.

3) sing supp $Au \subset$ sing supp u, $u \in \mathscr{E}'(\Omega)$. In other words, Au is C^∞ whenever u is. This property is referred to as the *pseudo-local property*.

We set

$$L^{-\infty}(\Omega) = \bigcap_{m \in \mathbf{R}} L^m(\Omega).$$

The next theorem characterizes the class $L^{-\infty}(\Omega)$.

6.5.1 Theorem. *The following three conditions are equivalent*:

(i) $A \in L^{-\infty}(\Omega)$.

(ii) *A is written in the form* (1) *with some* $a \in S^{-\infty}(\Omega \times \Omega \times \mathbf{R}^n)$.

(iii) *A is a regularizer, or equivalently, its kernel* K_A *is in* $C^\infty(\Omega \times \Omega)$.

Proof

(i) $\Rightarrow$ (iii): If $A \in L^{-\infty}(\Omega)$, then for every $m \in \mathbf{R}$ there exists a symbol $a \in S^m$ such that A can be written in the form (1). Then its kernel

$$K_A(x, y) = \int e^{i(x-y)\cdot\xi} a(x, y, \xi)\, đ\xi$$

is in $C^k(\Omega \times \Omega)$ for $k < -m - n$. This proves that K_A is in $C^\infty(\Omega \times \Omega)$.

(iii) $\Rightarrow$ (ii): If K_A is in $C^\infty(\Omega \times \Omega)$, we can write A in the form (1), by taking

$$a(x, y, \xi) = e^{-i(x-y)\cdot\xi} \theta(\xi) K_A(x, y)$$

with $\theta \in C_0^\infty(\mathbf{R}^n)$ and $\int \theta(\xi)\, đ\xi = 1$. This proves condition (ii), since $a \in S^{-\infty}$.

(ii) $\Rightarrow$ (i): This is trivial. ■

The next theorem states that every pseudo-differential operator can be written as the sum of a properly supported operator and a regularizer.

6.5.2 Theorem. *If $A \in L^m(\Omega)$, we have*

$$A = A_0 + R,$$

where $A_0 \in L^m(\Omega)$ is properly supported and $R \in L^{-\infty}(\Omega)$.

Proof. Choose a function $\rho \in C^\infty(\Omega \times \Omega)$ such that:

(a) $\rho = 1$ in a neighborhood of the diagonal $\{(x, x): x \in \Omega\}$ in $\Omega \times \Omega$;
(b) the restrictions to $\operatorname{supp} \rho$ of the projections $p_i: \Omega \times \Omega \ni (x_1, x_2) \mapsto x_i \in \Omega$ $(i = 1, 2)$ are proper mappings.

Then the operators A_0 and R, defined respectively by the kernels

$$K_{A_0} = \rho K_A,$$

$$K_R = (1 - \rho)K_A,$$

are the desired ones.

Symbols of a Pseudo-Differential Operator

First note that if $p(x, \xi) \in S^m(\Omega \times \mathbf{R}^n)$, then the operator $p(x, D)$, defined by

$$p(x, D)u(x) = \int e^{ix\cdot\xi} p(x, \xi)\hat{u}(\xi)\, d\!\!{}^-\xi, \qquad u \in C_0^\infty(\Omega), \tag{2}$$

is a pseudo-differential operator of order m on Ω, that is, $p(x, D) \in L^m(\Omega)$.

The next theorem asserts that every properly supported pseudo-differential operator can be reduced to the form (2).

6.5.3 Theorem. *If $A \in L^m(\Omega)$ is properly supported, then we have*

$$p(x, \xi) = e^{-ix\cdot\xi} A(e^{ix\cdot\xi}) \in S^m(\Omega \times \mathbf{R}^n),$$

and

$$A = p(x, D).$$

Furthermore, if $a(x, y, \xi) \in S^m(\Omega \times \Omega \times \mathbf{R}^n)$ is an amplitude for A, we have the asymptotic expansion

$$p(x, \xi) \sim \sum_{\alpha \geq 0} \frac{1}{\alpha!} \partial_\xi^\alpha D_y^\alpha(a(x, y, \xi))|_{y=x}.$$

The function $p(x, \xi)$ is called the *complete symbol* of A.

We extend the notion of a complete symbol to the whole space $L^m(\Omega)$. If $A \in L^m(\Omega)$, we choose a properly supported operator $A_0 \in L^m(\Omega)$ such that $A - A_0 \in L^{-\infty}(\Omega)$, and define

$$\sigma(A) = \text{the equivalence class of the complete symbol of } A_0 \text{ in } S^m(\Omega \times \mathbf{R}^n)/S^{-\infty}(\Omega \times \mathbf{R}^n).$$

In view of Theorems 6.5.1 and 6.5.3, it follows that $\sigma(A)$ does not depend on the operator A_0 chosen. The equivalence class $\sigma(A)$ is called the *complete symbol* of A. It is easy to see that the mapping

$$L^m(\Omega) \ni A \mapsto \sigma(A) \in S^m(\Omega \times \mathbf{R}^n)/S^{-\infty}(\Omega \times \mathbf{R}^n)$$

induces an isomorphism

$$L^m/L^{-\infty} \to S^m/S^{-\infty}.$$

Similarly, if $A \in L^m(\Omega)$, we define

$$\sigma_m(A) = \text{the equivalence class of the complete symbol of } A_0 \text{ in } S^m(\Omega \times \mathbf{R}^n)/S^{m-1}(\Omega \times \mathbf{R}^n).$$

The equivalence class $\sigma_m(A)$ is called the *principal symbol* of A. The mapping

$$L^m(\Omega) \ni A \mapsto \sigma_m(A) \in S^m(\Omega \times \mathbf{R}^n)/S^{m-1}(\Omega \times \mathbf{R}^n)$$

induces an isomorphism

$$L^m/L^{m-1} \to S^m/S^{m-1}.$$

We shall often identify the complete symbol $\sigma(A)$ with a representative in $S^m(\Omega \times \mathbf{R}^n)$ for notational convenience, and call any member of $\sigma(A)$ a complete symbol of A. We shall do the same for the principal symbol $\sigma_m(A)$.

A pseudo-differential operator $A \in L^m(\Omega)$ is said to be *classical* if its complete symbol $\sigma(A)$ has a representative in the class $S^m_{\mathrm{cl}}(\Omega \times \mathbf{R}^n)$.

We let

$$L^m_{\mathrm{cl}}(\Omega) = \text{the set of all classical pseudo-differential operators of order } m \text{ on } \Omega.$$

Then the mapping

$$L^m_{\mathrm{cl}}(\Omega) \ni A \mapsto \sigma(A) \in S^m_{\mathrm{cl}}(\Omega \times \mathbf{R}^n)/S^{-\infty}(\Omega \times \mathbf{R}^n)$$

induces an isomorphism

$$L^m_{\mathrm{cl}}/L^{-\infty} \to S^m_{\mathrm{cl}}/S^{-\infty}.$$

Also we have

$$L^{-\infty}(\Omega) = \bigcap_{m \in \mathbf{R}} L^m_{\mathrm{cl}}(\Omega).$$

If $A \in L^m_{\text{cl}}(\Omega)$, its principal symbol $\sigma_m(A)$ has a canonical representative $\sigma_A(x, \xi) \in C^\infty(\Omega \times (\mathbf{R}^n \setminus \{0\}))$ which is positively homogeneous of degree m in the variable ξ. The function $\sigma_A(x, \xi)$ is called the *homogeneous principal symbol* of a. We remark that

$$A \in L^{m-1}_{\text{cl}}(\Omega) \Leftrightarrow \sigma_A \equiv 0 \text{ on } \Omega \times (\mathbf{R}^n \setminus \{0\}).$$

The Algebra of Pseudo-Differential Operators

The next two theorems assert that the class of pseudo-differential operators forms an algebra closed under the operations of composition of operators and taking the transpose or adjoint of an operator.

6.5.4 Theorem. *If $A \in L^m(\Omega)$, then its transpose A' and its adjoint A^* are both in $L^m(\Omega)$, and the complete symbols $\sigma(A')$ and $\sigma(A^*)$ have respectively the following asymptotic expansions:*

$$\begin{cases} \sigma(A')(x, \xi) \sim \sum\limits_{\alpha \geq 0} \dfrac{1}{\alpha!} \partial_\xi^\alpha D_x^\alpha (\sigma(A)(x, -\xi)), \\[2ex] \sigma(A^*)(x, \xi) \sim \sum\limits_{\alpha \geq 0} \dfrac{1}{\alpha!} \partial_\xi^\alpha D_x^\alpha (\overline{\sigma(A)(x, \xi)}). \end{cases}$$

6.5.5 Theorem. *If $A \in L^m(\Omega)$ and $B \in L^{m'}(\Omega)$, and if one of them is properly supported, then the composition AB is in $L^{m+m'}(\Omega)$, and we have the asymptotic expansion*

$$\sigma(AB)(x, \xi) \sim \sum_{\alpha \geq 0} \frac{1}{\alpha!} \partial_\xi^\alpha(\sigma(A)(x, \xi))\, D_x^\alpha(\sigma(B)(x, \xi)).$$

Elliptic Pseudo-Differential Operators

A pseudo-differential operator $A \in L^m(\Omega)$ is said to be *elliptic* of order m if its complete symbol $\sigma(A)$ is elliptic of order m. In view of Corollary 6.1.4, it follows that a classical pseudo-differential operator $A \in L^m_{\text{cl}}(\Omega)$ is elliptic if and only if its homogeneous principal symbol $\sigma_A(x, \xi)$ does not vanish on the space $\Omega \times (\mathbf{R}^n \setminus \{0\})$.

The next theorem states that elliptic operators are the "invertible" elements in the algebra of pseudo-differential operators.

6.5.6 Theorem. *An operator $A \in L^m(\Omega)$ is elliptic if and only if there exists a properly supported operator $B \in L^{-m}(\Omega)$ such that*

$$\begin{cases} AB \equiv I & \mod L^{-\infty}(\Omega), \\ BA \equiv I & \mod L^{-\infty}(\Omega). \end{cases}$$

Such an operator B is called a *parametrix* for A. In other words, a parametrix for A is a two-sided inverse of A modulo $L^{-\infty}(\Omega)$. We observe that a parametrix is unique modulo $L^{-\infty}(\Omega)$.

Invariance of Pseudo-Differential Operators Under Change of Coordinates

We see what happens to a pseudo-differential operator under a change of coordinates.

6.5.7 Theorem. *Let Ω_1, Ω_2 be two open subsets of $\mathbf{R}^n$ and $\chi: \Omega_1 \to \Omega_2$ a C^∞ diffeomorphism. If $A \in L^m(\Omega_1)$, then the mapping*

$$A_\chi: C_0^\infty(\Omega_2) \to C^\infty(\Omega_2)$$
$$v \to A(v \circ \chi) \circ \chi^{-1}$$

is in $L^m(\Omega_2)$, and we have the asymptotic expansion

$$\sigma(A_\chi)(y, \eta) \sim \sum_{\alpha \geq 0} \frac{1}{\alpha!} (\partial_\xi^\alpha \sigma(A))(x, {}^t\chi'(x) \cdot \eta)\, D_z(e^{ir(x, z, \eta)})|_{z=x} \tag{3}$$

with

$$r(x, z, \eta) = \langle \chi(z) - \chi(x) - \chi'(x) \cdot (z - x), \eta \rangle.$$

Here $x = \chi^{-1}(y)$, $\chi'(x)$ is the derivative of χ at x and ${}^t\chi'(x)$ its transpose.

6.5.8 Remark. Formula (3) shows that

$$\sigma(A_\chi)(y, \eta) \equiv \sigma(A)(x, {}^t\chi'(x) \cdot \eta) \qquad \mod S^{m-1}.$$

Note that the mapping

$$\Omega_2 \times \mathbf{R}^n \ni (y, \eta) \mapsto (x, {}^t\chi'(x) \cdot \eta) \in \Omega_1 \times \mathbf{R}^n$$

is just a transition map of the cotangent bundle $T^*(\mathbf{R}^n)$. This implies that the principal symbol $\sigma_m(A)$ of $A \in L^m(\mathbf{R}^n)$ can be invariantly defined on $T^*(\mathbf{R}^n)$.

Pseudo-Differential Operators and Sobolev Spaces

A differential operator of order m with C^∞ coefficients on Ω is continuous on $H^s_{\mathrm{loc}}(\Omega)$ into $H^{s-m}_{\mathrm{loc}}(\Omega)$ for all $s \in \mathbf{R}$. This result extends to pseudo-differential operators:

6.5.9 Theorem. *Every operator $A \in L^m(\Omega)$ extends to a continuous linear operator*

$$A: H^s_{\mathrm{comp}}(\Omega) \to H^{s-m}_{\mathrm{loc}}(\Omega)$$

for all $s \in \mathbf{R}$. If in addition A is properly supported, it extends to continuous linear operators

$$A: H^s_{\mathrm{comp}}(\Omega) \to H^{s-m}_{\mathrm{comp}}(\Omega),$$

$$A: H^s_{\mathrm{loc}}(\Omega) \to H^{s-m}_{\mathrm{loc}}(\Omega)$$

for all $s \in \mathbf{R}$.

Combining Theorem 6.5.6 and Theorem 6.5.9, we obtain:

6.5.10 Theorem (the elliptic regularity theorem). *If $A \in L^m(\Omega)$ is properly supported and elliptic, then we have*:

(i) *A distribution $u \in \mathscr{D}'(\Omega)$ is in $H^{s+m}_{\mathrm{loc}}(\Omega)$ if and only if $Au \in H^s_{\mathrm{loc}}(\Omega)$.*
(ii) sing supp u = sing supp Au, $u \in \mathscr{D}'(\Omega)$.
In other words, u is C^∞ if and only if Au is.
(iii) *For every compact $K \subset \Omega$, $s \in \mathbf{R}$ and $t < s + m$, there exists a constant $C_{K,s,t} > 0$ such that*

$$\| u \|_{s+m} \leq C_{K,s,t}(\| Au \|_s + \| u \|_t), \qquad u \in C^\infty_K(\Omega).$$

Proof. Take a parametrix $B \in L^{-m}(\Omega)$ for A as in Theorem 6.5.6:

$$\begin{cases} AB = I + R_1, & R_1 \in L^{-\infty}(\Omega), \\ BA = I + R_2, & R_2 \in L^{-\infty}(\Omega). \end{cases}$$

Then parts (i) and (iii) follow from an application of Theorem 6.5.9. Further it follows that

$$\begin{aligned} u &= BAu - R_2 u \\ &\equiv BAu \qquad \mathrm{mod}\ C^\infty(\Omega), \end{aligned}$$

so that, by the pseudo-local property of B, we have

$$\begin{aligned}\text{sing supp } u &= \text{sing supp } B(Au) \\ &\subset \text{sing supp } Au.\end{aligned}$$

This proves part (ii), since the converse inclusion is simply the pseudo-local property of A. ■

6.6. Pseudo-Differential Operators on a Manifold

In this section we shall define the concept of a pseudo-differential operator on a manifold, and transfer all the machinery of pseudo-differential operators to manifolds.

Throughout this section, let M be an n-dimensional, *compact* C^∞ manifold without boundary.

Definition and Basic Properties

Theorem 6.5.7 leads us to the following:

6.6.1 Definition. A continuous linear operator $A: C^\infty(M) \to C^\infty(M)$ is called a *pseudo-differential operator* of order $m \in \mathbf{R}$ if it satisfies the following conditions:

(i) The kernel of A is C^∞ off the diagonal $\{(x, x); x \in M\}$ in $M \times M$.
(ii) For any chart (U, χ) on M, the mapping

$$\begin{aligned}A_\chi: C_0^\infty(\chi(U)) &\to C^\infty(\chi(U)) \\ u &\mapsto A(u \circ \chi) \circ \chi^{-1}\end{aligned}$$

belongs to $L^m(\chi(U))$ (cf. Section 4.7).

We let

$L^m(M) =$ the set of all pseudo-differential operators of order m on M,

and set

$$L^{-\infty}(M) = \bigcap_{m \in \mathbf{R}} L^m(M).$$

Some results about pseudo-differential operators on $\mathbf{R}^n$ stated in Section 6.5 are also true for pseudo-differential operators on M. In fact, pseudo-differential operators on M are defined to be locally pseudo-differential operators on $\mathbf{R}^n$.

For example, we have the following:

1. A pseudo-differential operator A extends to a continuous linear operator $A: \mathscr{D}'(M) \to \mathscr{D}'(M)$.
2. sing supp $Au \subset$ sing supp u, $u \in \mathscr{D}'(M)$.
3. A continuous linear operator $A: C^\infty(M) \to \mathscr{D}'(M)$ is a regularizer if and only if it is in $L^{-\infty}(M)$.
4. A pseudo-differential operator $A \in L^m(M)$ extends to a continuous linear operator $A: H^s(M) \to H^{s-m}(M)$ for all $s \in \mathbf{R}$.

Classical Pseudo-Differential Operators

A pseudo-differential operator $A \in L^m(M)$ is said to be *classical* if, for any chart (U, χ) on M, the mapping $A_\chi: C_0^\infty(\chi(U)) \to C^\infty(\chi(U))$ belongs to $L_{\mathrm{cl}}^m(\chi(U))$.

We let

$$L_{\mathrm{cl}}^m(M) = \text{the set of all classical pseudo-differential operators of order } m \text{ on } M.$$

We observe that

$$L^{-\infty}(M) = \bigcap_{m \in \mathbf{R}} L_{\mathrm{cl}}^m(M).$$

From now on, we only consider classical pseudo-differential operators that we often encounter in applications. For example, differential operators and parametrices for elliptic differential operators are classical pseudo-differential operators.

Let $A \in L_{\mathrm{cl}}^m(M)$. If (U, χ) is a chart on M, there is associated a homogeneous principal symbol $\sigma_{A_\chi} \in C^\infty(\chi(U) \times (\mathbf{R}^n \setminus \{0\}))$. In view of Remark 6.5.8, by smoothly patching together the functions σ_{A_χ}, we obtain a C^∞ function $\sigma_A(x, \xi)$ on $T^*(M) \setminus \{0\} = \{(x, \xi) \in T^*(M); \xi \neq 0\}$, which is positively homogeneous of degree m in the variable ξ. The function σ_A is called the *homogeneous principal symbol* of A. We remark that

$$A \in L_{\mathrm{cl}}^{m-1}(M) \quad \Leftrightarrow \quad \sigma_A \equiv 0 \text{ on } T^*(M) \setminus \{0\}.$$

The next theorem asserts that the class $L^m_{cl}(M)$ of classical pseudo-differential operators is stable under the operations of composition of operators and taking the transpose or adjoint of an operator.

6.6.2 Theorem

(i) *If $A \in L^m_{cl}(M)$, then its transpose A' and its adjoint A^* are both in $L^m_{cl}(M)$, and we have*

$$\begin{cases} \sigma_{A'}(x, \xi) = \sigma_A(x, -\xi), \\ \sigma_{A^*}(x, \xi) = \overline{\sigma_A(x, \xi)}. \end{cases} \tag{1}$$

(ii) *If $A \in L^m_{cl}(M)$ and $B \in L^{m'}_{cl}(M)$, then the composition AB is in $L^{m+m'}_{cl}(M)$, and we have*

$$\sigma_{AB}(x, \xi) = \sigma_A(x, \xi) \cdot \sigma_B(x, \xi). \tag{2}$$

Elliptic Pseudo-Differential Operators

A classical pseudo-differential operator $A \in L^m_{cl}(M)$ is said to be *elliptic* of order m if its homogeneous principal symbol $\sigma_A(x, \xi)$ does not vanish on the bundle $T^*(M) \setminus \{0\}$ of non-zero cotangent vectors.

The next theorem is a generalization of Theorem 6.5.6.

6.6.3 Theorem. *An operator $A \in L^m_{cl}(M)$ is elliptic if and only if there exists a parametrix $B \in L^{-m}_{cl}(M)$ for A:*

$$\begin{cases} AB \equiv I \quad \mod L^{-\infty}(M), \\ BA \equiv I \quad \mod L^{-\infty}(M). \end{cases}$$

6.7. Elliptic Pseudo-Differential Operators and their Indices

In this section, using the Riesz-Schauder theory, we shall prove some of the most important results about elliptic pseudo-differential operators on a manifold. These results will be useful for the study of boundary value problems in Chapter 8.

Throughout this section, let M be an n-dimensional, *compact* C^∞ manifold without boundary.

Pseudo-Differential Operators on Sobolev Spaces

Let $H^s(M)$ be the Sobolev space of order $s \in \mathbf{R}$ on M. Recall that

$$C^\infty(M) = \bigcap_{s \in \mathbf{R}} H^s(M),$$

$$\mathscr{D}'(M) = \bigcup_{s \in \mathbf{R}} H^s(M).$$

A linear operator $T\colon C^\infty(M) \to C^\infty(M)$ is said to be *of order* $m \in \mathbf{R}$ if it extends to a continuous linear operator on $H^s(M)$ into $H^{s-m}(M)$ for each $s \in \mathbf{R}$. For example, every pseudo-differential operator in $L^m(M)$ is of order m.

We say that $T\colon C^\infty(M) \to C^\infty(M)$ is *of order* $-\infty$ if it extends to a continuous linear operator on $H^s(M)$ into $C^\infty(M)$ for each $s \in \mathbf{R}$. This is equivalent to saying that T is a regularizer; hence we have

$$L^{-\infty}(M) = \text{the set of all operators of order } -\infty. \tag{1}$$

Let $T\colon H^s(M) \to H^t(M)$ be a linear operator with domain $D(T)$ everywhere dense in $H^s(M)$. Each element v of $H^{-t}(M)$ defines a linear functional G on $D(T)$ by the formula

$$G(u) = (Tu, v), \qquad u \in D(T),$$

where $(\ ,\ )$ on the right-hand side is the sesquilinear pairing of $H^t(M)$ and $H^{-t}(M)$. If this functional G is continuous everywhere on $D(T)$, then, applying Theorem 3.2.2, we obtain that G can be extended uniquely to a continuous linear functional $\tilde{G}$ on $\overline{D(T)} = H^s(M)$. Hence there exists a unique element v^* of $H^{-s}(M)$ such that

$$\tilde{G}(u) = (u, v^*), \qquad u \in H^s(M),$$

since the sesquilinear form $(\ ,\ )$ on the product space $H^s(M) \times H^{-s}(M)$ permits us to identify the strong dual space of $H^s(M)$ with $H^{-s}(M)$. In particular, we have

$$(Tu, v) = G(u) = (u, v^*), \qquad u \in D(T).$$

So we let

$D(T^*) =$ the totality of those $v \in H^{-t}(M)$ such that the mapping $u \mapsto (Tu, v)$ is continuous everywhere on $D(T)$,

and define

$$T^*v = v^*.$$

Therefore T^* is a linear operator from $H^{-t}(M)$ into $H^{-s}(M)$ with domain $D(T^*)$ such that

$$(Tu, v) = (u, T^*v), \qquad u \in D(T), \qquad v \in D(T^*). \tag{2}$$

The operator T^* is called the *adjoint* of T.

The *transpose* of T is a linear operator T' from $H^{-t}(M)$ into $H^{-s}(M)$ with domain $D(T')$ such that

$D(T')$ = the totality of those $v \in H^{-t}(M)$ such that the mapping $u \mapsto \langle Tu, v\rangle$ is continuous everywhere on $D(T)$,

and satisfies

$$\langle Tu, v\rangle = \langle u, T'v\rangle, \qquad u \in D(T), \qquad v \in D(T'). \tag{3}$$

Here $\langle\ ,\ \rangle$ on the left-hand (resp. right-hand) side is the bilinear pairing of $H^t(M)$ and $H^{-t}(M)$ (resp. $H^s(M)$ and $H^{-s}(M)$).

In view of formulas (2) and (3), it follows that:

(a) $v \in D(T') \Leftrightarrow \bar{v} \in D(T^*)$,
(b) $T'v = \overline{T^*\bar{v}}$, $v \in D(T')$

where $\overline{}$ denotes complex conjugation. Hence we have the following:

$$\begin{cases} \text{1. The ranges } R(T^*) \text{ and } R(T') \text{ are isomorphic.} \\ \text{2. The null spaces } N(T^*) \text{ and } N(T') \text{ are isomorphic.} \end{cases} \tag{4}$$

Now let $A \in L^m(M)$. Then the operator $A: C^\infty(M) \to C^\infty(M)$ extends uniquely to a continuous linear operator

$$A_s: H^s(M) \to M^{s-m}(M)$$

for all $s \in \mathbf{R}$, and hence to a continuous linear operator

$$\tilde{A}: \mathscr{D}'(M) \to \mathscr{D}'(M).$$

The adjoint A^* of A is also in $L^m(M)$; hence the operator $A^*: C^\infty(M) \to C^\infty(M)$ extends uniquely to a continuous linear operator

$$A_s^*: H^s(M) \to H^{s-m}(M)$$

for all $s \in \mathbf{R}$.

The next lemma states a fundamental relationship between the operators A_s and A_s^*.

6.7.1 Lemma. *If $A \in L^m(M)$ we have, for all $s \in \mathbf{R}$,*

$$\begin{cases} A_s^* = A^*_{-s+m}, \\ (A^*_{-s+m})^* = A_s. \end{cases} \tag{5}$$

Proof. If $u \in D(A_s) = H^s(M)$ and $v \in D(A^*_{-s+m}) = H^{-s+m}(M)$, there exist sequences $\{u_j\}$ and $\{v_j\}$ in $C^\infty(M)$ such that $u_j \to u$ in $H^s(M)$ and $v_j \to v$ in $H^{-s+m}(M)$, respectively. Then we have

$$\begin{cases} Au_j \to A_s u & \text{in } H^{s-m}(M), \\ A^*v_j \to A^*_{-s+m}v & \text{in } H^{-s}(M), \end{cases}$$

so that

$$\begin{aligned} (A_s u, v) &= \lim_j (Au_j, v_j) \\ &= \lim_j (u_j, A^*v_j) \\ &= (u, A^*_{-s+m}v). \end{aligned}$$

This proves formulas (5). ■

The Index of an Elliptic Pseudo-Differential Operator

In this subsection, we study the operators A_s when A is a classical *elliptic* pseudo-differential operator.

The next theorem is an immediate consequence of Theorem 6.6.3.

6.7.2 Theorem (the elliptic regularity theorem). *Let $A \in L^m_{\mathrm{cl}}(M)$ be elliptic. Then we have, for all $s \in \mathbf{R}$,*

$$u \in \mathscr{D}'(M),\ Au \in H^s(M) \quad \Rightarrow \quad u \in H^{s+m}(M).$$

In particular, we have

$$R(A_s) \cap C^\infty(M) = R(A), \tag{6}$$

$$N(A_s) = N(A). \tag{7}$$

Here

$$R(A_s) = \{A_s u;\ u \in H^s(M)\}, \qquad R(A) = \{Au;\ u \in C^\infty(M)\};$$

$$N(A_s) = \{u \in H^s(M);\ A_s u = 0\}, \qquad N(A) = \{u \in C^\infty(M);\ Au = 0\}.$$

The next theorem states that the operators A_s are Fredholm operators.

6.7.3 Theorem. *Let $A \in L^m_{cl}(M)$ be elliptic. Then the operator A_s: $H^s(M) \to H^{s-m}(M)$ is a Fredholm operator for all $s \in \mathbf{R}$.*

Proof. Take a parametrix $B \in L^{-m}_{cl}(M)$ for A:

$$\begin{cases} BA = I + P, & P \in L^{-\infty}(M), \\ AB = I + Q, & Q \in L^{-\infty}(M). \end{cases}$$

Then we have

$$\begin{cases} B_{s-m} \cdot A_s = I + P_s, \\ A_s \cdot B_{s-m} = I + Q_{s-m}. \end{cases}$$

Further, in view of assertion (1), it follows from Rellich's theorem that the operators P_s: $H^s(M) \to H^s(M)$ and Q_{s-m}: $H^{s-m}(M) \to H^{s-m}(M)$ are compact. Therefore, applying Theorem 3.7.2 to our situation, we obtain that A_s is a Fredholm operator. ■

6.7.4 Corollary. *Let $A \in L^m_{cl}(M)$ be elliptic. Then we have:*

(i) *The range $R(A)$ of A is a closed linear subspace of $C^\infty(M)$.*
(ii) *The null space $N(A)$ of A is a finite dimensional, closed linear subspace of $C^\infty(M)$.*

Proof.

(i) It follows from Theorem 6.7.3 that the range $R(A_s)$ of A_s is closed in $H^{s-m}(M)$; hence it is closed in $C^\infty(M)$, since the injection $C^\infty(M) \to H^{s-m}(M)$ is continuous. In view of formula (6), this proves part (i).
(ii) Similarly, in view of formula (7), it follows from Theorem 6.7.3 that $N(A)$ has finite dimension; so it is closed in each $H^s(M)$ and hence in $C^\infty(M) = \bigcap_{s \in \mathbf{R}} H^s(M)$. ■

The next theorem asserts that

$$\text{ind } A_s = \dim N(A_s) - \text{codim } R(A_s)$$

does not depend on $s \in \mathbf{R}$.

6.7.5 Theorem. *If $A \in L^m_{cl}(M)$ is elliptic then we have, for all $s \in \mathbf{R}$,*

$$\text{ind } A_s = \dim N(A) - \dim N(A^*). \tag{8}$$

Here

$$N(A^*) = \{v \in C^\infty(M); A^*v = 0\}.$$

Proof. Since the range $R(A_s)$ is closed in $H^{s-m}(M)$, applying the closed range theorem (Theorem 3.4.6) to our situation, we obtain that

$$\operatorname{codim} R(A_s) = \dim N(A_s').$$

But, in view of assertion (4), it follows that

$$\dim N(A_s') = \dim N(A_s^*).$$

Further we have, by formulas (5) and (7),

$$N(A_s^*) = N(A^*_{-s+m}) = N(A^*), \tag{9}$$

since $A^* \in L^m_{\mathrm{cl}}(M)$ is also elliptic (cf. formula (6.6.1)). Summing up, we obtain that

$$\operatorname{codim} R(A_s) = \dim N(A^*). \tag{10}$$

Therefore, formula (8) follows from formulas (7) and (10). ■

We give another useful expression for ind A_s. To do so, we need the following:

6.7.6 Lemma. *Let $A \in L^m_{\mathrm{cl}}(M)$ be elliptic. Then the spaces $N(A^*)$ and $R(A)$ are orthogonal complements of each other in $C^\infty(M)$ relative to the inner product of $L^2(M)$:*

$$C^\infty(M) = N(A^*) \oplus R(A). \tag{11}$$

Proof. Since the range $R(A_m)$ is closed in $L^2(M)$, applying the closed range theorem, we obtain that

$$L^2(M) = N(A_m^*) \oplus R(A_m). \tag{12}$$

But it follows from formula (9) that

$$N(A_m^*) = N(A^*). \tag{13}$$

Therefore, combining (12) and (13), we have

$$L^2(M) = N(A^*) \oplus R(A_m).$$

In view of formula (6), this implies the decomposition (11). ■

Now we can prove:

6.7.7 Theorem. *If $A \in L^m_{\mathrm{cl}}(M)$ is elliptic then we have, for all $s \in \mathbf{R}$,*

$$\operatorname{ind} A_s = \dim N(A) - \operatorname{codim} R(A). \tag{14}$$

Here

$$\operatorname{codim} R(A) = \dim C^\infty(M)/R(A).$$

Proof. The decomposition (11) tells us that

$$\dim N(A^*) = \operatorname{codim} R(A).$$

Hence formula (14) follows from formula (8). ■

We let

$$\begin{aligned} \operatorname{ind} A &= \dim N(A) - \dim N(A^*) \\ &= \dim N(A) - \operatorname{codim} R(A). \end{aligned} \tag{15}$$

The next theorem states that the index of an elliptic pseudo-differential operator depends only on its principal symbol.

6.7.8 Theorem. *If A, $B \in L^m_{\mathrm{cl}}(M)$ are elliptic and if they have the same homogeneous principal symbol, then we have*

$$\operatorname{ind} A = \operatorname{ind} B. \tag{16}$$

Proof. Since the difference $A - B$ is in $L^{m-1}_{\mathrm{cl}}(M)$, it follows from Rellich's theorem that the operator

$$A_s - B_s \colon H^s(M) \to H^{s-m}(M)$$

is compact. Hence, applying Theorem 3.7.4, we obtain that

$$\begin{aligned} \operatorname{ind} A_s &= \operatorname{ind}(B_s + (A_s - B_s)) \\ &= \operatorname{ind} B_s. \end{aligned}$$

In view of Theorem 6.7.7, this proves formula (16). ■

As for the product of elliptic pseudo-differential operators, we have:

6.7.9 Theorem. *If $A \in L^m_{\mathrm{cl}}(M)$ and $B \in L^{m'}_{\mathrm{cl}}(M)$ are elliptic, then we have*

$$\operatorname{ind} BA = \operatorname{ind} B + \operatorname{ind} A. \tag{17}$$

Proof. Observe that for each $s \in \mathbf{R}$ we have

$$(BA)_s = B_{s-m} \cdot A_s.$$

Hence, applying Theorem 3.7.3, we obtain that

$$\mathrm{ind}(BA)_s = \mathrm{ind}\, B_{s-m} + \mathrm{ind}\, A_s.$$

This proves formula (17), since BA is an elliptic operator in $L_{\mathrm{cl}}^{m+m'}(M)$ (cf. formula (6.6.2)). ■

As for the adjoints, we have:

6.7.10 Theorem. *If $A \in L_{\mathrm{cl}}^m(M)$ is elliptic, then we have*

$$\mathrm{ind}\, A^* = -\,\mathrm{ind}\, A. \tag{18}$$

In fact, it suffices to note that $A^{**} = A$.
We give some useful criteria for ind $A = 0$.

6.7.11 Theorem. *If $A \in L_{\mathrm{cl}}^m(M)$ is elliptic and if A and A^* have the homogeneous principal symbol, then we have*

$$\mathrm{ind}\, A = 0. \tag{19}$$

Proof. Theorem 6.7.8 tells us that ind A = ind A^*. But, in view of formula (18), this implies formula (19). ■

6.7.12 Corollary. *If $A \in L_{\mathrm{cl}}^m(\Omega)$ is elliptic and if its homogeneous principal symbol is real, then we have*

$$\mathrm{ind}\, A = 0.$$

Proof. In view of formula (6.6.1), it follows that A and A^* have the same homogeneous principal symbol. Hence Theorem 6.7.11 applies. ■

6.7.13 Theorem. *If $A \in L_{\mathrm{cl}}^m(\Omega)$ is elliptic and if $A^* = \lambda A$ for some $\lambda \in \mathbf{C}$, then we have*

$$|\lambda| = 1,$$

and

$$\mathrm{ind}\, A = 0.$$

Proof. First note that

$$A = A^{**} = (\lambda A)^* = \bar{\lambda} A^* = |\lambda|^2 A.$$

Hence we have $|\lambda| = 1$, and so

$$\begin{cases} N(\lambda A) = N(A), \\ R(\lambda A) = R(A). \end{cases}$$

Thus it follows from formula (18) that

$$\text{ind } A = \text{ind } \lambda A = \text{ind } A^* = -\text{ ind } A.$$

This implies that ind $A = 0$. ■

The next theorem describes conditions under which an elliptic pseudo-differential operator is invertible on Sobolev spaces.

6.7.14 Theorem. *Let $A \in L^m_{\text{cl}}(M)$ be elliptic. Suppose that*

$$\begin{cases} \text{ind } A = 0, \\ N(A) = \{0\}. \end{cases}$$

Then we have:

(i) *The operator A: $C^\infty(M) \to C^\infty(M)$ is bijective.*
(ii) *The operator A_s: $H^s(M) \to H^{s-m}(M)$ is an isomorphism for each $s \in \mathbf{R}$.*
(iii) *The inverse A^{-1} of A is in $L^{-m}_{\text{cl}}(M)$.*

Proof. (i) Since ind $A = 0$ and $N(A) = \{0\}$, it follows from formula (15) that $N(A^*) = \{0\}$. Hence the surjectivity of A follows from the decomposition (11).

(ii) Since $N(A_s) = N(A) = \{0\}$ and ind A_s = ind $A = 0$, it follows that the operator A_s: $H^s(M) \to H^{s-m}(M)$ is bijective for each $s \in \mathbf{R}$. Therefore, applying the closed graph theorem (Theorem 3.4.3) to our situation, we obtain that the inverse A_s^{-1}: $H^{s-m}(M) \to H^s(M)$ is continuous for each $s \in \mathbf{R}$.

(iii) Since we have

$$A^{-1} = A_s^{-1}|_{C^\infty(M)}$$

and each A_s^{-1}: $H^{s-m}(M) \to H^s(M)$ is continuous, it follows that the operator A^{-1}: $C^\infty(M) \to C^\infty(M)$ is continuous, and also it is of order $-m$.

To prove that $A^{-1} \in L_{\rm cl}^{-m}(M)$, take a parametrix $B \in L_{\rm cl}^{-m}(M)$ for A:

$$\begin{cases} AB = I + P, & P \in L^{-\infty}(M), \\ BA \equiv I + Q, & Q \in L^{-\infty}(M). \end{cases}$$

Then we have

$$\begin{aligned} A^{-1} - B &= (I - BA)A^{-1} \\ &= -Q \cdot A^{-1}. \end{aligned}$$

But, in view of assertion (1), it follows that the operator $Q \cdot A^{-1}$ is in $L^{-\infty}(M)$, since it is of order $-\infty$. This proves that $A^{-1} = B - Q \cdot A^{-1} \in L_{\rm cl}^{-m}(M)$. ■

The next theorem states that the Sobolev spaces $H^s(M)$ can be characterized in terms of elliptic pseudo-differential operators.

6.7.15 Theorem. *Let $A \in L_{\rm cl}^m(M)$ be elliptic with $m > 0$. Suppose that*

$$\begin{cases} A = A^*, \\ N(A) = \{0\}. \end{cases}$$

Then we have:

(i) *There exists a complete orthonormal system $\{\varphi_j\}$ of $L^2(M)$ consisting of eigenfunctions of A, and its corresponding eigenvalues $\{\lambda_j\}$ are real and $|\lambda_j| \to +\infty$.*

(ii) *A distribution $u \in \mathscr{D}'(M)$ belongs to $H^{mr}(M)$ for some integer r if and only if we have*

$$\sum_{j=1}^{\infty} \lambda_j^{2r} |(u, \varphi_j)|^2 < +\infty.$$

More precisely, the quantity

$$(u, v)_{mr} = \sum_{j=1}^{\infty} \lambda_j^{2r} (u, \varphi_j)\overline{(v, \varphi_j)} \tag{20}$$

is an admissible inner product for the space $H^{mr}(M)$.

Proof. (i) Since $A = A^*$, it follows from Theorem 6.7.13 that $\operatorname{ind} A = 0$. Hence, applying Theorem 6.7.14, we obtain that the operator $A\colon C^\infty(M) \to C^\infty(M)$ is bijective, and its inverse A^{-1} is an elliptic operator in $L_{\rm cl}^{-m}(M)$.

We let

$$\widetilde{A^{-1}} = \text{the composition of } (A^{-1})_0 \colon L^2(M) \to H^m(M) \text{ and the injection: } H^m(M) \to L^2(M).$$

Then it follows from Rellich's theorem that the operator $\widetilde{A^{-1}}: L^2(M) \to L^2(M)$ is compact. Further, since $C^\infty(M)$ is dense in $L^2(M)$ and $A = A^*$, we have

$$(\widetilde{A^{-1}}u, v) = (u, \widetilde{A^{-1}}\, v), \qquad u, v \in L^2(M),$$

where (,) is the inner product of $L^2(M)$. This implies that the operator $\widetilde{A^{-1}}$ is self-adjoint. Also, we have

$$N(\widetilde{A^{-1}}) = N((A^{-1})_0) = N(A^{-1}) = \{0\},$$

since $A^{-1} \in L_{\mathrm{cl}}^{-m}(M)$ is elliptic. Therefore, applying the Hilbert-Schmidt theorem (Theorem 3.8.14) to the operator $\widetilde{A^{-1}}$, we obtain that there exists a complete orthonormal system $\{\varphi_j\}$ of $L^2(M)$ consisting of eigenfunctions of $\widetilde{A^{-1}}$, and its corresponding eigenvalues $\{\mu_j\}$ are real and converge to zero.

Since the eigenvalues μ_j are all non-zero, it follows that

$$\varphi_j = \frac{1}{\mu_j}\widetilde{A^{-1}}\varphi_j = \frac{1}{\mu_j}(A^{-1})_0\varphi_j \in H^m(M).$$

But note that $(A^{-1})_0|_{H^m(M)} = (A^{-1})_m$ and that $(A^{-1})_m: H^m(M) \to H^{2m}(M)$. Hence we have

$$\varphi_j = \frac{1}{\mu_j}(A^{-1})_m\varphi_j \in H^{2m}(M).$$

Continuing this way, we obtain that

$$\varphi_j \in \bigcap_{k \in \mathbf{N}} H^{km}(M) = C^\infty(M).$$

Therefore we have

$$A^{-1}\varphi_j = \mu_j\varphi_j,$$

and hence

$$A\varphi_j = \lambda_j\varphi_j, \qquad \lambda_j = \frac{1}{\mu_j},$$

with

$$|\lambda_j| = \frac{1}{|\mu_j|} \to +\infty.$$

(ii) For each integer r, we let

$$A^r = \begin{cases} A^r & \text{if } r \geq 0, \\ (A^{-1})^{|r|} & \text{if } r < 0, \end{cases}$$

where $A^0 = I$. Then it follows that A^r is an elliptic operator in $L_{\text{cl}}^{mr}(M)$ and that

$$N(A^r) = \{0\}.$$

Further we have, by Theorem 6.7.9,

$$\text{ind } A^r = \begin{cases} r \text{ ind } A = 0 & \text{if } r \geq 0, \\ |r| \text{ ind}(A^{-1}) = 0 & \text{if } r < 0. \end{cases}$$

Therefore, applying Theorem 6.7.14, we obtain that the operator

$$(A^r)_{mr}\colon H^{mr}(M) \to L^2(M)$$

is an isomorphism. Thus the quantity

$$(u, v)_{mr} = ((A^r)_{mr} u, (A^r)_{mr} v)$$

is an admissible inner product for $H^{mr}(M)$. Further, since $\{\varphi_j\}$ is a complete orthonormal system of $L^2(M)$, we have, by Parseval's formula,

$$(u, v)_{mr} = \sum_{j=1}^{\infty} ((A^r)_{mr} u, \varphi_j)\overline{((A^r)_{mr} v, \varphi_j)}. \tag{21}$$

But we have, by formula (5),

$$((A^r)_{mr})^* = ((A^r)^*)_0 = (A^r)_0,$$

since $A = A^*$. Hence it follows that

$$\begin{aligned} ((A^r)_{mr} u, \varphi_j) &= (u, (A^r)_0 \varphi_j) \\ &= (u, A^r \varphi_j) \\ &= \lambda_j^r (u, \varphi_j). \end{aligned} \tag{22}$$

Consequently, formula (20) follows from formulas (21) and (22). ■

As one of the important applications of Theorem 6.7.15, we can obtain the following:

6.7.16 Theorem. *Let Δ_M be the Laplace-Beltrami operator on M, and let $\{\chi_j\}$ be the orthonormal system of $L^2(M)$ consisting of eigenfunctions of $-\Delta_M$ and $\{\lambda_j\}$ its corresponding eigenvalues:*

$$-\Delta_M \chi_j = \lambda_j \chi_j \qquad (\lambda_j \geq 0).$$

Then the functions χ_j span the Sobolev spaces $H^s(M)$, $s \in \mathbf{R}$. More precisely, the quantity

$$(u, v)_s = \sum_{j=1}^{\infty} (1 + \lambda_j)^s (u, \chi_j)\overline{(v, \chi_j)}$$

is an admissible inner product for the space $H^s(M)$.

6.8. Potentials and Pseudo-Differential Operators

The purpose of this section is to describe, in terms of pseudo-differential operators, the surface and volume potentials arising in boundary value problems for elliptic differential operators.

We give a formal description of a background. Let Ω be a bounded domain in $\mathbf{R}^n$ with C^∞ boundary. Its closure $\bar{\Omega}$ is an n-dimensional, compact C^∞ manifold with boundary. We may suppose that $\bar{\Omega}$ is the closure of a relatively compact open subset Ω of an n-dimensional, compact C^∞ manifold M without boundary in which Ω has a C^∞ boundary $\partial\Omega$ (cf. Figure 5-1). Let P be a differential operator of order m with C^∞ coefficients on M. Then we have the jump formula (5.6.3):

$$P(u^0) = (Pu)^0 + \tilde{P}\gamma u, \qquad u \in C^\infty(\bar{\Omega});$$

Here $\tilde{P}\gamma u$ is a distribution on M with support in $\partial\Omega$. If P admits an "inverse" Q, then the function u may be expressed as follows:

$$u = Q((Pu)^0)|_\Omega + Q(\tilde{P}\gamma u)|_\Omega.$$

The first term on the right-hand side is a volume potential and the second term is a surface potential with m "layers". For example, if P is the Laplacian, the first term is the classical Newtonian potential and the second term is the familiar combination of single and double layer potentials.

First we state a theorem which covers surface potentials.

6.8.1 Theorem. *Let $A \in L^m_{cl}(M)$ be properly supported. Suppose that:*

$$\text{Every term in the complete symbol } \sum_{j=0}^{\infty} a_j(x,\xi) \text{ of } A \text{ is a rational function of } \xi. \tag{1}$$

Then we have:

(i) *The operator*

$$H: v \mapsto A(v \otimes \delta)|_\Omega$$

is continuous on $C^\infty(\partial\Omega)$ into $C^\infty(\bar{\Omega})$. If $v \in \mathscr{D}'(\partial\Omega)$, the distribution Hv has sectional traces of any order on $\partial\Omega$.

(ii) *The operator*

$$S: C^\infty(\partial\Omega) \to C^\infty(\partial\Omega)$$
$$v \mapsto \gamma_0(Hv)$$

is in $L^{m+1}_{cl}(\partial\Omega)$. Furthermore, its homogeneous principal symbol is given by the following:

$$(x', \xi') \mapsto \frac{1}{2\pi} \int_\Gamma a_0(x', 0, \xi', \xi_n)\, d\xi_n, \tag{2}$$

where $a_0 \in C^\infty(T^*(M) \setminus \{0\})$ is the homogeneous principal symbol of A, and Γ is a circle in the plane $\{\xi_n \in \mathbf{C}; \operatorname{Im} \xi_n > 0\}$ which encloses the poles ξ_n of $a_0(x', 0, \xi', \xi_n)$ there.

(iii) *The operator H extends to a continuous linear operator*

$$H: H^s(\partial\Omega) \to H^{s-m-1/2}(\bar{\Omega})$$

for all $s \in \mathbf{R}$.

6.8.2 Remark. In view of Theorem 6.5.7, it follows that condition (1) is invariant under change of coordinates. Also it is easy to see that every parametrix for an elliptic differential operator satisfies condition (1).

The next theorem covers volume potentials.

6.8.3 Theorem. *Let $A \in L^m_{cl}(M)$ be as in Theorem 6.8.1. Then we have:*

(i) *The operator*

$$G: f \mapsto A(f^0)|_\Omega$$

is continuous on $C^\infty(\bar{\Omega})$ into itself.

(ii) *The operator G extends to a continuous linear operator*

$$G: H^s(\bar{\Omega}) \to H^{s-m}(\bar{\Omega})$$

for all $s > -1/2$.

6.9. The Sharp Gårding Inequality

Let Ω be an open subset of $\mathbf{R}^n$, and let A be a properly supported pseudo-differential operator of order m on Ω. In this section we are concerned with inequalities from below for A of the form

$$\operatorname{Re}(Au, u) \geq C_K \|u\|_s^2, \qquad u \in C_K^\infty(\Omega), \tag{1}$$

where K is a compact subset of Ω and $(\ ,\ )$ is the inner product of $L^2(\Omega)$. We remark that inequality (1) is always true for $s \geq m/2$, since we have

$$|(Au, u)| \leq C'_K \|u\|_{m/2}^2, \qquad u \in C_K^\infty(\Omega),$$

with a constant $C'_K > 0$.

In what follows we give sufficient conditions on A for inequality (1) to hold for $s < m/2$. These results will play an important role in deriving *a priori* estimates for (non-)elliptic boundary value problems in Chapter 10.

The next result, first proved by Gårding [1] for differential operators, is a milestone in the theory of elliptic boundary value problems.

6.9.1 Theorem. *Let A be a properly supported pseudo-differential operator of order m on Ω with principal symbol $a_m(x, \xi)$. Suppose that there exists a constant $a_0 > 0$ such that*

$$\operatorname{Re} a_m(x, \xi) \geq a_0 |\xi|^m, \qquad x \in \Omega, \qquad \xi \in \mathbf{R}^n.$$

Then, for every compact $K \subset \Omega$ and $s < m/2$, there exist constants $c_{K,s} > 0$ and $C_{K,s} > 0$ such that

$$\operatorname{Re}(Au, u) \geq c_{K,s} \|u\|_{m/2}^2 - C_{K,s} \|u\|_s^2, \qquad u \in C_K^\infty(\Omega).$$

This inequality is called *Gårding's inequality.*

A sharpened form of Gårding's inequality is given by Hörmander [2]:

6.9.2 Theorem (the sharp Gårding inequality). *Let $A \in L^m(\Omega)$ be as in Theorem 6.9.1. Suppose that*

$$\operatorname{Re} a_m(x, \xi) \geq 0, \qquad x \in \Omega,\ \xi \in \mathbf{R}^n.$$

Then, for every compact $K \subset \Omega$ and $s < (m-1)/2$, there exist constants $c_{K,s} > 0$ and $C_{K,s} > 0$ such that

$$\operatorname{Re}(Au, u) \geq -c_{K,s}\|u\|^2_{(m-1)/2} - C_{K,s}\|u\|^2_s, \qquad u \in C_K^\infty(\Omega).$$

6.9.3 Remark. Melin [1] goes further, giving a necessary and sufficient condition on A for the following inequality to hold for every $\varepsilon > 0$:

$$\operatorname{Re}(Au, u) \geq -\varepsilon\|u\|^2_{(m-1)/2} - C_{K,s,\varepsilon}\|u\|^2_s, \qquad u \in C_K^\infty(\Omega).$$

Fefferman-Phong [1] have proved some result of this nature for differential operators, which we now state.

Let M be an n-dimensional, *compact* C^∞ manifold without boundary, and let A be a second-order, degenerate elliptic differential operator with real coefficients on M such that in local coordinates

$$A = \sum_{i,j=1}^n a^{ij}(x)\frac{\partial^2}{\partial x_i\,\partial x_j} + \sum_{i=1}^n b^i(x)\frac{\partial}{\partial x_i} + c(x),$$

where:

1) The a^{ij} are the components of a C^∞ symmetric contravariant tensor of type $\binom{2}{0}$ on M, and

$$\sum_{i,j=1}^n a^{ij}(x)\xi_i\xi_j \geq 0, \qquad x \in M,\ \xi = \sum_{j=1}^n \xi_j\,dx_j \in T_x^*(M),$$

where $T_x^*(M)$ is the cotangent space of M at x. That is, the principal symbol $\sum_{i,j} a^{ij}(x)\xi_i\xi_j$ of $-A$ is non-negative on the cotangent bundle $T^*(M) = \bigcup_{x\in M} T_x^*(M)$.

2) $b^i \in C^\infty(M)$.

3) $c \in C^\infty(M)$.

A tangent vector $v = \sum_{j=1}^n v^j\,(\partial/\partial x_j) \in T_x(M)$ is *subunit* for the operator $A^0 = \sum_{i,j=1}^n a^{ij}\,(\partial^2/\partial x_i\,\partial x_j)$ if it satisfies

$$\left(\sum_{j=1}^n v^j\xi_j\right)^2 \leq \sum_{i,j=1}^n a^{ij}(x)\xi_i\xi_j, \qquad \xi = \sum_{j=1}^n \xi_j\,dx_j \in T_x^*(M).$$

If $\rho > 0$, we define a "non-Euclidean" ball $B_{A^0}(x, \rho)$ of radius ρ about x as follows:

$B_{A^0}(x, \rho) =$ the set of all points $y \in M$ which can be joined to x by a Lipschitz path $\gamma: [0, \rho] \to M$ for which the tangent vector $\dot{\gamma}(t)$ of M at $\gamma(t)$ is subunit for A^0 for almost every t.

Also we let

$$B_E(x, \rho) = \text{the ordinary Euclidean ball of radius } \rho \text{ about } x.$$

The next result is due to Fefferman-Phong [1] (cf. [1], Theorem 1).

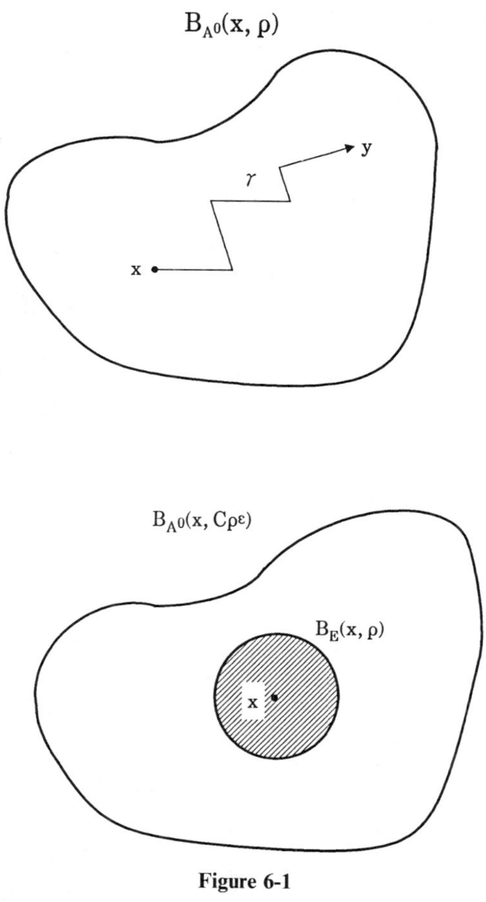

Figure 6-1

6.9.4 Theorem. *Let the differential operator A be as above. Then the following two conditions are equivalent*:

(i) *There exist constants $0 < \varepsilon \leq 1$ and $C > 0$ such that for all sufficiently small $\rho > 0$ we have*

$$B_E(x, \rho) \subset B_{A^0}(x, C\rho^\varepsilon), \qquad x \in M. \qquad \text{(cf. Figure 6-1)}$$

(ii) *There exist constants $C_0 > 0$ and $C_1 > 0$ such that*

$$-\operatorname{Re}(Au, u) \geq C_0 \|u\|_\varepsilon^2 - C_1 \|u\|_0^2, \qquad u \in C^\infty(M).$$

Here (,) is the inner product of $L^2(M)$.

6.10. Hypoelliptic Pseudo-Differential Operators

Let Ω be an open subset of $\mathbf{R}^n$. A properly supported pseudo-differential operator A on Ω is said to be *hypoelliptic* if it satisfies

$$\text{sing supp } u = \text{sing supp } Au, \qquad u \in \mathscr{D}'(\Omega). \tag{1}$$

For example, Theorem 6.5.10 tells us that elliptic operators are hypoelliptic. It is easy to see that condition (1) is equivalent to the following:

$$\begin{cases} \text{For any open subset } \Omega_1 \text{ of } \Omega, \text{ we have:} \\ u \in \mathscr{D}'(\Omega),\ Au \in C^\infty(\Omega_1) \quad \Rightarrow \quad u \in C^\infty(\Omega_1). \end{cases} \tag{1'}$$

We say that A is *globally hypoelliptic* if it satisfies the weaker condition:

$$u \in \mathscr{D}'(\Omega),\ Au \in C^\infty(\Omega) \quad \Rightarrow \quad u \in C^\infty(\Omega).$$

We remark that these two notions can be transferred to manifolds.

In this section we describe two classes of hypoelliptic pseudo-differential operators of Hörmander [6] and Melin-Sjöstrand [1], which arise in the study of elliptic boundary value problems.

1) Let A be a properly supported, classical pseudo-differential operator of order m on $\Omega \subset \mathbf{R}^n$ such that the complete symbol $\sigma(A)(x, \xi)$ has an asymptotic expansion

$$\sigma(A)(x, \xi) \sim a_m(x, \xi) + a_{m-1}(x, \xi) + \cdots,$$

where $a_j(x, \xi)$ is positively homogeneous of degree j in the variable ξ. For simplicity, we suppose that

$$a_m(x, \xi) \geq 0 \text{ on the cotangent bundle } T^*(\Omega) = \Omega \times \mathbf{R}^n. \tag{2}$$

We let

$$\Sigma = \{(x, \xi) \in \Omega \times (\mathbf{R}^n \setminus \{0\}); a_m(x, \xi) = 0\}.$$

The set Σ is called the *characteristic set* of A.

Now let u be an arbitrary tangent vector of $T^*(\Omega)$ at a point ρ of Σ. Then we choose a C^∞ vector field v on $T^*(\Omega)$ equal to u at ρ, and define a quadratic form $Q_\rho(u, u)$ on the product space $T_\rho(T^*(\Omega)) \times T_\rho(T^*(\Omega))$ by the formula

$$Q_\rho(u, u) = (v^2 a_m)(\rho).$$

In view of inequality (2), it is easy to verify that $Q_\rho(u, u)$ is independent of the vector field v chosen. The form Q_ρ is called the *Hessian* of a_m at ρ.

Let $\tilde{T}_\rho(T^*(\Omega))$ be the complexification of the tangent space $T_\rho(T^*(\Omega))$. We consider the symplectic form $\sigma = \sum_{j=1}^n d\xi_j \wedge dx_j$ and the quadratic form Q_ρ as bilinear forms on the product space $\tilde{T}_\rho(T^*(\Omega)) \times \tilde{T}_\rho(T^*(\Omega))$. Since the form σ is non-degenerate, one can define a linear map

$$F_\rho\colon \tilde{T}_\rho(T^*(\Omega)) \to \tilde{T}_\rho(T^*(\Omega))$$

by the formula

$$\sigma(u, F_\rho v) = Q_\rho(u, v), \qquad u, v \in \tilde{T}_\rho(T^*(\Omega)).$$

The map F_ρ is called the *Hamilton map* of Q_ρ. It is easy to see that the eigenvalues of F_ρ are situated on the imaginary axis, symmetrically around the origin.

We let

$$a'_{m-1}(x, \xi) = a_{m-1}(x, \xi) + \frac{\sqrt{-1}}{2} \sum_{j=1}^n \frac{\partial^2 a_m}{\partial x_j\, \partial \xi_j}(x, \xi).$$

The function $a'_{m-1}(x, \xi)$ is invariantly defined at the points of Σ, and is called the *subprincipal symbol* of A.

The following criterion for hypoellipticity is due to Hörmander [6] (cf. [6], Theorem 5.9).

6.10.1 Theorem. *Let $A \in L^m_{\mathrm{cl}}(\Omega)$ be properly supported. Suppose that $a_m \geq 0$ on $T^*(\Omega)$ and that the range of $-a'_{m-1}$ on Σ belongs to a closed angle which intersects with the positive real axis only at the origin. Then the following two conditions are equivalent:*

(i) *For every compact $K \subset \Omega$, $s \in \mathbf{R}$ and $t < s + m - 1$, there exists a constant $C_{K,s,t} > 0$ such that*

$$\| u \|^2_{s+m-1} \leq C_{K,s,t}(\| Au \|^2_s + \| u \|^2_t), \qquad u \in C^\infty_K(\Omega).$$

(ii) *At every point* ρ *of* Σ, *either the subprincipal symbol* $a'_{m-1}(\rho)$ *is non-zero or else the Hamilton map* F_ρ *of the Hessian* Q_ρ *of* a_m *is not nilpotent.*

Furthermore, each of conditions (i) *and* (ii) *implies that:*

$$u \in \mathscr{D}'(\Omega),\ Au \in H^s_{\mathrm{loc}}(\Omega) \quad \Rightarrow \quad u \in H^{s+m-1}_{\mathrm{loc}}(\Omega). \tag{3}$$

6.10.2 Remark. Regularity result (3) involves a loss of one derivative compared with the elliptic regularity theorem (Theorem 6.5.10). We express this by saying that A is hypoelliptic, with loss of one derivative.

2) Let M be an n-dimensional, *compact* C^∞ manifold without boundary, and let A be a classical pseudo-differential operator of first order on M such that

$$A(x, D) = \beta(x, D) + \sqrt{-1}\, B(x, D), \tag{4}$$

where:

1. $\beta(x, \partial) = \sqrt{-1}\,\beta(x, D)$ is a real C^∞ vector field on M.
2. $B \in L^1_{\mathrm{cl}}(M)$ and its homogeneous principal symbol $b_1(x, \xi)$ is real.

We remark that the homogeneous principal symbol $\beta(x, \xi)$ of $\beta(x, D)$ is a *polynomial* of degree one in the variable ξ.

The following criterion for global hypoellipticity is due to Melin-Sjöstrand [1] (cf. [1], Introduction).

6.10.3 Theorem. *Let* $A \in L^1_{\mathrm{cl}}(M)$ *be of the form* (4). *Suppose that:*

(a) *The symbol* $b_1(x, \xi)$ *does not change sign on the cotangent bundle* $T^*(M)$, *that is,* $b_1(x, \xi) \geq 0$ *or* $b_1(x, \xi) \leq 0$ *on* $T^*(M)$.

(b) *The vector field* β *is non-zero on the set* $K = \{x \in M;\ b_1(x, \xi) = 0$ *for some* $(x, \xi) \in T^*(M),\ \xi \neq 0\}$.

(c) *Any maximal integral curve of* β *is not* entirely *contained in* K.

Then we have, for all $s \in \mathbf{R}$,

$$u \in \mathscr{D}'(M),\ Au \in H^s(M) \quad \Rightarrow \quad u \in H^s(M).$$

Furthermore, for any $t < s$, *there exists a constant* $C_{s,t} > 0$ *such that*

$$\|u\|_s \leq C_{s,t}(\|Au\|_s + \|u\|_t).$$

Thus the operator A is globally hypoelliptic, with loss of one derivative.

Notes

Our treatment of pseudo-differential operators follows the exposition of Chazarain-Piriou [1]. For detailed studies of pseudo-differential operators, the reader is referred to Kumano-go [1] and Taylor [1].

Section 6.1: The symbol classes S^m were first introduced by Hörmander [4].

Section 6.4: For the theory of Fourier integral operators, see Hörmander [5], Duistermaat-Hörmander [1] and Duistermaat [1].

Section 6.7: Our treatment of index theory of elliptic operators is adapated from Palais [1]. To prove Theorem 6.7.16, one needs an interpolation argument. See Lions-Magenes [1] or Taylor [1].

Section 6.8: Theorem 6.8.1 is due to Hörmander [2]; see also Seeley [2] and Vaĭnberg-Grušin [1]. Theorem 6.8.3 is due to Boutet de Monvel [1].

Section 6.10: The notion of hypoellipticity was introduced by Schwartz (cf. Schwartz [1]). Hypoelliptic second-order differential operators have been studied in detail by Hörmander [3], Frediĭ [1], Oleĭnik-Radkevič [1] and many others.

A properly supported pseudo-differential operator $A \in L^m(\Omega)$ is said to be *subelliptic* if there exists a constant $0 < \varepsilon < 1$ such that, for every compact $K \subset \Omega$, $s \in \mathbf{R}$ and $t < s + m - \varepsilon$, we have

$$\| u \|_{s+m-\varepsilon} \leq C_{K,s,t}(\| Au \|_s + \| u \|_t), \qquad u \in C_K^\infty(\Omega).$$

It is known (cf. Hörmander [2]) that subelliptic operators are hypoelliptic, with loss of ε derivatives. Egorov [1] and Hörmander [7] have obtained necessary and sufficient conditions that a properly supported, classical pseudo-differential operator $A \in L^m_{\text{cl}}(\Omega)$ be subelliptic.

7 Maximum Principles for Degenerate Elliptic Operators

In this chapter we prove various maximum principles for degenerate elliptic differential operators of second order, and reveal the underlying analytical mechanism of propagation of maximums. In Chapter 8 the results will be applied to questions of uniqueness for elliptic boundary value problems. Furthermore, the mechanism of propagation of maximums plays an important role in the interpretation and study of Markov processes in terms of partial differential equations, as will be seen in Chapter 10.

7.1. Maximum Principles

Let D be a bounded domain in $\mathbf{R}^N$ with boundary ∂D, and let A be a second-order, degenerate elliptic differential operator with real coefficients such that

$$A = \sum_{i,j=1}^{N} a^{ij}(x) \frac{\partial^2}{\partial x_i \, \partial x_j} + \sum_{i=1}^{N} b^i(x) \frac{\partial}{\partial x_i} + c(x),$$

where:

1. $a^{ij} \in C(\mathbf{R}^N)$, $a^{ij} = a^{ji}$ and
$$\sum_{i,j=1}^{N} a^{ij}(x)\xi_i\xi_j \geq 0, \qquad x \in \mathbf{R}^N,\ \xi \in \mathbf{R}^N.$$
2. $b^i \in C(\mathbf{R}^N)$.
3. $c \in C(\mathbf{R}^N)$ and $c \leq 0$ in D.

First we prove the following:

7.1.1 Theorem (the weak maximum principle). *Suppose that a function $u \in C(\bar{D}) \cap C^2(D)$ satisfies either*

$$Au \geq 0 \text{ and } c < 0 \text{ in } D \tag{1}$$

or

$$Au > 0 \text{ and } c \leq 0 \text{ in } D. \tag{2}$$

Then the function u may take its positive *maximum only on the boundary ∂D.*

Proof. Assume to the contrary that:

The function u takes its positive maximum at a point x_0 of D. (3)

Without loss of generality, we may choose a local coordinate system $(y_1, \ldots, y_N)$ in a neighborhood of x_0 such that:

$$\begin{cases} 1.\ x_0 = \text{the origin.} \\ 2.\ A = \displaystyle\sum_{j,k=1}^{N} \alpha^{jk}(y) \frac{\partial^2}{\partial y_j\, \partial y_k} + \sum_{k=1}^{N} \beta^k(y) \frac{\partial}{\partial y_k} + c(y) \\ \quad \text{with } (\alpha^{jk}(0)) = \begin{pmatrix} E_r & 0 \\ 0 & 0 \end{pmatrix}. \end{cases}$$

Here $r = \text{rank}\,(a^{ij}(x_0))$ and E_r is the $r \times r$ unit matrix. Then assumption (3) implies that

$$\frac{\partial u}{\partial y_k}(0) = 0, \qquad \frac{\partial^2 u}{\partial y_k^2}(0) \leq 0,$$

so that we have

$$Au(x_0) = Au(0) \leq c(0)u(0) \begin{cases} < 0 & \text{if } c(x_0) < 0, \\ \leq 0 & \text{if } c(x_0) \leq 0. \end{cases}$$

This contradicts hypothesis (1) or (2). ■

As an application of the weak maximum principle, we can obtain a pointwise estimate for solutions of the inhomogeneous equation $Au = f$:

7.1.2 Theorem. *Suppose that*

$$c < 0 \qquad \text{on } \bar{D} = D \cup \partial D.$$

Then we have, for all $u \in C(\bar{D}) \cap C^2(D)$,

$$\max_{\bar{D}} |u| \leq \max\left\{\sup_D \left|\frac{Au}{c}\right|, \max_{\partial D} |u|\right\}. \tag{4}$$

Proof. We let

$$M = \max\left\{\sup_D \left|\frac{Au}{c}\right|, \max_{\partial D} |u|\right\},$$

and consider the functions

$$v_{\pm} = M \pm u.$$

Then it follows that

$$Av_{\pm} = cM \pm Au \leq 0 \qquad \text{in } D.$$

Hence, applying Theorem 7.1.1 to the functions $-v_{\pm}$, we obtain that the functions $v_{\pm}$ may take their negative minimums only on the boundary ∂D. But we have

$$v_{\pm} = M \pm u \geq 0 \qquad \text{on } \partial D,$$

so that

$$v_{\pm} \geq 0 \qquad \text{on } \bar{D} = D \cup \partial D.$$

This proves estimate (4). ■

7.1.3 Remark. In the case when $Au = 0$ in D, estimate (4) can be replaced by the following equality:

$$\max_{\bar{D}} |u| = \max_{\partial D} |u|. \tag{4'}$$

We consider the case when the operator A is *elliptic* on $\bar{D}$, that is, there exists a constant $a_0 > 0$ such that

$$\sum_{i,j=1}^{N} a^{ij}(x)\xi_i\xi_j \geq a_0|\xi|^2, \qquad x \in \bar{D},\ \xi \in \mathbf{R}^N. \tag{5}$$

Then we have:

7.1.4 Theorem. *Suppose that A is elliptic on $\bar{D}$ and $c \equiv 0$ in D. If $u \in C^2(D) \cap C(\bar{D})$ and $Au \geq 0$ in D, then we have*

$$\max_{\bar{D}} u = \max_{\partial D} u. \tag{6}$$

Proof. Taking $\xi = (1, 0, \ldots, 0)$ in inequality (5), we obtain that

$$a^{11} \geq a_0 \qquad \text{on } \bar{D}.$$

Hence we can find a constant $\alpha > 0$, so large that

$$Ae^{\alpha x_1} = (\alpha^2 a^{11} + \alpha b^1) e^{\alpha x_1} > 0 \qquad \text{on } \bar{D}.$$

Then we have, for all $\varepsilon > 0$,

$$A(u + \varepsilon e^{\alpha x_1}) \geq \varepsilon A e^{\alpha x_1} > 0 \qquad \text{in } D.$$

Thus, arguing as in the proof of Theorem 7.1.1, we obtain that the function $u + \varepsilon e^{\alpha x_1}$ may take its maximum only on the boundary ∂D. This implies that

$$\max_{\bar{D}}(u + \varepsilon e^{\alpha x_1}) = \max_{\partial D}(u + \varepsilon e^{\alpha x_1}). \tag{7}$$

Equality (6) follows by letting $\varepsilon \downarrow 0$ in equality (7). ■

7.1.5 Corollary. *Suppose that A is elliptic on $\bar{D}$. If $u \in C^2(D) \cap C(\bar{D})$ and $Au = 0$ in D, then we have*

$$\max_{\bar{D}} |u| = \max_{\partial D} |u|.$$

Proof. Replacing u by $-u$ if necessary, we may assume that

$$\max_{\bar{D}} u > 0.$$

We let

$$D^+ = \{x \in D;\, u(x) > 0\}.$$

Then we have

$$\sum_{i,j=1}^{N} a^{ij} \frac{\partial^2 u}{\partial x_i \, \partial x_j} + \sum_{i=1}^{N} b^i \frac{\partial u}{\partial x_i} = -cu \geq 0 \qquad \text{in } D^+.$$

Hence, applying Theorem 7.1.4 with $D = D^+$, we obtain that

$$\max_{\overline{D^+}} u = \max_{\partial D^+} u.$$

But, since $u = 0$ on $\partial D^+ \cap D$, this implies that

$$\max_{\overline{D}} u = \max_{\overline{D^+}} u = \max_{\partial D^+} u = \max_{\partial D} u.$$

The proof of Corollary 7.1.5 is complete. ■

Now we study the interior normal derivative $\partial u/\partial \mathbf{n}$ of u at a point where the function u takes its non-negative maximum. In what follows let D be a domain of class C^2.

We let

$$\rho(x) = \operatorname{dist}(x, \partial D), \qquad x \in \mathbf{R}^N.$$

Then it follows that:

$$\begin{cases} \rho \in C^1(\mathbf{R}^N), \\ x \in \partial D \Leftrightarrow \rho(x) = 0, \\ \operatorname{grad} \rho = \text{the unit interior normal } \mathbf{n} \text{ to } \partial D. \end{cases}$$

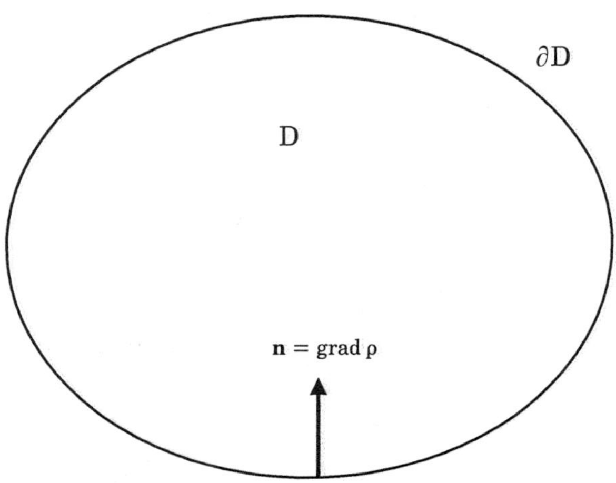

Figure 7-1

We define a subset Σ_3 of the boundary ∂D by

$$\Sigma_3 = \left\{x' \in \partial D;\ \sum_{i,j=1}^{N} a^{ij}(x')n_i n_j > 0\right\},$$

where $\mathbf{n} = (n_1, \ldots, n_N)$. In other words, Σ_3 is the set of non-characteristic points with respect to the operator A.

The next lemma justifies the definition of the set Σ_3.

7.1.6 Lemma. *The set Σ_3 is invariant under C^2 diffeomorphisms preserving normal vectors.*

Proof. Let x'_0 be an arbitrary point of Σ_3 and consider, in a neighborhood U of x'_0, a C^2 diffeomorphism

$$y = F(x) = (F^1(x), \ldots, F^N(x))$$

which preserves normal vectors. Then it follows that

$$\partial D \cap U = \{y \in U;\ \Phi(y) = 0\}, \qquad \Phi = \rho \circ F^{-1},$$

and also

$$n_i = \frac{\partial \rho}{\partial x_i} = \sum_{\ell=1}^{N} \frac{\partial F^\ell}{\partial x_i}\frac{\partial \Phi}{\partial y_\ell}.$$

Further the operator A is written in the form

$$\begin{aligned} A &= \sum_{i,j=1}^{N} a^{ij}\frac{\partial^2}{\partial x_i\,\partial x_j} + \sum_{i=1}^{N} b^i \frac{\partial}{\partial x_i} + c \\ &= \sum_{\ell,m=1}^{N}\left(\sum_{i,j=1}^{N} a^{ij}\frac{\partial F^\ell}{\partial x_i}\frac{\partial F^m}{\partial x_j}\right)\frac{\partial^2}{\partial y_\ell\,\partial y_m} \\ &\quad + \sum_{\ell=1}^{N}\left(\sum_{i=1}^{N} b^i \frac{\partial F^\ell}{\partial x_i} + \sum_{i,j=1}^{N} a^{ij}\frac{\partial^2 F^\ell}{\partial x_i\,\partial x_j}\right)\frac{\partial}{\partial y_\ell} + c. \end{aligned}$$

But we have

$$\begin{aligned} \sum_{i,j=1}^{N} a^{ij}\frac{\partial \rho}{\partial x_i}\frac{\partial \rho}{\partial x_j} &= \sum_{i,j=1}^{N} a^{ij}\left(\sum_{\ell,m=1}^{N}\frac{\partial F^\ell}{\partial x_i}\frac{\partial F^m}{\partial x_j}\frac{\partial \Phi}{\partial y_\ell}\frac{\partial \Phi}{\partial y_m}\right) \\ &= \sum_{\ell,m=1}^{N}\left(\sum_{i,j=1}^{N} a^{ij}\frac{\partial F^\ell}{\partial x_i}\frac{\partial F^m}{\partial x_j}\right)\frac{\partial \Phi}{\partial y_\ell}\frac{\partial \Phi}{\partial y_m}. \end{aligned}$$

This proves the invariance of the set Σ_3, since the diffeomorphism F preserves normal vectors and so grad Φ has the same direction as the interior normal. ■

The next lemma will be useful in Chapter 9.

7.1.7 Lemma (the boundary point lemma). *Suppose that a function* $u \in C(\bar{D}) \cap C^2(D)$ *satisfies*

$$Au \geq 0 \qquad \text{in } D, \tag{8}$$

and that there exists a point x_0' *of the set* Σ_3 *such that*

$$\begin{cases} u(x_0') = \max_{x \in \bar{D}} u(x) \geq 0, & \text{(9.a)} \\ u(x) < u(x_0'), \qquad x \in D. & \text{(9.b)} \end{cases}$$

Then the interior normal derivative $\partial u/\partial \mathbf{n}(x_0')$ *of* u *at* x_0', *if it exists, satisfies*

$$\frac{\partial u}{\partial \mathbf{n}}(x_0') < 0. \tag{10}$$

Proof. By virtue of Lemma 7.1.6, we may choose a local coordinate system $(y_1, \ldots, y_N)$ in a neighborhood of x_0' such that

$$\begin{cases} x_0' = \text{the origin}, \\ \rho = y_N. \end{cases}$$

Suppose that the operator A is written in the form

$$A = \sum_{j,k=1}^{N} \alpha^{jk}(y) \frac{\partial^2}{\partial y_j \, \partial y_k} + \sum_{k=1}^{N} \beta^k(y) \frac{\partial}{\partial y_k} + c(y).$$

Note that

$$\alpha^{NN}(0) > 0, \tag{11}$$

since $0 \in \Sigma_3$ and $\mathbf{n} = (0, \ldots, 0, 1)$.

Now we consider the function

$$v(y) = \alpha \sum_{i=1}^{N-1} y_i^2 - \beta y_N - y_N^2,$$

where α, β are positive constants to be chosen later on. Then we have

$$Av(0) = 2\alpha \sum_{i=1}^{N-1} \alpha^{ii}(0) - 2\alpha^{NN}(0) - \beta \cdot \beta^N(0).$$

In view of inequality (11), it follows that there exists a neighborhood V of 0 such that

$$Av < 0 \qquad \text{in } V, \tag{12}$$

if the constants α and β are chosen sufficiently small.

We let

$E =$ the domain surrounded by the hypersurface $\{v = 0\}$ and the hyperplane $\{y_N = \eta\}$.

Here η is a positive constant to be chosen small enough so that $E \subset V$ (cf. Figure 7-2). Further we let

$$w(y) = \varepsilon v(y) - u(y) + u(0),$$

where ε is a positive constant to be chosen later on. Then it follows from inequalities (8), (9.a) and (12) that

$$Aw = \varepsilon Av - Au + cu(0) \le \varepsilon Av < 0 \qquad \text{in } E.$$

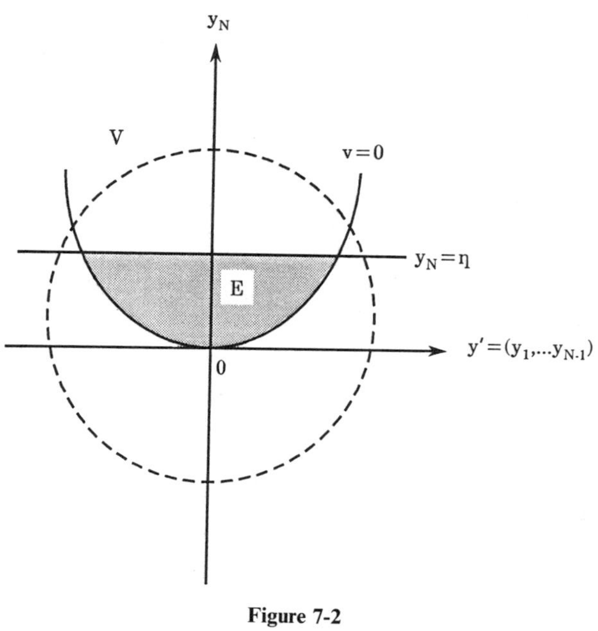

Figure 7-2

Thus, applying Theorem 7.1.1 to the function $-w$, we find that the function w may take its negative minimum only on the boundary ∂E of E. But, condition (9.b) implies that

$$w(y) = \varepsilon v(y) + (u(0) - u(y)) \geq 0, \qquad y \in \partial E,$$

if ε is chosen sufficiently small. Hence it follows that

$$w(y) = \varepsilon v(y) + u(0) - u(y) \geq 0, \qquad y \in E \cup \partial E.$$

Therefore, taking $y = (0, y_N)$ with $0 < y_N < \eta$, we have

$$\varepsilon\left(\frac{v(0, y_N) - v(0, 0)}{y_N}\right) \geq \frac{u(0, y_N) - u(0, 0)}{y_N}. \tag{13}$$

If the derivative $\partial u/\partial \mathbf{n}$ exists at x'_0, we can let $y_N \downarrow 0$ in inequality (13) to obtain that

$$\frac{\partial u}{\partial \mathbf{n}}(x'_0) = \frac{\partial u}{\partial y_N}(0) \leq \varepsilon \frac{\partial v}{\partial y_N}(0) = -\varepsilon\beta.$$

This proves inequality (10). ■

7.2. Propagation of Maximums

Let D be a connected open subset of $\mathbf{R}^N$. The following result is well-known by the name of the *strong maximum principle* for the Laplacian $\Delta = \sum_{i=1}^N \partial^2/\partial x_i^2$:

If $u \in C^2(D)$, $\Delta u \geq 0$ in D and u takes its maximum at a point of D, then u is a constant.

The purpose of this section is to reveal the underlying *analytical* mechanism of propagation of maximums for *degenerate* elliptic differential operators of second order, explaining the above result. The mechanism of propagation of maximums is closely related to the diffusion phenomenon of Markovian particles.

Let A be a second-order, degenerate elliptical differential operator with real coefficients such that

$$A = \sum_{i,j=1}^N a^{ij}(x) \frac{\partial^2}{\partial x_i \, \partial x_j} + \sum_{i=1}^N b^i(x) \frac{\partial}{\partial x_i},$$

where:

1. The a^{ij} are C^2 functions on $\mathbf{R}^N$ all of whose derivatives of order ≤ 2 are bounded in $\mathbf{R}^N$, $a^{ij} = a^{ji}$ and

$$\sum_{i,j=1}^{N} a^{ij}(x)\xi_i\xi_j \geq 0, \qquad x \in \mathbf{R}^N,\ \xi \in \mathbf{R}^N.$$

2. The b^i are C^1 functions on $\mathbf{R}^N$ with bounded derivatives in $\mathbf{R}^N$.

In this section we shall consider the following:

Problem. *Let D be a connected open subset of $\mathbf{R}^N$ and x a point of D. Then determine the largest connected, relatively closed subset $D(x)$ of D, containing x, such that:*

$$\begin{cases} \text{If } u \in C^2(D),\ Au \geq 0 \text{ in } D,\ \sup_D u = M < +\infty \text{ and } u(x) = M, \text{ then} \\ u \equiv M \text{ throughout } D(x). \end{cases} \tag{$*$}$$

The set $D(x)$ is called the *propagation set* of x in D (cf. Figure 7-3).

We shall give a coordinate-free description of the propagation set $D(x)$ in terms of subunit vectors, introduced by Fefferman-Phong [1] (cf. Section 6.9, Theorem 6.9.4).

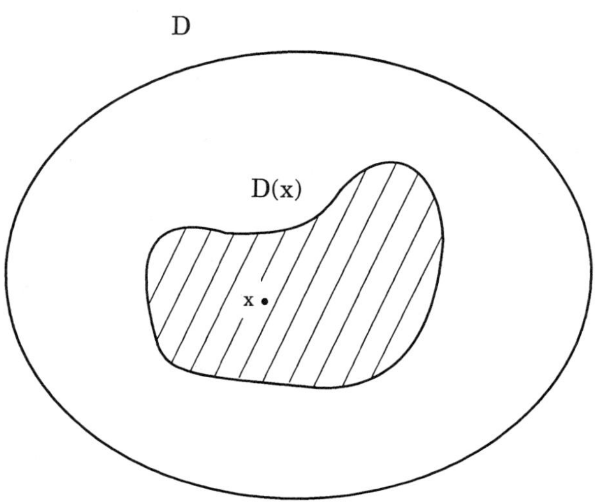

Figure 7-3

Statement of Results

Following Fefferman-Phong [1], we say that a tangent vector $X = \sum_{j=1}^{N} \gamma^j(\partial/\partial x_j)$ at $x \in D$ is *subunit* for the operator $A^0 = \sum_{i,j=1}^{N} a^{ij}(\partial^2/\partial x_i\, \partial x_j)$ if it satisfies

$$\left(\sum_{j=1}^{N} \gamma^j \eta_j\right)^2 \leq \sum_{i,j=1}^{N} a^{ij}(x)\eta_i \eta_j, \qquad \eta = \sum_{j=1}^{N} \eta_j\, dx_j \in T_x^*(D),$$

where $T_x^*(D)$ is the cotangent space of D at x. We remark that this notion is coordinate-free. So we rotate the coordinate axes so that the matrix (a^{ij}) is diagonalized at x:

$$(a^{ij}(x)) = (\lambda_i \delta_{ij}), \qquad \lambda_1 > 0, \ldots, \lambda_r > 0,\ \lambda_{r+1} = \cdots = \lambda_N = 0;$$

here $r = \operatorname{rank}(a^{ij}(x))$. Then it is easy to see that the vector X is subunit for A^0 if and only if it is contained in the following ellipsoid of dimension r (cf. Figure 0-5):

$$\frac{(\gamma^1)^2}{\lambda_1} + \cdots + \frac{(\gamma^r)^2}{\lambda_r} \leq 1, \qquad \gamma^{r+1} = \cdots = \gamma^N = 0. \tag{1}$$

A *subunit trajectory* is a Lipschitz path $\gamma\colon [t_1, t_2] \to D$ such that the tangent vector $\dot{\gamma}(t) = (d/dt)(\gamma(t))$ is subunit for A^0 at $\gamma(t)$ for almost every t. We remark that if $\dot{\gamma}(t)$ is subunit for A^0, so is $-\dot{\gamma}(t)$; hence subunit trajectories are not oriented.

We let

$$X_0 = \sum_{i=1}^{N} \left(b^i - \sum_{j=1}^{N} \frac{\partial a^{ij}}{\partial x_j}\right) \frac{\partial}{\partial x_i}.$$

The vector field X_0 is called the drift vector field in probability theory, while it is the so-called subprincipal part of the operator A in terms of the theory of partial differential equations (cf. Section 6.10, the function a'_{m-1}).

A *drift trajectory* is a curve $\theta\colon [t_1, t_2] \to D$ such that

$$\dot{\theta}(t) = X_0(\theta(t)) \qquad \text{on } [t_1, t_2],$$

and this curve is oriented in the direction of increasing t.

Now we can state our main result:

7.2.1 Theorem. *The propagation set $D(x)$ of x in D contains the closure $D'(x)$ in D of all points $y \in D$ which can be joined to x by a finite number of subunit and drift trajectories.*

Theorem 7.2.1 tells us that if the matrix (a^{ij}) is non-degenerate at x, that is, if $r = \text{rank}\,(a^{ij}(x)) = N$, then the maximum propagates in a neighborhood of x; but if the matrix (a^{ij}) is degenerate at x, then the maximum propagates only in a "thin" ellipsoid of dimension r (cf. formula (1)) and in the direction of X_0. Now we see the reason why the strong maximum principle holds for the Laplacian Δ.

In [2], Stroock and Varadhan characterized the *support* of the diffusion process corresponding to the operator A (which is the closure of the collection of all possible trajectories of a Markovian particle, starting at x, with generator A) and, as one of its applications, they gave a (not coordinate-free) description of the propagation set $D(x)$.

The next theorem asserts that our propagation set $D'(x)$ coincides with that of Stroock-Varadhan [2]:

7.2.2 Theorem. *The propagation set $D'(x)$ of Theorem 7.2.1 coincides with the closure in D of the points $\phi(t)$, $t \geq 0$, where $\phi\colon [0, t] \to D$ is a path for which there exists a piecewise C^1 function $\psi\colon [0, t] \to \mathbf{R}^N$ such that*

$$\phi^i(s) = x_i + \int_0^s \sum_{j=1}^N a^{ij}(\phi(\tau))\psi^j(\tau)\,d\tau$$

$$+ \int_0^s \left(b^i(\phi(\tau)) - \sum_{j=1}^N \frac{\partial a^{ij}}{\partial x_j}(\phi(\tau)) \right) d\tau \qquad (1 \leq i \leq N). \tag{2}$$

7.2.3 Remark. By Theorem 4.1 of Stroock-Varadhan [2], we see that our propagation set $D'(x)$ is the largest subset of D having property $(*)$ in some weak sense (see also Ikeda-Watanabe [1], Chapter VI, Theorem 8.3).

In the case when the operator A is written as the sum of squares of vector fields, Hill [1] gave another (coordinate-free) description of a propagation set, although his proof was not complete. Hill's result is completely proved and extended to the non-linear case by Redheffer [1] (cf. Bony [1]). As a byproduct of Theorem 7.2.1, we can prove that our propagation set $D'(x)$ coincides with that of Hill [1].

Now suppose that the operator A is written as the sum of squares of vector fields:

$$A = \sum_{k=1}^r Y_k^2 + Y_0; \tag{3}$$

here the Y_k are real C^2 vector fields on $\mathbf{R}^N$ and Y_0 is a real C^1 vector field on $\mathbf{R}^N$. *Hill's diffusion trajectory* is a curve β: $[t_1, t_2] \to D$ such that

$$\dot{\beta}(t) = Y_k(\beta(t)), \qquad \dot{\beta}(t) \neq 0 \text{ on } [t_1, t_2].$$

Hill's diffusion trajectories are not oriented; they may be traversed in either direction. *Hill's drift trajectories* are defined similarly, with Y_k replaced by Y_0, but they are oriented in the direction of increasing t.

We can prove the following:

7.2.4 Theorem. *Suppose that the operator A is written in the form* (3). *Then the propagation set $D'(x)$ of Theorem* 7.2.1 *coincides with the closure in D of all points $y \in D$ which can be joined to x by a finite number of Hill's diffusion and drift trajectories.*

7.2.5 Remark. Theorem 7.2.4 is implicitly proved by Stroock-Varadhan (cf. [1], Theorem 5.2; [2], Theorem 3.2), since the support of the diffusion process corresponding to the operator A does not depend on the expression of A.

Theorem 7.2.1 may be reformulated in various ways. For example, we have:

7.2.1′ Theorem. *Let c be a continuous function on D such that $c \leq 0$ in D. If $u \in C^2(D)$, $(A + c)u \geq 0$ in D and if u attains its* positive *maximum M at a point x of D, then $u \equiv M$ throughout $D'(x)$.*

Preliminaries

First we prove the *weak maximum principle* (cf. Theorem 7.1.1):

7.2.6 Theorem. *Suppose that $u \in C^2(D)$, $Au > 0$ in D and* $\sup_D u = M < +\infty$. *Then the function u takes its maximum M only on the boundary ∂D.*

Proof. Assume to the contrary that:

The function u takes its maximum M at a point x_0 of D.

Without loss of generality, we may choose a local coordinate system $(y_1, \ldots, y_N)$ in a neighborhood of x_0 such that:

$$\begin{cases} 1.\ x_0 = \text{the origin}, \\ 2.\ A = \displaystyle\sum_{j,k=1}^{N} \alpha^{jk}(y) \frac{\partial^2}{\partial y_j\, \partial y_k} + \sum_{k=1}^{N} \beta^k(y) \frac{\partial}{\partial y_k} \\ \quad \text{with } (\alpha^{jk}(0)) = \begin{pmatrix} E_r & 0 \\ 0 & 0 \end{pmatrix}. \end{cases} \tag{1}$$

Here $r = \operatorname{rank}\, (a^{ij}(x_0))$ and E_r is the $r \times r$ unit matrix. Since the function u takes its maximum M at x_0, it follows from (1) that

$$\frac{\partial u}{\partial y_k}(0) = 0, \qquad \frac{\partial^2 u}{\partial y_k^2}(0) \leq 0, \qquad 1 \leq k \leq N,$$

and hence

$$Au(x_0) = \sum_{k=1}^{r} \frac{\partial^2 u}{\partial y_k^2}(0) \leq 0.$$

This contradicts the hypothesis: $Au > 0$ in D. ■

Next we prove two elementary lemmas on non-negative functions.

7.2.7 Lemma. *Let f be a non-negative C^2 function on $\mathbf{R}$ such that*

$$\sup_{x \in \mathbf{R}} |f''(x)| \leq C \tag{2}$$

for some constant $C > 0$. Then we have

$$|f'(x)| \leq \sqrt{2C}\sqrt{f(x)} \qquad \text{on } \mathbf{R}. \tag{3}$$

Proof. In view of Taylor's formula, it follows that

$$0 \leq f(y) = f(x) + f'(x)(y - x) + \frac{f''(\xi)}{2}(y - x)^2,$$

where ξ is between x and y. Thus, letting $z = x - y$, we obtain from estimate (2) that

$$\begin{aligned} 0 &\leq f(x) + f'(x)z + \frac{f''(\xi)}{2} z^2 \\ &\leq f(x) + f'(x)z + \frac{C}{2} z^2, \end{aligned}$$

so that

$$\frac{C}{2} z^2 + f'(x)z + f(x) \geq 0 \qquad \text{for all } z \in \mathbf{R}.$$

Therefore we have

$$f'(x)^2 - 2Cf(x) \leq 0.$$

This proves inequality (3). ∎

7.2.8 Lemma. *Let f be a non-negative C^2 function on $\mathbf{R}$ such that*

$$\sup_{x \in \mathbf{R}} |f''(x)| \leq 1. \tag{4}$$

Then we have

$$\frac{1}{3}(y^2 + f(0)) \leq y^2 + f(y) \leq 2(y^2 + f(0)) \qquad \text{on } \mathbf{R}. \tag{5}$$

Proof. Since the derivative of the function $f(y) + f(-y)$ vanishes at $y = 0$, using the Taylor expansion, we obtain that

$$\begin{aligned} y^2 + f(y) &\leq y^2 + f(y) + f(-y) \\ &\leq y^2 + \left[2f(0) + \sup_{x \in \mathbf{R}} |f''(x)| \cdot y^2 \right]. \end{aligned}$$

By virtue of estimate (4), this yields the inequality on the right-hand side of (5).

On the other hand, we have, by the mean value theorem,

$$\begin{aligned} \frac{|f(0) + f(2y) - 2f(y)|}{y^2} &= \frac{1}{|y|} \left| \left(\frac{f(2y) - f(y)}{y} \right) - \left(\frac{f(y) - f(0)}{y} \right) \right| \\ &= \frac{1}{|y|} |f'(z) - f'(w)| \\ &= \left| \frac{z - w}{y} \right| \cdot \left| \frac{f'(z) - f'(w)}{z - w} \right| \\ &\leq \left| \frac{z - w}{y} \right| \sup_{x \in \mathbf{R}} |f''(x)|. \end{aligned} \tag{6}$$

Here z is between y and $2y$, and w is between 0 and y, and so

$$|z - w| \leq 2|y|. \tag{7}$$

Therefore, in view of (4) and (7), it follows from inequality (6) that

$$\begin{aligned} f(0) &\leq 2y^2 + 2f(y) - f(2y) \\ &\leq 2y^2 + 2f(y). \end{aligned}$$

This yields the inequality on the left-hand side of (5).

Lemma 7.2.8 is proved. ■

As one of the applications of Lemma 7.2.7, we obtain the following lemmas on positive semi-definite quadratic forms:

7.2.9 Lemma. *Let a^{ij} be bounded continuous functions on $\mathbf{R}^N$, and suppose that*

$$\sum_{i,j=1}^{N} a^{ij}(x)\xi_i\xi_j \geq 0, \qquad x \in \mathbf{R}^N,\ \xi \in \mathbf{R}^N.$$

Then we have, for $1 \leq j \leq N$,

$$\left|\sum_{i=1}^{N} a^{ij}(x)\xi_i\right|^2 \leq a^{jj}(x)\left(\sum_{k,\ell=1}^{N} a^{k\ell}(x)\xi_k\xi_\ell\right), \qquad x \in \mathbf{R}^N,\ \xi \in \mathbf{R}^N. \tag{8}$$

Proof. Inequality (8) is an immediate consequence of inequality (3) if we apply Lemma 7.2.7 to the function

$$\mathbf{R} \ni \xi_j \mapsto \frac{1}{2} \sum_{k,\ell=1}^{N} a^{k\ell}(x)\xi_k\xi_\ell. \qquad ■$$

7.2.10 Lemma. *Suppose that a^{ij} are C^2 functions on $\mathbf{R}^N$ all of whose second derivatives are bounded on $\mathbf{R}^N$, and that*

$$\sum_{i,j=1}^{N} a^{ij}(x)\xi_i\xi_j \geq 0, \qquad x \in \mathbf{R}^N,\ \xi \in \mathbf{R}^N.$$

Then we have, for $1 \leq k \leq N$,

$$\begin{aligned} \left|\sum_{i,j=1}^{N} \frac{\partial a^{ij}}{\partial x_k}(x)\lambda_i\mu_j\right| \leq C\Bigg\{|\mu|\left(\sum_{i,j=1}^{N} a^{ij}(x)\lambda_i\lambda_j\right)^{1/2} \\ + |\lambda|\left(\sum_{i,j=1}^{N} a^{ij}(x)\mu_i\mu_j\right)^{1/2}\Bigg\}, \end{aligned} \tag{9}$$

$$x \in \mathbf{R}^N,\ \lambda,\ \mu \in \mathbf{R}^N,$$

where $C > 0$ is a constant depending only on the bounds on the second derivatives of a^{ij}.

Proof. Let x_0 be an arbitrary point of $\mathbf{R}^N$. We may assume that the matrix $(a^{ij}(x_0))$ is diagonal. Then, applying Lemma 7.2.7 to the function

$$\mathbf{R} \ni x_k \mapsto \sum_{i,j=1}^{N} a^{ij}(x)\lambda_i\lambda_j,$$

we obtain that

$$\left(\sum_{i,j=1}^{N} \frac{\partial a^{ij}}{\partial x_k}(x)\lambda_i\lambda_j\right)^2 \leq 2 \sup_{\substack{x\in\mathbf{R}^N \\ 1\leq \ell, m\leq N}} \left|\sum_{i,j=1}^{N} \frac{\partial^2 a}{\partial x_\ell\, \partial x_m}(x)\lambda_i\lambda_j\right| \cdot \left(\sum_{i,j=1}^{N} a^{ij}(x)\lambda_i\lambda_j\right). \tag{10}$$

Thus, taking $x = x_0$ and $\lambda = \mathbf{e}_i$ in inequality (10), where $\mathbf{e}_i$ is the i-th coordinate vector, we have

$$\left|\frac{\partial a^{ii}}{\partial x_k}(x_0)\right| \leq C_1(a^{ii}(x_0))^{1/2}. \tag{11}$$

Further, taking $\lambda = \mathbf{e}_i + \mathbf{e}_j$ $(i \neq j)$, we have

$$\left|\frac{\partial a^{ii}}{\partial x_k}(x_0) + 2\frac{\partial a^{ij}}{\partial x_k}(x_0) + \frac{\partial a^{jj}}{\partial x_k}(x_0)\right| \leq C_2(a^{ii}(x_0) + a^{jj}(x_0))^{1/2},$$

so that

$$\begin{aligned}\left|2\frac{\partial a^{ij}}{\partial x_k}(x_0)\right| &\leq C_2(a^{ii}(x_0) + a^{jj}(x_0))^{1/2} + \left|\frac{\partial a^{ii}}{\partial x_k}(x_0) + \frac{\partial a^{jj}}{\partial x_k}(x_0)\right| \\ &\leq C_2(a^{ii}(x_0) + a^{jj}(x_0))^{1/2} + C_1((a^{ii}(x_0))^{1/2} + (a^{jj}(x_0))^{1/2}) \\ &\leq C_3(a^{ii}(x_0) + a^{jj}(x_0))^{1/2}. \end{aligned} \tag{12}$$

Here C_1, C_2, C_3 are positive constants depending only on the bounds on the second derivatives of a^{ij}.

Therefore, it follows from inequalities (11) and (12) that

$$\begin{aligned}
&\left|\sum_{i,j=1}^{N} \frac{\partial a^{ij}}{\partial x_k}(x_0)\lambda_i\mu_j\right| \\
&\leq \sum_{i,j=1}^{N} \left|\frac{\partial a^{ij}}{\partial x_k}(x_0)\right| |\lambda_i||\mu_j| \\
&\leq C_4 \sum_{i,j=1}^{N} (a^{ii}(x_0) + a^{jj}(x_0))^{1/2} |\lambda_i||\mu_j| \\
&\leq NC_4\left[\sum_{i,j=1}^{N} (a^{ii}(x_0) + a^{jj}(x_0))\lambda_i^2\mu_j^2\right]^{1/2} \\
&= NC_4\left[\left(\sum_{i=1}^{N} a^{ii}(x_0)\lambda_i^2\right)\left(\sum_{j=1}^{N} \mu_j^2\right)\right. \\
&\quad \left. + \left(\sum_{j=1}^{N} a^{jj}(x_0)\mu_j^2\right)\left(\sum_{i=1}^{N} \lambda_i^2\right)\right]^{1/2} \\
&\leq NC_4\left\{\left(\sum_{i,j=1}^{N} a^{ij}(x_0)\lambda_i\lambda_j\right)^{1/2} |\mu|\right. \\
&\quad \left. + \left(\sum_{i,j=1}^{N} a^{ij}(x_0)\mu_i\mu_j\right)^{1/2} |\lambda|\right\},
\end{aligned} \tag{13}$$

since the matrix $(a^{ij}(x_0))$ is diagonal. Here $C_4 = \max(C_1, C_3/2)$. Inequality (13) proves inequality (9) with $C = NC_4$. ■

We prove an approximation theorem for integral curves of vector fields. To do so, we need two elementary lemmas.

7.2.11 Lemma (Gronwall). *Suppose that $y(t)$ is an absolutely continuous function on* **R**, *and that there exist two continuous functions f and g on* **R** *such that*

$$\dot{y}(t) + f(t)y(t) \leq g(t) \qquad \text{a.e. in } \mathbf{R}. \tag{14}$$

Then we have

$$y(t) \leq e^{-\int_0^t f(\tau)\,d\tau}\left\{y(0) + \int_0^t g(\tau)e^{\int_0^\tau f(\sigma)\,d\sigma}\,d\tau\right\} \qquad \text{on } \mathbf{R}. \tag{15}$$

Proof. It follows from inequality (14) that

$$\frac{d}{dt}\left(y(t)e^{\int_0^t f(\tau)\,d\tau}\right) \le g(t)e^{\int_0^t f(\tau)\,d\tau} \qquad \text{a.e. in } \mathbf{R}.$$

Hence, integrating with respect to t, we obtain inequality (15). ■

7.2.12 Lemma. *Let Z be a Lipschitz continuous vector field on $\mathbf{R}^N$ and let ω be a bounded continuous function on $\mathbf{R}$. Suppose that $x(t)$ is the unique solution of the initial-value problem*

$$\begin{cases} \dot{x}(t) = Z(x(t)) & \text{in } \mathbf{R}, \\ x(0) = x_0 \in \mathbf{R}^N, \end{cases} \tag{16}$$

and that $y(t)$ is a piecewise C^1 function on $\mathbf{R}$ satisfying

$$\begin{cases} \dot{y}(t) = Z(y(t)) + \omega(t) & \text{a.e. in } \mathbf{R}, \\ y(0) = x_0. \end{cases} \tag{17}$$

Then we have

$$|x(t) - y(t)| \le \frac{\varepsilon}{K}(e^{Kt} - 1) \qquad \text{on } \mathbf{R}, \tag{18}$$

where $\varepsilon = \sup_{t \in \mathbf{R}} |\omega(t)|$ and K is a Lipschitz constant for the vector field Z.

Proof. We let

$$u(t) = |x(t) - y(t)|.$$

We observe that the function $u(t)$ is absolutely continuous, since $x(t)$ and $y(t)$ are piecewise C^1 functions. Thus, in view of (16) and (17), it follows that

$$\begin{aligned} \dot{u}(t) &\le |\dot{x}(t) - \dot{y}(t)| \\ &\le |Z(x(t)) - Z(y(t))| + |\omega(t)| \\ &\le K|x(t) - y(t)| + \varepsilon \\ &= Ku(t) + \varepsilon \qquad \text{a.e. in } \mathbf{R}. \end{aligned}$$

Therefore, inequality (18) follows from an application of Lemma 7.2.11, since $u(0) = 0$. ■

Now we can prove an approximation theorem for integral curves of vector fields, essentially due to Bony [1]:

7.2.13 Theorem. *Let $X_1, \ldots, X_m$ be Lipschitz continuous vector fields on $\mathbf{R}^N$, and let $Z = \sum_{k=1}^{m} \lambda_k X_k$ where the λ_k are real-valued C^1 functions on $\mathbf{R}^N$. Then each integral curve of the vector field Z can be approximated uniformly by piecewise differentiable curves, of which each differentiable arc is an integral curve of one of the vector fields X_k.*

Proof. It suffices to prove the theorem in the case $m = 2$:

$$Z = \lambda_1 X_1 + \lambda_2 X_2.$$

We consider a piecewise differentiable curve $x(t)$ defined by the following:

$$\begin{cases} x(0) = x_0 \in \mathbf{R}^N; \\ \dot{x}(t) = \lambda_1(x(2k\theta))X_1(x(t)), & 2k\theta \le t \le (2k+1)\theta; \\ \dot{x}(t) = \lambda_2(x(2k\theta))X_2(x(t)), & (2k+1)\theta \le t \le (2k+2)\theta, \end{cases} \tag{19}$$

where θ is a positive parameter and k ranges over all integers. Further we let $y(t)$ be a polygonal line defined by the following (cf. Figure 7-4):

$$y(t) = x(2k\theta) + \frac{t - k\theta}{\theta}(x((2k+2)\theta) - x(2k\theta)), \; k\theta \le t \le (k+1)\theta. \tag{20}$$

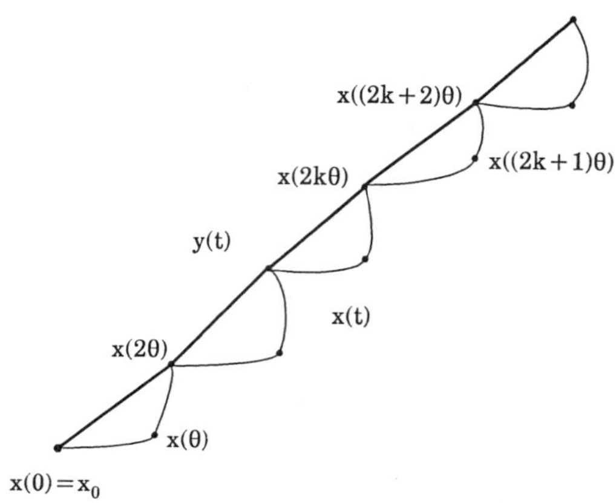

Figure 7-4

Then, by virtue of the Taylor expansion, it follows from equations (20) and (19) that for $k\theta \leq t \leq (k+1)\theta$ we have

$$\begin{aligned}\dot{y}(t) &= \frac{1}{\theta}(x((2k+2)\theta) - x(2k\theta)) \\ &= \frac{x((2k+1)\theta) - x(2k\theta)}{\theta} + \frac{x((2k+2)\theta) - x((2k+1)\theta)}{\theta} \\ &= \dot{x}(2k\theta) + \dot{x}((2k+1)\theta) + \text{an error term of order } \theta \\ &= \lambda_1(x(2k\theta))X_1(x(2k\theta)) + \lambda_2(x(2k\theta))X_2(x((2k+1)\theta) \\ &\quad + \text{an error term of order } \theta. \end{aligned} \tag{21}$$

But, by the mean value theorem, we have, for $k\theta \leq t \leq \; < k + 1)\theta$,

$$\begin{aligned}|y(t) - x(2k\theta)| &= \frac{t - k\theta}{\theta}|x((2k+2)\theta) - x(2k\theta)| \\ &\leq |x((2k+2)\theta) - x(2k\theta)| \\ &\leq |x((2k+2)\theta) - x((2k+1)\theta)| + |x((2k+1)\theta) - x(2k\theta)| \\ &= \text{a term of order } \theta, \end{aligned} \tag{22}$$

and also

$$\begin{aligned}|y(t) - x((2k+1)\theta)| &\leq |y(t) - x(2k\theta)| + |x(2k\theta) - x((2k+1)\theta)| \\ &= \text{a term of order } \theta. \end{aligned} \tag{23}$$

Therefore, combining (21), (22) and (23), we find that

$$\dot{y}(t) = Z(y(t)) + \text{an error term of order } \theta \qquad \text{a.e. in } \mathbf{R}.$$

In view of Lemma 7.2.12, this implies that, as $\theta \downarrow 0$, the polygonal line $y(t)$ converges uniformly to the integral curve of Z issuing from x_0.

Since the distance between $x(t)$ and $y(t)$ tends to zero as $\theta \downarrow 0$, it follows that, as $\theta \downarrow 0$, the piecewise differentiable curve $x(t)$, defined by (19), converges uniformly to the integral curve of Z issuing from x_0.

Theorem 7.2.13 is proved. ■

Finally we study the behavior of integral curves of vector fields with small initial data.

7.2.14 Lemma. *Suppose that* $X = \sum_{i=1}^N a^i(\partial/\partial x_i)$ *is a* C^1 *vector field on* $\mathbf{R}^N$ *such that*

$$X = \frac{\partial}{\partial x_1} \qquad \text{at } x = 0.$$

Let $x(t, y) = (x_1(t, y), \ldots, x_N(t, y))$ *be the unique solution of the initial-value problem*

$$\begin{cases} \dot{x}(t, y) = X(x(t, y)), \\ x(0, y) = y \in \mathbf{R}^N. \end{cases} \tag{24}$$

Then we have, as $|t| + |y| \to 0$,

$$\begin{cases} x_1(t, y) = y_1 + t + \dfrac{1}{2}\dfrac{\partial a^1}{\partial x_1}(0)t^2 + \displaystyle\sum_{j=1}^N \dfrac{\partial a^1}{\partial x_j}(0)y_j t + o(|t|^2 + |y|^2), \\ x_i(t, y) = y_i + \dfrac{1}{2}\dfrac{\partial a^i}{\partial x_1}(0)t^2 + \displaystyle\sum_{j=1}^N \dfrac{\partial a^i}{\partial x_j}(0)y_j t + o(|t|^2 + |y|^2), \\ \qquad\qquad\qquad\qquad\qquad\qquad\qquad 2 \le i \le N. \end{cases} \tag{25}$$

Proof. We let

$$w_i(t, y) = x_i(t, y) - y_i - \frac{1}{2}\frac{\partial a^i}{\partial x_1}(0)t^2 - \sum_{j=1}^N \frac{\partial a^i}{\partial x_j}(0)y_j t, \qquad 1 \le i \le N.$$

Then it follows from (24) that

$$w_i(0, y) = 0, \tag{26}$$

and that

$$\begin{aligned} \dot{w}_i(t, y) &= a^i(x(t, y)) - \frac{\partial a^i}{\partial x_1}(0)t - \sum_{j=1}^N \frac{\partial a^i}{\partial x_j}(0)y_j \\ &= \delta_{i1} + \left(a^i(x(t, y)) - a^i(0) - \sum_{j=1}^N \frac{\partial a^i}{\partial x_j}(0)x_j(t, y)\right) \\ &\quad + \frac{\partial a^i}{\partial x_1}(0)\left(\frac{x_1(t, y) - x_1(0, y)}{t} - 1\right)t + \sum_{j=2}^N \frac{\partial a^i}{\partial x_j}(0)(x_j(t, y) - y_j), \end{aligned} \tag{27}$$

since $a^i(0) = \delta_{i1}$ and $x_1(0, y) = y_1$.

Using the mean value theorem, we can estimate each term of the right-hand side of (27) as follows:

1) $$a^i(x(t, y)) - a^i(0) - \sum_{j=1}^{N} \frac{\partial a^i}{\partial x_j}(0)x_j(t, y) = O(|x(t, y)|^2)$$
$$= o(|t| + |y|),$$

since $x_j(t, y) = y_j + ta^j(x(s, y))$ for some s between 0 and t.

2) $$\left(\frac{x_1(t, y) - x_1(0, y)}{t} - 1\right)t = (a^1(x(s, y)) - 1)t$$
$$= o(|t| + |y|),$$

since $a^1(0) = 1$.

3) $$x_j(t, y) - y_j = ta^j(x(s, y))$$
$$= o(|t| + |y|), \qquad 2 \leq j \leq N,$$

since $a^j(0) = 0$.

Summing up, we can rewrite formula (27) as

$$\dot{w}_i(t, y) = \delta_{i1} + o(|t| + |y|). \tag{27'}$$

Hence it follows from (26) and (27′) that

$$w_i(t, y) = \int_0^t \dot{w}_i(s, y)\, ds$$
$$= \delta_{i1}t + o(|t| + |y|), \qquad 1 \leq i \leq N.$$

This proves formula (25). ■

Proof of Theorem 7.2.1

We shall use a modification of the techniques originally introduced by E. Hopf [1] for elliptic operators and later adapted by Bony [1] for degenerate elliptic ones (cf. Hill [1], Redheffer [1], Oleĭnik-Radkevič [1], Amano [1]). Before the proof of Theorem 7.2.1, we summarize these techniques in the form of lemmas (Lemma 7.2.15 and Lemma 7.2.17 below).

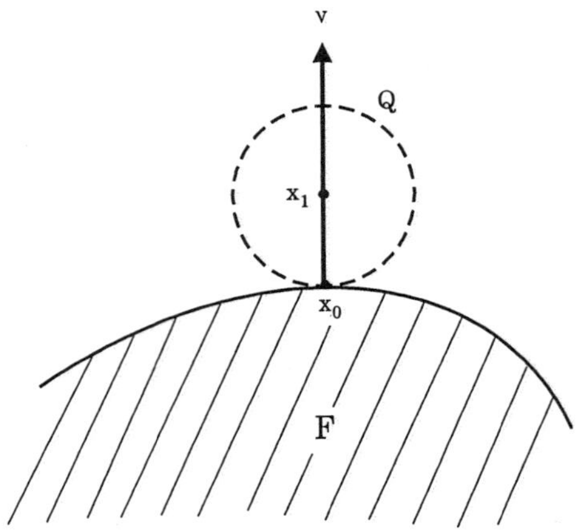

Figure 7-5

Let F be a (relatively) closed subset of D. Following Bony [1], we say that a vector v is *normal* to the set F at one of its points, x_0, if there exists an open ball Q contained in the set $D \setminus F$, centered at x_1, such that (cf. Figure 7-5):

$$\begin{cases} 1. \text{ The point } x_0 \text{ is on the boundary of the ball } Q; \\ 2. \; v = s(x_1 - x_0) \text{ with } s > 0. \end{cases}$$

The next lemma, essentially due to Bony [1], will play a fundamental role in the proof of Theorem 7.2.1.

7.2.15 Lemma. *Let X be a Lipschitz continuous vector field on $\mathbf{R}^N$, and let $x(t)$ be an integral curve of X. Suppose that:*

At each point x_0 of the set F, the inner product $\langle X(x_0), v\rangle$ is non-positive for any vector v normal to F at x_0: (1)

$$\langle X(x_0), v\rangle \leq 0.$$

Then, if $x(t_0) \in F$ for some t_0, it follows that $x(t) \in F$ for all $t \geq t_0$.

7.2.16 Remark. If $\langle X(x_0), v\rangle = 0$ for any vector v normal to F at x_0, then we can replace t by $-t$ and deduce that $x(t) \in F$ for all t, not just for $t \geq t_0$.

Proof of Lemma 7.2.15. We let

$$\delta(t) = \inf_{z \in F} |x(t) - z|,$$

and assume to the contrary that

$$\delta(t) > 0, \qquad t_0 < t \le t_1. \tag{2}$$

Here recall that the set F is a (relatively) closed subset of D.

1) First we show that

$$\liminf_{h \downarrow 0} \frac{\delta(t-h)^2 - \delta(t)^2}{h} \ge -2K\delta(t)^2, \qquad t_0 < t \le t_1, \tag{3}$$

where K is a Lipschitz constant for the vector field X.

Let $\{h_n\}$ be a sequence, $h_n \downarrow 0$, such that

$$\lim_{n\to\infty} \frac{\delta(t-h_n)^2 - \delta(t)^2}{h_n} = \liminf_{h\downarrow 0} \frac{\delta(t-h)^2 - \delta(t)^2}{h}, \tag{4}$$

and let y_n be the projection on the set F of the point $x(t - h_n)$ (cf. Figure 7-6):

$$|x(t-h_n) - y_n| = \inf_{z\in F} |x(t-h_n) - z| = \delta(t-h_n). \tag{5}$$

Now we remark that one can choose a sufficiently small open ball B, centered at $x(t)$, such that its closure $\bar{B}$ in $\mathbf{R}^N$ is contained in D. Since the set $\bar{B} \cap F$ is compact and $x(t - h_n) \in B$ for sufficiently large n, by passing to a subsequence we may assume that the sequence $\{y_n\}$ converges to some point y of F. Then it follows from formula (5) that

$$|x(t) - y| = \inf_{z\in F} |x(t) - z| = \delta(t). \tag{6}$$

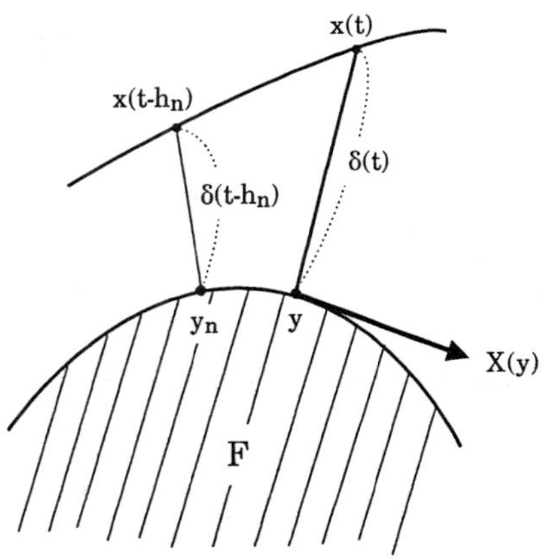

Figure 7-6

In other words, the limit point y is the projection on F of the point $x(t)$. Thus we have, by the mean value theorem,

$$\begin{aligned}
\frac{\delta(t-h_n)^2 - \delta(t)^2}{h_n} &= \frac{|x(t-h_n) - y_n|^2 - |x(t) - y|^2}{h_n} \\
&\geq \frac{|x(t-h_n) - y_n|^2 - |x(t) - y_n|^2}{h_n} \\
&= -2\langle x(t_n) - y_n, X(x(t_n))\rangle, \qquad (7)
\end{aligned}$$

where $t - h_n < t_n < t$. By virtue of formula (4), we can let $n \to \infty$ in inequality (7) to obtain that

$$\begin{aligned}
\liminf_{h \downarrow 0} \frac{\delta(t-h)^2 - \delta(t)^2}{h} &\geq -2\langle x(t) - y, X(x(t))\rangle \\
&= -2\langle x(t) - y, X(y)\rangle \\
&\quad - 2\langle x(t) - y, X(x(t)) - X(y)\rangle. \qquad (8)
\end{aligned}$$

But we have, by hypothesis (1),

$$\langle x(t) - y, X(y)\rangle \leq 0,$$

since formula (6) implies that the vector $x(t) - y$ is normal to F at y (cf. Figure 7-6). Hence, using Schwarz's inequality, we obtain from inequality (8) that

$$\begin{aligned}
\liminf_{h \downarrow 0} \frac{\delta(t-h)^2 - \delta(t)^2}{h} &\geq -2\langle x(t) - y, X(x(t)) - X(y)\rangle \\
&\geq -2|x(t) - y| \cdot |X(x(t)) - X(y)| \\
&\geq -2K|x(t) - y|^2 \\
&= -2K\delta(t)^2.
\end{aligned}$$

This proves inequality (3).

2) Next we show that:

If f is a continuous function on the closed interval $[t_0, t_1]$ such that

$$\begin{cases} \liminf_{h \downarrow 0} \dfrac{f(t-h) - f(t)}{h} \geq -C, & t_0 < t \leq t_1, \\ f(t_0) = 0, & \end{cases} \qquad (9)$$

then we have

$$f(t) \le C(t - t_0) \qquad \text{on } [t_0, t_1]. \tag{10}$$

Here C is a non-negative constant.

Assume to the contrary that:

There exists a point $s \in [t_0, t_1]$ such that

$$f(s) > C(s - t_0). \tag{11}$$

We remark that $s \ne t_0$, since $f(t_0) = 0$, and that $f(s) > 0$, since $C \ge 0$. We let

$$\Phi(t) = f(t) - \frac{f(s)}{s - t_0}(t - t_0),$$

and let s_0 be a point of $[t_0, s]$ at which the function Φ attains its non-negative maximum on $[t_0, s]$. We may take $s_0 \ne t_0$, since $\Phi(s) = 0$. Then we have, by (9) and (11),

$$\begin{aligned} 0 &\ge \liminf_{h \downarrow 0} \frac{\Phi(s_0 - h) - \Phi(s_0)}{h} \\ &= \liminf_{h \downarrow 0} \frac{f(s_0 - h) - f(s_0)}{h} + \frac{f(s)}{s - t_0} \\ &\ge -C + \frac{f(s)}{s - t_0} \\ &> 0. \end{aligned}$$

This is a contradiction.

3) Now the proof of Lemma 7.2.15 is easy. We let

$$\theta = \min\left(t_1 - t_0, \frac{1}{4K}\right).$$

Then we find from inequality (3) that the function $f(t) = \delta(t)^2$ satisfies hypothesis (9) with

$$t_1 = t_0 + \theta, \quad C = 2K \max_{t_0 \le s \le t_0 + \theta} \delta(s)^2.$$

Thus it follows from inequality (10) that for $t_0 \leq t \leq t_0 + \theta$ we have

$$\begin{aligned}\delta(t)^2 &\leq 2K \max_{t_0 \leq s \leq t_0 + \theta} \delta(s)^2 \cdot (t - t_0) \\ &\leq 2K\theta \max_{t_0 \leq s \leq t_0 + \theta} \delta(s)^2 \\ &\leq \frac{1}{2} \max_{t_0 \leq s \leq t_0 + \theta} \delta(s)^2,\end{aligned}$$

so that

$$\delta(t) \equiv 0, \qquad t_0 \leq t \leq t_0 + \theta.$$

This contradicts assumption (2).

The proof of Lemma 7.2.15 is complete. ■

Next we prove a lemma on "barriers":

7.2.17 Lemma. *Suppose that* $u \in C^2(D)$, $Au \geq 0$ *in* D, $\sup_D u = M < +\infty$ *and* $u(x_0) = M$ *for some point* z_0 *of* D. *If there exists a* C^2 *function* v *on* D *such that* $v(x_0) = 0$, grad $v(x_0) \neq 0$ *and* $Av(x_0) > 0$, *then, for any sufficiently small neighborhood* U *of* x_0, *the function* u *attains its maximum* M *at some point of the set* $\{x \in \partial U; v(x) > 0\}$. *Here* ∂U *denotes the boundary of* U.

Proof. Since $v(x_0) = 0$ and grad $v(x_0) \neq 0$, we can construct a C^2 function V from v such that

$$\begin{cases} V(x_0) = 0, \\ \text{grad } V(x_0) \neq 0, \\ AV(x_0) > 0, \\ \{x \in D;\, V(x) \geq 0\} \setminus \{x_0\} \subset \{x \in D;\, v(x) > 0\}. \end{cases}$$

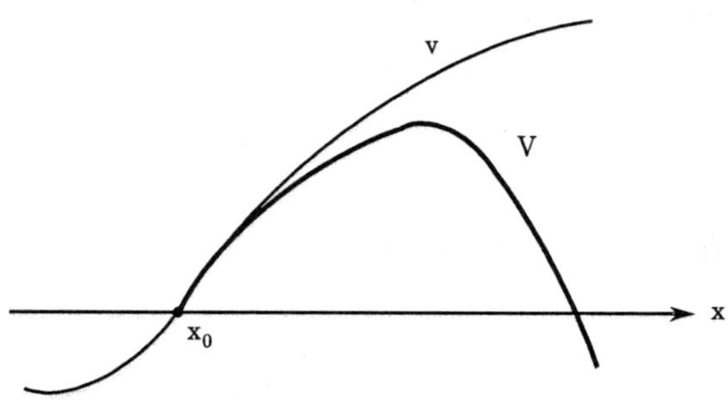

Figure 7-7

Thus, in order to prove the lemma, it suffices to show the following:

If we choose a sufficiently small open ball B centered at x_0, then, for any neighborhood U of x_0 contained in B, we have (12)

$$\sup\{u(x);\, x \in \partial U,\, V(x) \geq 0\} = M.$$

Assume to the contrary that:

> For any open ball B centered at x_0, there exists a neighborhood U of x_0, contained in B, such that
>
> $$\sup\{u(x);\, x \in \partial U,\, V(x) \geq 0\} < M.$$

Then we can find a neighborhood σ of the set $\partial U \cap \{V \geq 0\}$ and a sufficiently small constant $\varepsilon > 0$ such that (cf. Figure 7-8)

$$u(x) + \varepsilon V(x) < M, \qquad x \in \sigma. \tag{13}$$

Also, since there exists a constant $\delta > 0$ such that

$$V(x) \leq -\delta, \qquad x \in \partial U \setminus \sigma,$$

it follows that

$$u(x) + \varepsilon V(x) < M, \qquad x \in \partial U \setminus \sigma. \tag{14}$$

Therefore we obtain from inequalities (13) and (14) that

$$u(x) + \varepsilon V(x) < M \qquad \text{on } \partial U.$$

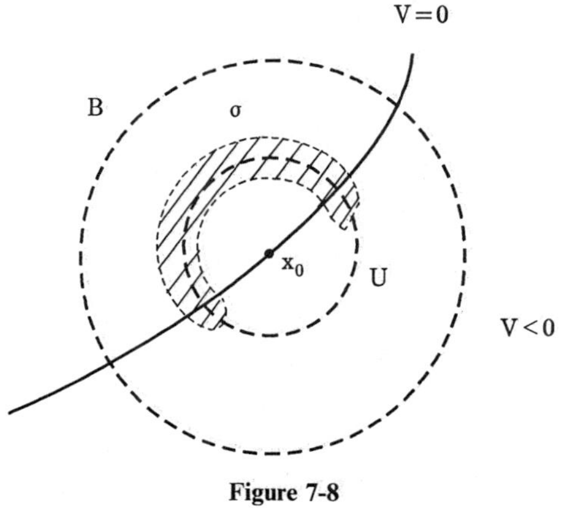

Figure 7-8

On the other hand, since $AV(x_0) > 0$, choosing the ball B sufficiently small, we may assume that

$$AV > 0 \qquad \text{in } U.$$

Then we have

$$A(u + \varepsilon V) \geq \varepsilon AV > 0 \qquad \text{in } U.$$

Therefore, applying Theorem 7.2.6 to the function $u + \varepsilon V$, we obtain that the function $u + \varepsilon V$ may take its maximum only on the boundary ∂U. But this is a contradiction, since we have

$$\begin{cases} u(x_0) + \varepsilon V(x_0) = u(x_0) = M, \\ \quad u(x) + \varepsilon V(x) < M \qquad \text{on } \partial U. \end{cases}$$

This contradiction proves assertion (12) and hence Lemma 7.2.17. ■

Proof of Theorem 7.2.1. Suppose that $u \in C^2(D)$, $Au \geq 0$ in D, $\sup_D u = M < +\infty$ and $u(x) = M$ for some point x of D. We let

$$F = \{y \in D; u(y) = M\}.$$

We remark that, by the continuity of u, the set F is a (relatively) closed subset of D.

I) First we prove that the maximum M propagates along *subunit trajectories*. To do so, in view of Lemma 7.2.15 and Remark 7.2.16, it suffices to show the following:

7.2.18 Lemma. *Let x_0 be a point of the set F, and let γ be a subunit vector for the operator $A^0 = \sum_{i,j=1}^{N} a^{ij} (\partial^2/\partial x_i \, \partial x_j)$ at x_0. Then we have*

$$\langle \gamma, v \rangle = 0$$

for all vectors v normal to F at x_0.

Proof. Assume to the contrary that:

For some vector v normal to F at x_0, we have

$$\langle \gamma, v \rangle = s \sum_{j=1}^{N} \gamma^j (x_1^j - x_0^j) \neq 0,$$

where $v = s(x_1 - x_0)$, $s > 0$ and x_1 is the center of some open ball Q contained in $D \setminus F$, and x_0 is on the boundary of Q (cf. Figure 7-5).

Then, since the vector v is subunit for A^0 at x_0, it follows that

$$0 < \left(\sum_{j=1}^{N} \gamma^j (x_1^j - x_0^j) \right)^2 \leq \sum_{i,j=1}^{N} a^{ij}(x_0)(x_1^i - x_0^i)(x_1^j - x_0^j). \tag{15}$$

Now we consider the function

$$v(x) = e^{-q|x-x_1|^2} - e^{-q|x_1-x_0|^2},$$

where q is a positive constant to be chosen later on. Clearly we have

$$\begin{cases} v(x_0) = 0, \\ \text{grad}\, v(x_0) = -2q \cdot e^{-q|x_1-x_0|^2} \cdot (x_0 - x_1) \neq 0, \end{cases}$$

and further

$$Av(x_0) = e^{-q|x_1-x_0|^2} \left\{ 4q^2 \sum_{i,j=1}^{N} a^{ij}(x_0)(x_1^i - x_0^i)(x_1^j - x_0^j) - 2q \sum_{k=1}^{N} (a^{kk}(x_0) + b^k(x_0)(x_0^k - x_1^k)) \right\}.$$

Hence it follows from inequality (15) that

$$Av(x_0) > 0,$$

if we choose the constant q sufficiently large.

Therefore, applying Lemma 7.2.17, we obtain that, for any sufficiently small neighborhood U of x_0, the function u attains its maximum M at some point of the set $\{x \in \partial U; v(x) > 0\} = \partial U \cap Q$. This contradicts the assumption that $Q \subset D \setminus F$. ▼

II) Next we prove that the maximum M propagates along *drift trajectories*. To do so, in view of Lemma 7.2.15, it suffices to show the following:

7.2.19 Lemma. *The drift vector field*

$$X_0 = \sum_{i=1}^{N} \left(b^i - \sum_{j=1}^{N} \frac{\partial a^{ij}}{\partial x_j} \right) \frac{\partial}{\partial x_i}$$

satisfies condition (1) *in Lemma 7.2.15, that is,*

At each point x_0 of the set F, the inner product $\langle X_0(x_0), v\rangle$ is non-positive for any vector v normal to F at x_0:

$$\langle X_0(x_0), v\rangle \leq 0. \tag{16}$$

Proof. We divide the proof into four steps.

1) Assume to the contrary that:

For some vector v normal to F at x_0, we have

$$\langle X_0(x_0), v\rangle = s\sum_{i=1}^{N}\left(b^i(x_0) - \sum_{j=1}^{N}\frac{\partial a^{ij}}{\partial x_j}(x_0)\right)(x_1^i - x_0^i) > 0, \tag{17}$$

where $v = s(x_1 - x_0)$, $s > 0$ as in the proof of Lemma 7.2.18 (cf. Figure 7-5).

First it follows from inequality (8) in Lemma 7.2.9 that the tangent vector

$$\left(\frac{\sum_{j=1}^{N} a^{kj}(x_0)(x_1^j - x_0^j)}{(\sum_{i,j=1}^{N} a^{ij}(x_0)(x_1^i - x_0^i)(x_1^j - x_0^j))^{1/2}}\right)_{1\leq k\leq N} \tag{18}$$

is well defined. Further, using Schwarz's inequality, we find that the tangent vector (18) is subunit for the operator $A^0 = \sum_{i,j=1}^{N} a^{ij}\,(\partial^2/\partial x_i\,\partial x_j)$ at x_0. Hence we have, by Lemma 7.2.18,

$$\sum_{k,j=1}^{N} a^{kj}(x_0)(x_1^k - x_0^k)(x_1^j - x_0^j) = 0. \tag{19}$$

Therefore, without loss of generality, we may choose a local coordinate system $(y_1, \ldots, y_N)$ in a neighborhood of x_0 such that:

$$\begin{cases} x_0 = \text{the origin}, \\ x_1 - x_0 = (0, \ldots, 0, 1), \\ (a^{ij}(x_0)) = \begin{pmatrix} E_r & 0 \\ 0 & 0 \end{pmatrix}, \end{cases} \tag{20}$$

where $r = \text{rank}\,(a^{ij}(x_0)) < N$ by (19), and E_r is the $r \times r$ unit matrix. In fact, we have only to choose coordinates so that $x_0 = 0$ and so that the vector $x_1 - x_0$ is directed along the positive x_N-axis, and then rotate the coordinates, keeping the x_N-axis fixed, so that the matrix $(a^{ij}(x_0))$ is diagonalized.

Then assumption (17) is expressed as

$$b^N(0) - \sum_{j=1}^{N} \frac{\partial a^{Nj}}{\partial y_j}(0) > 0. \tag{17'}$$

But we have, by inequality (9) in Lemma 7.2.10,

$$\sum_{i,j=r+1}^{N} \frac{\partial a^{ij}}{\partial y_k}(0)\lambda_i \mu_j = 0, \qquad \lambda_i, \mu_j \in \mathbf{R},\ 1 \le k \le N,$$

so that

$$\frac{\partial a^{Nj}}{\partial y_k}(0) = 0, \qquad r+1 \le j \le N,\ 1 \le k \le N. \tag{21}$$

Thus assumption (17′) is expressed as follows:

$$b^N(0) - \sum_{j=1}^{r} \frac{\partial a^{Nj}}{\partial y_j}(0) > 0. \tag{17''}$$

2) Now we consider the function

$$\begin{aligned} v(y) = y_N - \Bigg(&\frac{1}{2}\sum_{i=1}^{r} \frac{\partial a^{Ni}}{\partial y_i}(0) y_i^2 + \sum_{i=1}^{r}\sum_{j=i+1}^{N} \frac{\partial a^{Ni}}{\partial y_j}(0) y_i y_j \\ &+ c\sum_{i=1}^{r} y_i^2 + C\sum_{i=r+1}^{N} y_i^2\Bigg), \end{aligned}$$

where c and C are positive constants to be chosen later on. Then we have

$$\begin{cases} v(0) = 0, \\ \operatorname{grad} v(0) = (0, \ldots, 0, 1) \neq 0, \end{cases}$$

and further, by (20) and (17″),

$$Av(0) = -2rc + \left(b^N(0) - \sum_{i=1}^{r} \frac{\partial a^{Ni}}{\partial y_i}(0)\right) > 0,$$

if we choose the constant c sufficiently small.

Therefore, applying Lemma 7.2.17, we obtain that, for any sufficiently small $\varepsilon > 0$, the function u attains its maximum M at some point z of the set $\{y \in D;\ |y| < \varepsilon,\ v(y) > 0)$; thus $z \in F$. Since $v(z) > 0$, it follows that the point $z = (z_1, \dots, z_N)$ satisfies

$$z_N > \frac{1}{2} \sum_{i=1}^{r} \frac{\partial a^{Ni}}{\partial y_i}(0) z_i^2 + \sum_{i=1}^{r} \sum_{j=i+1}^{N} \frac{\partial a^{Ni}}{\partial y_j}(0) z_i z_j + c \sum_{i=1}^{r} z_i^2 + C \sum_{i=r+1}^{N} z_i^2. \tag{22}$$

3) We let

$$X_i = \sum_{j=1}^{N} a^{ij} \frac{\partial}{\partial y_j}, \qquad 1 \le i \le N,$$

and consider a chain of integral curves

$$y^{(i)}(t) = (y_1^{(i)}(t), \dots, y_N^{(i)}(t)), \qquad 1 \le i \le r,$$

defined by the following (cf. Figure 7-9):

$$\begin{cases} \dot{y}^{(1)}(t) = X_1(y^{(1)}(t)), & y^{(1)}(0) = z; \\ \dot{y}^{(2)}(t) = X_2(y^{(2)}(t)), & y^{(2)}(0) = y^{(1)}(-z_1); \\ \dot{y}^{(3)}(t) = X_3(y^{(3)}(t)), & y^{(3)}(0) = y^{(2)}(-y_2^{(2)}(0)); \\ \quad \vdots & \quad \vdots \\ \dot{y}^{(r)}(t) = X_r(y^{(r)}(t)), & y^{(r)}(0) = y^{(r-1)}(-y_{r-1}^{(r-1)}(0)). \end{cases}$$

For simplicity, we write

$$y(z) = (y_1(z), y_2(z), \dots, y_N(z)) = y^{(r)}(-y_r^{(r)}(0)).$$

First we show that

$$y(z) \in F. \tag{23}$$

In view of inequality (8) in Lemma 7.2.9, it follows that the vector fields

$$\frac{X_i}{(a^{ii})^{1/2}} = \sum_{j=1}^{N} \frac{a^{ij}}{(a^{ii})^{1/2}} \frac{\partial}{\partial y_j}, \qquad 1 \le i \le r,$$

are subunit for A^0. Therefore we find from Theorem 7.2.13 that a chain of the integral curves $y^{(i)}(t)$ can be approximated uniformly by subunit trajectories; so that assertion (23) is obtained from step 1), since $z \in F$.

Next we show that:

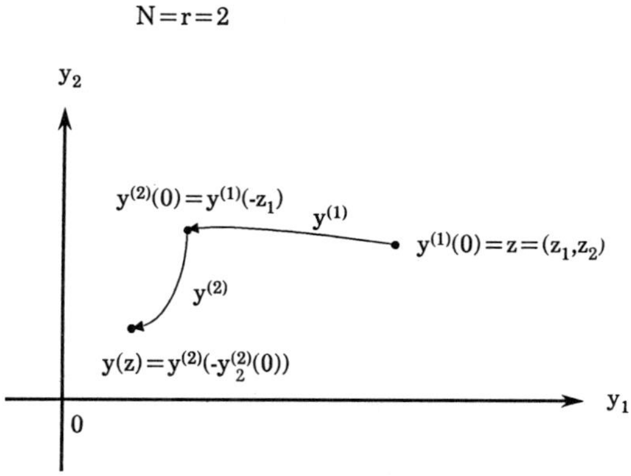

Figure 7-9

For any $\alpha > 0$, if we choose a constant $\varepsilon > 0$ sufficiently small and the constant C sufficiently large, then the point $y(z)$, $|z| < \varepsilon$, is contained in the open ball of radius α about $(0, \ldots, 0, \alpha)$ (cf. Figure 7-10):

$$\sum_{i=1}^{N-1} y_i(z)^2 + (y_N(z) - \alpha)^2 < \alpha^2, \qquad |z| < \varepsilon. \tag{24}$$

Since we have, by (20),

$$X_i(0) = \frac{\partial}{\partial y_i}, \qquad 1 \le i \le r,$$

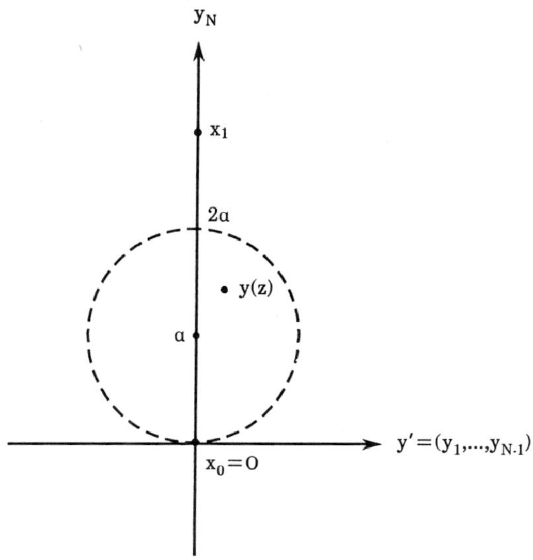

Figure 7-10

it follows from an application of Lemma 7.2.14 with $X = X_1$ that as $|z| \to 0$ we have

$$y_1^{(1)}(-z_1) = -\frac{1}{2}\frac{\partial a^{11}}{\partial y_1}(0)z_1^2 - \sum_{j=2}^{N}\frac{\partial a^{11}}{\partial y_j}(0)z_1 z_j + o(|z|^2),$$

$$y_i^{(1)}(-z_1) = z_i - \frac{1}{2}\frac{\partial a^{1i}}{\partial y_1}(0)z_1^2 - \sum_{j=2}^{N}\frac{\partial a^{1i}}{\partial y_j}(0)z_1 z_j + o(|z|^2), \quad 2 \le i \le N.$$

Further, replacing X_1 by X_2, we have, as $|z| \to 0$,

$$\begin{aligned} y_1^{(2)}(-y_2^{(1)}(-z_1)) = &-\frac{1}{2}\frac{\partial a^{11}}{\partial y_1}(0)z_1^2 - \frac{1}{2}\frac{\partial a^{21}}{\partial y_2}(0)z_2^2 \\ &- \sum_{j=2}^{N}\frac{\partial a^{11}}{\partial y_j}(0)z_1 z_j - \sum_{j=3}^{N}\frac{\partial a^{21}}{\partial y_j}(0)z_2 z_j + o(|z|^2), \end{aligned}$$

$$y_2^{(2)}(-y_2^{(1)}(-z_1)) = -\frac{1}{2}\frac{\partial a^{22}}{\partial y_2}(0)z_2^2 - \sum_{j=3}^{N}\frac{\partial a^{22}}{\partial y_j}(0)z_2 z_j + o(|z|^2),$$

$$\begin{aligned} y_i^{(2)}(-y_2^{(1)}(-z_1)) = z_i &- \frac{1}{2}\frac{\partial a^{1i}}{\partial y_1}(0)z_1^2 - \frac{1}{2}\frac{\partial a^{2i}}{\partial y_2}(0)z_2^2 \\ &- \sum_{j=2}^{N}\frac{\partial a^{1i}}{\partial y_j}(0)z_1 z_j - \sum_{j=3}^{N}\frac{\partial a^{2i}}{\partial y_j}(0)z_2 z_j + o(|z|^2), \end{aligned}$$

$$3 \le i \le N.$$

Continuing this process, we have, after r steps,

$$\begin{cases} \displaystyle\sum_{i=1}^{N-1} y_i(z)^2 \le K\left(\sum_{i=1}^{r} z_i^4 + \sum_{i=r+1}^{N} z_i^2\right), & (25) \\ \displaystyle y_N(z) = z_N - \frac{1}{2}\sum_{i=1}^{r}\frac{\partial a^{Ni}}{\partial y_i}(0)z_i^2 - \sum_{i=1}^{r}\sum_{j=i+1}^{N}\frac{\partial a^{Ni}}{\partial y_j}(0)z_i z_j + o(|z|^2), & (26) \end{cases}$$

where K is a positive constant independent of z. Combining formula (26) and inequality (22), we obtain that

$$y_N(z) > (c + o(1))\sum_{j=1}^{r} z_i^2 + (C + o(1))\sum_{i=r+1}^{N} z_i^2. \tag{27}$$

Therefore, for any $\alpha > 0$, if we choose the constant $\varepsilon > 0$ sufficiently small and the constant C sufficiently large, we conclude from (26), (27) and (25) that

$$\begin{aligned} y_N(z)(2\alpha - y_N(z)) &\geq \alpha y_N(z) \\ &> \alpha(c + o(1)) \sum_{i=1}^{r} z_i^2 + \alpha(C + o(1)) \sum_{i=r+1}^{N} z_i^2 \\ &\geq K\left(\sum_{i=1}^{r} z_i^4 + \sum_{i=r+1}^{N} z_i^2 \right) \\ &\geq \sum_{i=1}^{N-1} y_i(z)^2, \qquad |z| < \varepsilon. \end{aligned}$$

This proves assertion (24).

4) In view of (20), assertions (23) and (24) imply that the vector $v = s(x_1 - x_0)$ is *not* normal to F at x_0. This contradiction proves inequality (16) and hence Lemma 7.2.19. ▼

Now the proof of Theorem 7.2.1 is complete. ■

Proof of Theorem 7.2.1′. The proof of Theorem 7.2.1′ is essentially the same as that of Theorem 7.2.1. In fact, it is easy to see that, in the proof of Theorem 7.2.1, the assumption that $Au \geq 0$ is needed only in a sufficiently small neighborhood U of a point x_0 where $u(x_0) = M$. But, if $M > 0$, we may assume that $u > 0$ in U, and hence

$$Au \geq -cu \geq 0 \qquad \text{in } U.$$

Therefore the proof goes through as before. ■

Proof of Theorem 7.2.2

First we prove that:

Each trajectory $\phi(t)$ of the form (2) *in Theorem* 7.2.2 *can be approximated uniformly by a finite number of subunit and drift trajectories.* (1)

Let φ be a path: $[0, \rho] \to D$ for which there exists a piecewise C^1 function ψ: $[0, \rho] \to \mathbf{R}^N$ such that

$$\dot{\varphi}^i(t) = \sum_{j=1}^{N} a^{ij}(\varphi(t))\psi_j(t), \qquad 1 \leq i \leq N.$$

Then we have:

The path φ is a subunit trajectory if we change the scale of time. (2)

In fact, we let

$$\begin{cases} \|a\| = \sup\limits_{\substack{x \in \bar{D} \\ |\xi| = 1}} \sum\limits_{i,j=1}^{N} a^{ij}(x)\xi_i\xi_j, \\ \|\psi(\rho)\| = \sup\limits_{0 \le t \le \rho} \left(\sum\limits_{j=1}^{N} \psi_j(t)^2 \right)^{1/2}, \end{cases}$$

and define a path

$$\gamma(t) = \varphi(c_\rho t), \qquad 0 \le t \le \frac{\rho}{c_\rho},$$

where

$$c_\rho = \|a\|^{-1/2} \|\psi(\rho)\|^{-1}.$$

Then it follows that

$$\begin{aligned} \dot{\gamma}^i(t) &= c_\rho \dot{\varphi}^i(c_\rho t) \\ &= c_\rho \sum_{j=1}^{N} a^{ij}(\varphi(c_\rho t))\psi_j(c_\rho t) \\ &= c_\rho \sum_{j=1}^{N} a^{ij}(\gamma(t))\psi_j(c_\rho t). \end{aligned}$$

Hence it follows from Schwarz's inequality that we have, for all $\eta = \sum_{i=1}^{N} \eta_i \, dx_i \in T^*_{\gamma(t)}(D)$,

$$\begin{aligned} \left(\sum_{i=1}^{N} \dot{\gamma}^i(t)\eta_i \right)^2 &= \left(\sum_{i,j=1}^{N} c_\rho a^{ij}(\gamma(t))\psi_j(c_\rho t)\eta_i \right)^2 \\ &\le c_\rho^2 \left(\sum_{i,j=1}^{N} a^{ij}(\gamma(t))\psi_i(c_\rho t)\psi_j(c_\rho t) \right) \left(\sum_{i,j=1}^{N} a^{ij}(\gamma(t))\eta_i\eta_j \right) \\ &\le c_\rho^2 \|a\| \cdot \|\psi(\rho)\|^2 \left(\sum_{i,j=1}^{N} a^{ij}(\gamma(t))\eta_i\eta_j \right) \\ &= \sum_{i,j=1}^{N} a^{ij}(\gamma(t))\eta_i\eta_j. \end{aligned}$$

This implies that the path $\gamma(t) = \varphi(c_\rho t)$ is a subunit trajectory.

Therefore assertion (1) is obtained from fact (2) and an application of Theorem 7.2.13. ▼

The proof of the converse of assertion (1) is based on the following proposition, essentially due to Fefferman-Phong (cf. [1], the proof of Lemma 1):

7.2.20 Proposition. *Suppose that a point $y \in D$ can be joined to a point $x \in D$ by a Lipschitz path v: $[0, \rho] \to D$, for which the tangent vector $\dot{v}(t)$ is subunit for A^0 at $v(t)$ for almost every t. Then one can join x to y by a Lipschitz path $\tilde{\gamma}$: $[0, \tilde{C}\rho] \to D$ of the form*

$$\tilde{\gamma}^i(t) = x_i + \sum_{j=1}^{N} \int_0^t a^{ij}(\tilde{\gamma}(t))\xi_j(s)\, ds, \qquad 0 \le t \le \tilde{C}\rho, \tag{3}$$

where $\tilde{C}$ is a positive constant and $\xi(t) = \sum_{i=1}^N \xi_i(t)\, dx_i$ is a piecewise C^1 covector field (cf. Figure 7-11).

Granting Proposition 7.2.20 for the moment, we shall prove the converse of assertion (1).

1) First we remark that the trajectories $\phi(t)$ of the form (2) in Theorem 7.2.2 contain drift trajectories as the particular case $\psi_j \equiv 0$.

2) Let $\tilde{\gamma}$ be a Lipschitz path of the form (3). If n is a positive integer, we define a path ϕ_n of the form (2) in Theorem 7.2.2 as

$$\phi_n^i(t) = x_i + \sum_{j=1}^{N} \int_0^t a^{ij}(\phi_n(s)) n\xi_j(ns)\, ds$$
$$+ \int_0^t \left(b^i(\phi_n(s)) - \sum_{j=1}^{N} \frac{\partial a^{ij}}{\partial x_j}(\phi_n(s)) \right) ds, \qquad 0 \le t \le \frac{\tilde{C}\rho}{n},$$

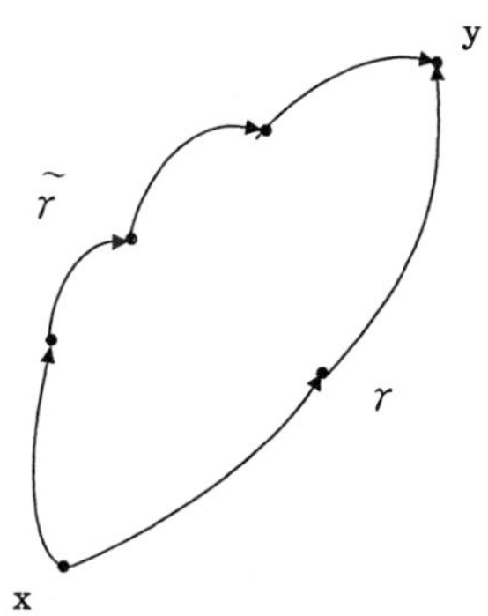

Figure 7-11

and let

$$\varphi_n(t) = \phi_n\left(\frac{t}{n}\right), \qquad 0 \le t \le \tilde{C}\rho.$$

Then we have

$$\begin{aligned}\varphi_n^i(t) &= x_i + \sum_{j=1}^N \int_0^{t/n} a^{ij}(\phi_n(s)) n\xi_j(ns)\,ds \\ &\quad + \int_0^{t/n} \left(b^i(\phi_n(s)) - \sum_{j=1}^N \frac{\partial a^{ij}}{\partial x_j}(\phi_n(s)) \right) ds \\ &= x_i + \sum_{j=1}^N \int_0^t a^{ij}(\varphi_n(\tau))\xi_j(\tau)\,d\tau \\ &\quad + \frac{1}{n}\int_0^t \left(b^i(\varphi_n(\tau)) - \sum_{j=1}^N \frac{\partial a^{ij}}{\partial x_j}(\varphi_n(\tau)) \right) d\tau.\end{aligned}$$

Thus it follows from an application of Lemma 7.2.12 that the path $\tilde{\gamma}$ can be approximated uniformly by the paths φ_n and hence by the paths ϕ_n. Therefore, combining this fact and Proposition 7.2.20, we find that the subunit trajectories can be approximated uniformly by paths of the form (2) in Theorem 7.2.3.

3) Summing up, we conclude that any finite number of subunit and drift trajectories can be approximated uniformly by a finite number of trajectories of the form (2) in Theorem 7.2.2. ▼

Theorem 7.2.2 is proved, apart from the proof of Proposition 7.2.20.

Proof of Proposition 7.2.20. Our proof mimics that of Lemma 1 of Fefferman-Phong [1]. We divide the proof into three steps.

I) Let x be a point of D and $\rho > 0$. With the differential operator $A^0 = \sum_{i,j=1}^N a^{ij}(\partial^2/\partial x_i\,\partial x_j)$, we associate a "non-Euclidean" ball $B_{A^0}(x, \rho)$ of radius ρ about x as follows (cf. Figure 6-1):

$B_{A^0}(x, \rho) =$ the set of all points $y \in D$ which can be joined to x by a Lipschitz path v: $[0, \rho] \to D$, for which the tangent vector $\dot{v}(t)$ is subunit for A^0 at $v(t)$ for almost every t.

Also we let

$$B_E(x, \rho) = \text{the ordinary Euclidean ball of radius } \rho \text{ about } x.$$

We remark that if the operator A^0 is the usual Laplacian $\Delta = \sum_{i=1}^N (\partial^2/\partial x_i^2)$, then the two balls $B_{A^0}(x, \rho)$ and $B_E(x, \rho)$ coincide.

Since we have $(a^{ij}) \leq (a^{ij} + \rho^8\delta_{ij})$ as matrices, it follows that

$$B_{A^0}(x, \rho) \subset B_{A^0 - \rho^8\Delta}(x, \rho).$$

Here the term $\rho^8\Delta$ is a technicality. Let (g_{ij}) be the inverse matrix of $(g^{ij}) = (a^{ij} + \rho^8\delta_{ij})$:

$$(g_{ij}) = (a^{ij} + \rho^8\delta_{ij})^{-1}.$$

Then, for any point $y \in B_{A^0}(x, \rho)$, we may join x to y by a *geodesic* $\gamma: [0, \rho] \to D$ in the metric

$$ds^2 = \sum_{i,j=1}^{N} g_{ij}(x)\, dx_i\, dx_j.$$

First we have:

7.2.21 Lemma. *If we parametrize the geodesic γ by its arc-length t, then the tangent vector $\dot{\gamma}(t)$ is a subunit vector for the operator $A^0 - \rho^8\Delta$:*

$$\left(\sum_{j=1}^{N} \dot{\gamma}^j(t)\eta_j\right)^2 \leq \sum_{i,j=1}^{N} (a^{ij}(\gamma(t)) + \rho^8\delta_{ij})\eta_i\eta_j,$$

$$\eta = \sum_{i=1}^{N} \eta_i\, dx_i \in T^*_{\gamma(t)}(D), \tag{4}$$

where $\dot{\gamma}(t) = \sum_{j=1}^{N} \dot{\gamma}^j(t) \dfrac{\partial}{\partial x_j} \in T_{\gamma(t)}(D).$

Proof. Since inequality (4) is independent of the particular local chart, we rotate the coordinate axes so that the matrix $(a^{ij}(\gamma(t)))$ is diagonalized:

$$(a^{ij}(\gamma(t))) = (\lambda_i\delta_{ij}), \qquad \lambda_i \geq 0.$$

Then it follows from Schwarz's inequality that

$$\begin{aligned}
\left(\sum_{j=1}^{N} \dot{\gamma}^j(t)\eta_j\right)^2 &\leq \left(\sum_{j=1}^{N} \frac{\dot{\gamma}^j(t)^2}{\lambda_j + \rho^8}\right)\left(\sum_{j=1}^{N} (\lambda_j + \rho^8)\eta_j^2\right) \\
&= \left(\sum_{1,j=1}^{N} g_{ij}(\gamma(t))\dot{\gamma}^i(t)\dot{\gamma}^j(t)\right)\left(\sum_{i,j=1}^{N} (a^{ij}(\gamma(t)) + \rho^8\delta_{ij})\eta_i\eta_j\right) \\
&= \sum_{i,j=1}^{N} (a^{ij}(\gamma(t)) + \rho^8\delta_{ij})\eta_i\eta_j,
\end{aligned}$$

since we have $\sum_{i,j=1}^{N} g_{ij}(\gamma(t))\dot{\gamma}^i(t)\dot{\gamma}^j(t) = 1.$ ▼

7.2.22 Corollary. *We have, for all sufficiently small* $\rho > 0$,

$$B_E(x, \rho^2) \subset B_{A^0 - \rho^8\Delta}(x, \rho). \tag{5}$$

Proof. Every point $y \in B_E(x, \rho^2)$ may be joined to the point x by a geodesic γ in the metric $ds^2 = \sum_{i,j=1}^N g_{ij}(x)\, dx_i\, dx_j$ where $(g_{ij}) = (a^{ij} + \rho^8\delta_{ij})^{-1} = ((\lambda_i + \rho^8)^{-1}\delta_{ij})$. So conclusion (5) is an immediate consequence of Lemma 7.2.21. In fact we have, for all sufficiently small $\rho > 0$,

$$\begin{aligned} \rho\, ds &= \rho\left(\sum_{i=1}^N \frac{(dx_i)^2}{\lambda_i + \rho^8}\right)^{1/2} \\ &\geq \frac{\rho}{(\max_{1 \leq i \leq N} \lambda_i + \rho^8)^{1/2}}\left(\sum_{i=1}^N (dx_i)^2\right)^{1/2} \\ &\geq \rho^2\left(\sum_{i=1}^N (dx_i)^2\right)^{1/2}. \end{aligned}$$ ▼

II) Let γ be the geodesic as in step I):

$$\begin{cases} \gamma(0) = x, \qquad \gamma(\rho) = y, \\ \text{the tangent vector } \dot{\gamma}(t) \text{ is subunit for } A^0 - \rho^8\Delta. \end{cases}$$

We shall perturb the path $\gamma(t)$ to a broken path $\gamma_{\#}(t)$, which starts at x and ends very near y, so that $c\dot{\gamma}_{\#}(t)$ is a subunit vector for A^0 for some constant $c > 0$.

II-a) First we need a lemma on perturbation of tangent vectors.

7.2.23 Lemma. *Let* $X = \sum_{j=1}^N \gamma^j(\partial/\partial x_j)$ *be a subunit vector for* $A^0 - \rho^8\Delta$ *at a point* $x^0 \in D$ *and let* x^1 *be an arbitrary point of* D *such that* $|x^1 - x^0| \leq c_1\rho^4$ *for a constant* $c_1 > 0$. *Then there exist a tangent vector* $Y = \sum_{j=1}^N \delta^j(\partial/\partial x_j)$ *at* x^1 *and a cotangent vector* $\zeta = \sum_{j=1}^N \zeta_j\, dx_j$ *at* x^1 *which satisfy the following conditions:*

(i) $|\delta^j - \gamma^j| \leq C\rho^4$;

(ii) *The vector* cY *is subunit for* A^0 *at* x^1;

(iii) $\delta^i = \sum_{j=1}^N a^{ij}(x^1)\zeta_j$ *and* $|\zeta| \leq C\rho^{-4}$;

here C *and* c *are positive constants depending only on* c_1 *and the bounds on the second derivatives of* a^{ij}.

Proof. Since the tangent vector $X = \sum_{j=1}^N \gamma^j(\partial/\partial x_j)$ is subunit for $A^0 - \rho^8\Delta$ at x^0, it follows that

$$\left(\sum_{j=1}^N \gamma^j\eta_j\right)^2 \leq \sum_{i,j=1}^N (a^{ij}(x^0) + \rho^8\delta_{ij})\eta_i\eta_j, \; \eta = \sum_{i=1}^N \eta_i \, dx_i \in T^*_{x^0}(D). \tag{6}$$

To estimate the right-hand side of (6), we let

$$g(t) = \frac{\sum_{i,j=1}^N (a^{ij}(x^1 + t(x^0 - x^1)) + \rho^8\delta_{ij})\eta_i\eta_j}{|\eta|^2}, \qquad t \in \mathbf{R}, \; \eta \neq 0.$$

Then we have

$$\begin{cases} g(t) \geq 0 & \text{on } \mathbf{R}, \\ |g''(t)| \leq K|x^1 - x^0|^2 & \text{on } \mathbf{R}, \end{cases}$$

where K is a positive constant depending only on the bounds on the second derivatives of a^{ij}. Applying Lemma 7.2.8 to the function

$$f(t) = \frac{g(t)}{K|x^1 - x^0|^2},$$

we obtain that

$$\frac{g(1)}{K|x^1 - x^0|^2} + 1 \leq 2\left(1 + \frac{g(0)}{K|x^1 - x^0|^2}\right),$$

so that

$$\frac{\sum_{i,j=1}^N (a^{ij}(x^0) + \rho^8\delta_{ij})\eta_i\eta_j}{|\eta|^2} \leq 2\,\frac{\sum_{i,j=1}^N (a^{ij}(x^1) + \rho^8\delta_{ij})\eta_i\eta_j}{|\eta|^2} + K|x^1 - x^0|^2.$$

Thus we have

$$\sum_{i,j=1}^N (a^{ij}(x^0) + \rho^8\delta_{ij})\eta_i\eta_j \leq K'\left(\sum_{i,j=1}^N (a^{ij}(x^1) + \rho^8\delta_{ij})\eta_i\eta_j + |x^1 - x^0|^2|\eta|^2\right), \tag{7}$$

where $K' = \max(K, 2)$.

Hence, combining inequalities (6) and (7), we have, for $|x^1 - x^0| \leq c_1\rho^4$,

$$\left(\sum_{j=1}^N \gamma^j\eta_j\right)^2 \leq K'(1 + c_1^2) \sum_{i,j=1}^N (a^{ij}(x^1) + \rho^8\delta_{ij})\eta_i\eta_j.$$

Therefore, letting

$$\tilde{X} = \sum_{j=1}^{N} \gamma^j \frac{\partial}{\partial x_j} \qquad \text{at } x^1,$$

$$\tilde{c} = (K'(1 + c_1^2))^{-1/2},$$

we obtain that:

$$\text{The vector } \tilde{c}\tilde{X} \text{ is subunit for } A^0 - \rho^8 \Delta \text{ at } x^1. \tag{8}$$

Since conditions (i), (ii) and (iii) are unaffected by the rotation of coordinate axes, we may assume that

$$(a^{ij}(x^1)) = (\lambda_i \delta_{ij}), \qquad \lambda_i \geq 0.$$

Then assertion (8) can be restated as follows:

$$\left(\sum_{i=1}^{N} \tilde{c}\gamma^i \eta_i\right)^2 \leq \sum_{i=1}^{N} (\lambda_i + \rho^8)\eta_i^2, \qquad \eta \in \mathbf{R}^N.$$

In particular, we have

$$\tilde{c}^2(\gamma^i)^2 \leq \lambda_i + \rho^8, \qquad 1 \leq i \leq N. \tag{9}$$

Now we define a tangent vector $Y = \sum_{i=1}^{N} \delta^i(\partial/\partial x_i)$ at x^1 as

$$\delta^i = \begin{cases} \gamma^i & \text{if } \lambda_i \geq \rho^8, \\ 0 & \text{if } 0 \leq \lambda_i < \rho^8, \end{cases}$$

and a cotangent vector $\zeta = \sum_{i=1}^{N} \zeta_i \, dx_i$ at x^1 as

$$\zeta_i = \begin{cases} \dfrac{\gamma^i}{\lambda_i} & \text{if } \lambda_i \geq \rho^8, \\ 0 & \text{if } 0 \leq \lambda_i < \rho^8. \end{cases}$$

Then we can verify conditions (i), (ii) and (iii) as follows:

(i) First, using inequality (9), we have

$$|\delta^i - \gamma^i|^2 = \begin{cases} 0 & \text{if } \lambda_i \geq \rho^8, \\ (\gamma^i)^2 \leq \dfrac{\lambda_i + \rho^8}{\tilde{c}^2} \leq \dfrac{2}{\tilde{c}^2}\rho^8 & \text{if } 0 \leq \lambda_i < \rho^8, \end{cases}$$

so that

$$|\delta^i - \gamma^i| \leq C\rho^4$$

with

$$C = \sqrt{2}/\tilde{c} = (2/K'(1 + c_1^2))^{1/2}.$$

(ii) Secondly we have

$$\left(\sum_{i=1}^{N} \delta^i \eta_i\right)^2 = \left(\sum_{\substack{1 \le i \le N \\ \lambda_i \ge \rho^8}} \gamma^i \eta_i\right)^2$$

$$\le N \sum_{\substack{1 \le i \le N \\ \lambda_i \ge \rho^8}} (\gamma^i)^2 \eta_i^2$$

$$\le N \sum_{\substack{1 \le i \le N \\ \lambda_i \ge \rho^8}} \frac{\lambda_i + \rho^8}{\tilde{c}^2} \eta_i^2$$

$$\le \frac{2N}{\tilde{c}^2} \sum_{i=1}^{N} \lambda_i \eta_i^2.$$

Hence, letting

$$c = \tilde{c}/\sqrt{2N} = (2NK'(1 + c_1^2))^{-1/2}, \tag{10}$$

we obtain that:

The vector cY is subunit for A^0 at x^1.

(iii) Finally we have

$$\sum_{j=1}^{N} a^{ij}(x^1)\zeta_j = \lambda_i \zeta_i$$

$$= \begin{cases} \gamma^i & \text{if } \lambda_i \ge \rho^8, \\ 0 & \text{if } 0 \le \lambda_i < \rho^8, \end{cases}$$

$$= \delta^i,$$

and also

$$|\zeta|^2 = \sum_{\substack{1 \le i \le N \\ \lambda_i \ge \rho^8}} \frac{(\gamma^i)^2}{\lambda_i^2}$$

$$\le \sum_{\substack{1 \le i \le N \\ \lambda_i \ge \rho^8}} \frac{\lambda_i + \rho^8}{\tilde{c}^2 \lambda_i^2}$$

$$\le \sum_{\substack{1 \le i \le N \\ \lambda_i \ge \rho^8}} \frac{2}{\tilde{c}^2} \frac{1}{\lambda_i}$$

$$\le \frac{2N}{\tilde{c}^2} \rho^{-8}.$$

This proves that

$$|\zeta| \leq C\rho^{-4}$$

with

$$C = \sqrt{2N/\tilde{c}} = (2NK'(1 + c_1^2))^{1/2}.$$

Lemma 7.2.23 is proved. ▼

II-b) Next we need a lemma on estimates for the second derivatives of Hamiltonian paths.

We let

$$H(x, \xi) = \frac{1}{2} \sum_{i,j=1}^{N} a^{ij}(x)\xi_i\xi_j,$$

and consider the *Hamiltonian equations*:

$$\begin{cases} \dot{x}_i(t) = \sum_{j=1}^{N} a^{ij}(x(t))\xi_j(t), & x(0) = x_0; \\ \dot{\xi}_\ell(t) = -\frac{1}{2} \sum_{i,j=1}^{N} \frac{\partial a^{ij}}{\partial x_\ell}(x(t))\xi_i(t)\xi_j(t), & \xi(0) = \xi_0. \end{cases} \tag{11}$$

Then we obtain the following:

7.2.24 Lemma. *Let $|\xi_0| \leq C\rho^{-4}$ for some constant $C > 0$. If we flow for time $0 \leq t \leq c_1\rho^4$ where $c_1 > 0$ is a sufficiently small constant, then we have*

$$\begin{cases} |\xi(t)| \leq C'\rho^{-4}, \\ |\ddot{x}(t)| \leq C''\rho^{-8}, \end{cases} \tag{12}$$

along the path. Here C' and C'' are positive constants depending only on C, c_1 and the bounds on the a^{ij} and their first derivatives.

Proof. First it follows from equations (11) that

$$|\dot{x}(t)| \leq K|\xi(t)|, \qquad |\dot{\xi}(t)| \leq K|\xi(t)|^2, \tag{13}$$

and so

$$|\ddot{x}(t)| \leq K|\xi(t)|^2, \tag{14}$$

since we have

$$\ddot{x}_i(t) = \sum_{j=1}^{N} a^{ij}(x(t))\dot{\xi}_j(t) + \sum_{j,\ell=1}^{N} \frac{\partial a^{ij}}{\partial x_\ell}(x(t))\dot{x}_\ell(t)\xi_j(t).$$

Here K is a positive constant depending only on the bounds on the a^{ij} and their first derivatives. By the mean value theorem and estimates (13), it follows that for $0 \le t \le c_1\rho^4$ we have

$$\begin{aligned}|\xi(t)| &\le |\xi_0| + \sup_{0\le s\le c_1\rho^4} |\dot{\xi}(s)|\cdot t \\ &\le C\rho^{-4} + K \sup_{0\le s\le c_1\rho^4} |\xi(s)|^2 \cdot c_1\rho^4,\end{aligned}$$

so that

$$Kc_1\rho^4\Big(\sup_{0\le s\le c_1\rho^4} |\xi(s)|\Big)^2 - \sup_{0\le s\le c_1\rho^4} |\xi(s)| + C\rho^{-4} \ge 0.$$

Thus, if the constant c_1 is sufficiently small so that

$$1 - 4c_1KC > 0,$$

we obtain that

$$\sup_{0\le s\le c_1\rho^4} |\xi(s)| \le \left(\frac{1-\sqrt{1-4c_1KC}}{2Kc_1}\right)\rho^{-4}. \tag{15}$$

In fact, it suffices to note that as $c_1 \downarrow 0$ we have

$$\frac{1+\sqrt{1-4c_1KC}}{2Kc_1} \to +\infty,$$

$$\sup_{0\le s\le c_1\rho^4} |\xi(s)| \to |\xi_0|.$$

Therefore, estimates (12) follow from estimates (15) and (14), with

$$\begin{cases} C' = \dfrac{1-\sqrt{1-4c_1KC}}{2Kc_1}, \\ C'' = KC'^2. \end{cases}$$

The proof of Lemma 7.2.24 is complete. ▼

III-c) Now we construct the broken path $\gamma_{\#}(t)$ mentioned above.

1) Suppose that:

We have constructed a path $\gamma_{\#}(t)$ for $0 \leq t \leq \tau_k$ such that

$$\begin{cases} \gamma_{\#}(0) = \gamma(0) = x, \\ |\gamma_{\#}(t) - \gamma(t)| \leq C_{+}\tau_k \rho^4, \qquad 0 \leq t \leq \tau_k. \end{cases} \tag{16}$$

The large constant C_{+} and the division points τ_k will be picked later on (see (32), (33) below) so that for all sufficiently small $\rho > 0$ we have

$$C_{+}\tau_k \leq C_{+}\rho \leq c_1, \tag{17}$$

where c_1 is the same constant as in Lemma 7.2.24. We remark that assertion (16) is vacuous for $\tau_0 = 0$.

By virtue of (16) and (17), we can apply Lemma 7.2.23 with $x^0 = \gamma(\tau_k)$, $x^1 = \gamma_{\#}(\tau_k)$ and $X = \sum_{j=1}^{N} \dot{\gamma}^j(\tau_k)(\partial/\partial x_j)$ to obtain that:

There exist a tangent vector $Y = \sum_{j=1}^{N} \delta^j(\partial/\partial x_j)$ at $\gamma_{\#}(\tau_k)$ and a cotangent vector $\zeta = \sum_{j=1}^{N} \zeta_j \, dx_j$ at $\gamma_{\#}(\tau_k)$ which satisfy the following conditions:

(i) $|\delta^j - \dot{\gamma}^j(\tau_k)| \leq C\rho^4$;
(ii) The vector cY is subunit for A^0 at $\gamma_{\#}(\tau_k)$; (18)
(iii) $\delta^i = \sum_{j=1}^{N} a^{ij}(\gamma_{\#}(\tau_k))\zeta_j$ and $|\zeta| \leq C\rho^{-4}$.

Now we define a path $\gamma_{\#}(t)$ for $\tau_k \leq t \leq \tau_{k+1}$ as the projection onto the x-coordinate of the Hamiltonian curve for $H(x, \xi) = \frac{1}{2}\sum_{i,j=1}^{N} a^{ij}(x)\xi_i\xi_j$ starting at $(\gamma_{\#}(\tau_k), \zeta)$ for $t = \tau_k$:

$$\begin{cases} \dot{\gamma}^i_{\#}(t) = \sum_{j=1}^{N} a^{ij}(\gamma_{\#}(t))\xi_j(t), \qquad \gamma_{\#}(\tau_k) = \gamma_{\#}(\tau_k); \\ \dot{\xi}_\ell(t) = -\dfrac{1}{2}\sum_{i,j=1}^{N} \dfrac{\partial a^{ij}}{\partial x_\ell}(\gamma_{\#}(t))\xi_i(t)\xi_j(t), \qquad \xi(\tau_k) = \zeta. \end{cases} \tag{19}$$

Then it follows from conditions (iii) and (ii) of (18) that

$$c^2\left(\sum_{i,j=1}^{N} a^{ij}(\gamma_{\#}(\tau_k))\zeta_i\zeta_j\right)^2 = \left(\sum_{i=1}^{N} c\delta^i\zeta_i\right)^2$$
$$\leq \sum_{i,j=1}^{N} a^{ij}(\gamma_{\#}(\tau_k))\zeta_i\zeta_j,$$

so that

$$\sum_{i,j=1}^{N} a^{ij}(\gamma_{\#}(\tau_k))\zeta_i\zeta_j \leq \frac{1}{c^2}. \tag{20}$$

On the other hand, in view of Schwarz's inequality, it follows from the initial-value problem (19) that we have, for all $\eta = \sum_{i=1}^{N} \eta_i \, dx_i \in T^*_{\gamma\#(t)}(D)$,

$$\begin{aligned}\left(\sum_{i=1}^{N} \dot{\gamma}^i_{\#}(t)\eta_i\right)^2 &= \left(\sum_{i,j=1}^{N} a^{ij}(\gamma_{\#}(t))\eta_i\xi_j(t)\right)^2 \\ &\leq \left(\sum_{i,j=1}^{N} a^{ij}(\gamma_{\#}(t))\xi_i(t)\xi_j(t)\right)\left(\sum_{i,j=1}^{N} a^{ij}(\gamma_{\#}(t))\eta_i\eta_j\right) \\ &= \left(\sum_{i,j=1}^{N} a^{ij}(\gamma_{\#}(\tau_k))\zeta_i\zeta_j\right)\left(\sum_{i,j=1}^{N} a^{ij}(\gamma_{\#}(t))\eta_i\eta_j\right),\end{aligned} \tag{21}$$

since the function $H(\gamma_{\#}(t), \xi(t))$ is conserved along the path.

Therefore, combining inequalities (21) and (20), we obtain that:

The tangent vector $c\dot{\gamma}_{\#}(t)$ is subunit for A^0 throughout $\tau_k \leq t \leq \tau_{k+1}$.

Here we remark that the constant c is independent of the division points τ_k, depending essentially on the constant c_1 (see formula (10)).

Moreover, since $|\xi(\tau_k)| = |\zeta| \leq C\rho^{-4}$, it follows from an application of Lemma 7.2.24 that

$$|\ddot{\gamma}_{\#}(t)| \leq C''\rho^{-8}, \qquad \tau_k \leq t \leq \tau_{k+1}, \tag{22}$$

provided

$$\tau_{k+1} - \tau_k \leq c_1\rho^4.$$

2) We estimate the second derivative $\ddot{\gamma}(t)$ of the geodesic $\gamma(t)$. First recall that the geodesic $\gamma(t)$ satisfies the equations

$$\frac{d^2\gamma^i}{dt^2} + \sum_{j,k=1}^{N} \Gamma^i_{jk} \frac{d\gamma^j}{dt}\frac{d\gamma^k}{dt} = 0, \qquad 1 \leq i \leq N, \tag{23}$$

where

$$\Gamma^i_{jk} = \frac{1}{2}\sum_{\ell=1}^{N} g^{i\ell}\left(\frac{\partial g_{\ell j}}{\partial x_k} + \frac{\partial g_{\ell k}}{\partial x_j} - \frac{\partial g_{jk}}{\partial x_\ell}\right) \tag{24}$$

are the Christoffel symbols. Thus, to estimate the second derivative $\ddot{\gamma}(t)$, it suffices to estimate the Γ^i_{jk} and the first derivative $\dot{\gamma}(t)$.

Differentiating the identity

$$\sum_{j=1}^{N} g^{ij} g_{jk} = \delta^i_k$$

with respect to the variable x_ℓ, and then multiplying it to the left by g_{im} and summing over i, we obtain that

$$\frac{\partial g_{mk}}{\partial x_\ell} = - \sum_{i,j=1}^{N} g_{im} \frac{\partial g^{ij}}{\partial x_\ell} g_{jk}.$$

Substituting these into the right-hand side of (24), we have

$$\begin{aligned}
\Gamma^i_{jk} &= -\frac{1}{2}\left(\sum_{n=1}^{N} \frac{\partial g^{in}}{\partial x_k} g_{nj} + \sum_{n=1}^{N} \frac{\partial g^{in}}{\partial x_j} g_{nk} \right. \\
&\qquad \left. - \sum_{\ell,m,n=1}^{N} g^{i\ell} g_{mj} g_{nk} \frac{\partial g^{mn}}{\partial x_\ell} \right) \\
&= -\frac{1}{2}\left(\sum_{n=1}^{N} \frac{\partial a^{in}}{\partial x_k} g_{nj} + \sum_{n=1}^{N} \frac{\partial a^{in}}{\partial x_j} g_{nk} \right. \\
&\qquad \left. - \sum_{\ell,m,n=1}^{N} (a^{i\ell} + \rho^8 \delta_{i\ell}) g_{mj} g_{nk} \frac{\partial a^{mn}}{\partial x_\ell} \right).
\end{aligned} \tag{25}$$

We estimate the terms of the right-hand side of (25). Since the bounds on $\ddot{\gamma}(t)$ are unaffected by the rotation of coordinate axes, one may assume that

$$(a^{ij}(\gamma(t))) = (\lambda_i \delta_{ij}), \qquad \lambda_i \geq 0;$$

so that

$$\begin{cases} (g^{ij}(\gamma(t))) = ((\lambda_i + \rho^8)\delta_{ij}), \\ (g_{ij}(\gamma(t))) = ((\lambda_i + \rho^8)^{-1}\delta_{ij}). \end{cases} \tag{26}$$

Then, since the tangent vector $\dot{\gamma}(t)$ is subunit for $A^0 - \rho^8 \Delta$, it follows that

$$\left(\frac{d\gamma^j}{dt}(t) \right)^2 \leq \lambda_j + \rho^8, \qquad 1 \leq j \leq N. \tag{27}$$

Hence we have, by (26) and (27),

$$\left|\sum_{n=1}^{N} \frac{\partial a^{in}}{\partial x_k} g_{nj} \frac{d\gamma^j}{dt}\frac{d\gamma^k}{dt}\right| \leq \left|\frac{\partial a^{ij}}{\partial x_k}\right| \frac{(\lambda_j + \rho^8)^{1/2}(\lambda_k + \rho^8)^{1/2}}{\lambda_j + \rho^8}$$

$$\leq K\rho^{-4}, \tag{28}$$

where K is a positive constant depending only on the bounds on the first derivatives of a^{ij}.

Similarly, we have

$$\begin{cases} \left|\sum_{n=1}^{N} \frac{\partial a^{in}}{\partial x_j} g_{nk} \frac{d\gamma^j}{dt}\frac{d\gamma^k}{dt}\right| \leq K\rho^{-4}, \\ \left|\sum_{\ell, m, n=1}^{N} (a^{i\ell} + \rho^8 \delta_{i\ell}) g_{mj} g_{nk} \frac{\partial a^{mn}}{\partial x_\ell} \frac{d\gamma^j}{dt}\frac{d\gamma^k}{dt}\right| \leq K\rho^{-8}. \end{cases} \tag{29}$$

Therefore, in view of estimates (28) and (29), it follows from equations (23) and (25) that

$$\left|\frac{d^2\gamma^i}{dt^2}\right| = \left| - \sum_{j,k=1}^{N} \Gamma^i_{jk} \frac{d\gamma^j}{dt}\frac{d\gamma^k}{dt}\right|$$

$$\leq C'''\rho^{-8},$$

so that

$$|\ddot{\gamma}(t)| \leq C'''\rho^{-8}. \tag{30}$$

Here C''' is a positive constant depending only on the bounds on the a^{ij} and their first derivatives.

3) Now we pick the constants C_+ and τ_k so that

$$|\gamma_\#(t) - \gamma(t)| \leq C_+ \tau_{k+1} \rho^4, \qquad \tau_k \leq t \leq \tau_{k+1}.$$

First, it follows from conditions (iii) and (i) of (18) that

$$|\dot{\gamma}^i_\#(\tau_k) - \dot{\gamma}^1(\tau_k)| = \left|\sum_{j=1}^{N} a^{ij}(\gamma_\#(\tau_k))\xi_j(\tau_k) - \dot{\gamma}^i(\tau_k)\right|$$

$$= |\delta^i - \dot{\gamma}^i(\tau_k)|$$

$$\leq C\rho^4. \tag{31}$$

We assume that

$$\tau_{k+1} - \tau_k = \rho^{12}.$$

Then, by the mean value theorem, it follows from estimates (31), (22) and (30) that

$$\begin{aligned} |\dot{\gamma}_{\#}(t) - \dot{\gamma}(t)| &\leq C\rho^4 + (C'' + C''')\rho^{-8}(t - \tau_k) \\ &\leq (C + C'' + C''')\rho^4, \qquad \tau_k \leq t \leq \tau_{k+1}, \end{aligned}$$

so that, by induction hypothesis (16), we have

$$\begin{aligned} |\gamma_{\#}(t) - \gamma(t)| &\leq C_+ \tau_k \rho^4 + (C + C'' + C''')\rho^4(t - \tau_k) \\ &\leq C_+ \tau_{k+1}\rho^4, \qquad \tau_k \leq t \leq \tau_{k+1}. \end{aligned}$$

provided we pick

$$C_+ \geq C + C'' + C'''. \tag{32}$$

Our construction of the path $\gamma_{\#}(t)$ is now complete with

$$\tau_k = k\rho^{12}. \tag{33}$$

Summing up, we have constructed a broken path $\gamma_{\#}: [0, \rho] \to D$ such that:

$$\begin{cases} \text{(i) } \gamma_{\#}(0) = \gamma(0) = x; \\ \text{(ii) } |\gamma_{\#}(t) - \gamma(t)| \leq C_+(k+1)\rho^{16} \textit{ for } k\rho^{12} \leq t \leq (k+1)\rho^{12}; \\ \text{(iii) } \textit{The tangent vector } c\dot{\gamma}_{\#}(t) \textit{ is subunit for } A^0 \textit{ throughout} \\ \qquad k\rho^{12} \leq t \leq (k+1)\rho^{12}. \quad \blacktriangledown \end{cases} \tag{34}$$

III) *End of the proof of Proposition* 7.2.20 Given any point $y \in B_{A^0}(x, \rho)$, let $\gamma: [0, \rho] \to D$ be a geodesic in the metric $ds^2 = \sum_{i,j=1}^N g_{ij}(x)\, dx_i\, dx_j$, where $(g_{ij}) = (a^{ij} + \rho^8\delta_{ij})^{-1}$, such that γ joins $x = \gamma(0)$ to $y = \gamma(\rho)$.

Let $\gamma_{\#}: [0, \rho] \to D$ be a broken path satisfying conditions (34). Then it follows from condition (ii) of (34) that

$$|\gamma_{\#}(\rho) - y| = |\gamma_{\#}(\rho) - \gamma(\rho)| \leq C_+(\rho^5 + \rho^{16}) \leq \rho^4, \tag{35}$$

since $k\rho^{12} \leq t = \rho$ and ρ is sufficiently small. Thus we have

$$y \in B_E(y_{\#}^1, \rho^4), \qquad y_{\#}^1 = \gamma_{\#}(\rho).$$

Applying Corollary 7.2.22 with $x = y_{\#}^1$ and $\rho = \rho^2$, we obtain that

$$B_E(y_{\#}^1, \rho^4) \subset B_{A^0 - (\rho^2)^8\Delta}(y_{\#}^1, \rho^2). \tag{36}$$

Now we define a path

$$\tilde{\gamma}_{\#}(t) = \gamma_{\#}(ct), \qquad 0 \le t \le \frac{\rho}{c}, \tag{37}$$

where c is the same constant as in condition (iii) of (34). Then we have the following:

1. $\tilde{\gamma}_{\#}(0) = \gamma_{\#}(0) = x.$
2. $\tilde{\gamma}_{\#}\left(\frac{\rho}{c}\right) = \gamma_{\#}(\rho) = y_{\#}^{1}.$
3. The tangent vector $\dot{\tilde{\gamma}}_{\#} = c\dot{\gamma}_{\#}$ is subunit for A^0.

This proves that there is a broken path $\gamma_{\#}^{0} = \tilde{\gamma}_{\#}: [0, \rho/c] \to D$ with tangent vectors subunit for A^0 such that $\gamma_{\#}^{0}$ joins $x = \gamma_{\#}^{0}(0)$ to $y_{\#}^{1} = \gamma_{\#}^{0}(\rho/c)$. Also it follows from estimate (35) that $|y - y_{\#}^{1}| \le \rho^4$.

By virtue of (36), we can repeat the above process, replacing x and ρ by $y_{\#}^{1}$ and ρ^2, respectively, to obtain that there is a broken path $\gamma_{\#}^{1}: [0, \rho^2/c] \to D$ with tangent vectors subunit for A^0 such that $\gamma_{\#}^{1}$ joins $y_{\#}^{1} = \gamma_{\#}^{1}(0)$ to $y_{\#}^{2} = y_{\#}^{1}(\rho^2/c)$, and $|y - y_{\#}^{2}| \le (\rho^2)^4$.

Repeating the process yields a sequence of paths $\gamma_{\#}^{\mu}: [0, t_{\mu+1}] \to D$ ($\mu = 0, 1, 2, \ldots$) with tangent vectors subunit for A^0 such that $\gamma_{\#}^{0}$ joins $x = \gamma_{\#}^{0}(0)$ to $y_{\#}^{1} = \gamma_{\#}^{0}(t_1)$, $\gamma_{\#}^{\mu}$ joins $y_{\#}^{\mu}$ to $y_{\#}^{\mu+1}$, $y_{\#}^{\mu} \to y$ as $\mu \to \infty$, and $\sum_{\mu=0}^{\infty} t_{\mu+1} \le \tilde{C}\rho$ for some constant $\tilde{C} > 0$ (cf. Figure 7-12).

Combining a sequence of paths $\gamma_{\#}^{\mu}$ into a single Lipschitz path $\tilde{\gamma}: [0, \tilde{C}\rho] \to D$, we see by formulas (19) and (37) that the path $\tilde{\gamma}$ is of the form (3) and joins x to y, as desired.

Now the proof of Proposition 7.2.20 and hence that of Theorem 7.2.2 is complete. ■

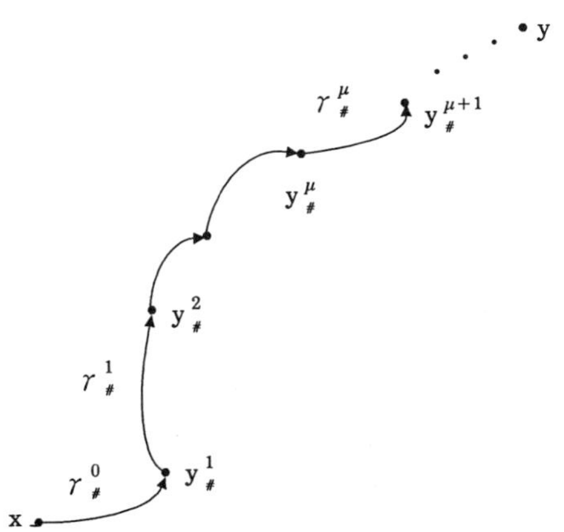

Figure 7-12

Proof of Theorem 7.2.4

1) We let

$$\begin{cases} Y_k = \sum_{i=1}^{N} b^{ik} \dfrac{\partial}{\partial x_i}, & 1 \leq k \leq r; \\ Y_0 = \sum_{i=1}^{N} c^i \dfrac{\partial}{\partial x_i}. \end{cases}$$

Then we have

$$\begin{aligned} A &= \sum_{k=1}^{r} Y_k^2 + Y_0 \\ &= \sum_{i,j=1}^{N} a^{ij} \frac{\partial^2}{\partial x_i \, \partial x_j} + \sum_{i=1}^{N} b^i \frac{\partial}{\partial x_i}, \end{aligned}$$

where

$$\begin{cases} a^{ij} = \sum_{k=1}^{r} b^{ik} b^{jk}, \\ b^i = c^i + \sum_{k=1}^{r} \sum_{j=1}^{N} b^{jk} \dfrac{\partial b^{ik}}{\partial x_j}. \end{cases} \tag{1}$$

Thus it follows that the vector fields Y_k are subunit for the operator $A^0 = \sum_{i,j=1}^{N} a^{ij}(\partial^2/\partial x_i \, \partial x_j)$; thus Hill's diffusion trajectories (integral curves of Y_k) are subunit trajectories.

Moreover it follows from formula (1) that the vector field Y_0 is expressed as

$$\begin{aligned} Y_0 &= \sum_{i=1}^{N} c^i \frac{\partial}{\partial x_i}. \\ &= \sum_{i=1}^{N} \left(b^i - \sum_{k=1}^{r} \sum_{j=1}^{N} b^{jk} \frac{\partial b^{ik}}{\partial x_j} \right) \frac{\partial}{\partial x_i} \\ &= \sum_{i=1}^{N} \left(b^i - \sum_{j=1}^{N} \frac{\partial a^{ij}}{\partial x_j} \right) \frac{\partial}{\partial x_i} + \sum_{k=1}^{r} \left(\sum_{j=1}^{N} \frac{\partial b^{jk}}{\partial x_j} \right) \left(\sum_{i=1}^{N} b^{ik} \frac{\partial}{\partial x_i} \right) \\ &= X_0 + \sum_{k=1}^{r} \left(\sum_{j=1}^{N} \frac{\partial b^{jk}}{\partial x_j} \right) Y_k. \end{aligned} \tag{2}$$

Therefore, in view of Theorem 7.2.13, we obtain that Hill's drift trajectories (integral curves of Y_0) can be approximated uniformly by piecewise differentiable curves, of which each differentiable arc is a subunit or drift trajectory.

Summing up, we have proved that any finite number of Hill's diffusion and drift trajectories can be approximated uniformly by a finite number of subunit and drift trajectories.

2) To prove the converse, we remark that, by Proposition 7.2.20, the subunit trajectories can be replaced by integral curves of the vector fields $X_i = \sum_{j=1}^N a^{ij}(\partial/\partial x_j)$. Also it follows from formula (1) that the vector fields X_i are expressed as

$$\begin{aligned} X_i &= \sum_{j=1}^{N} \left(\sum_{k=1}^{r} b^{ik} b^{jk} \right) \frac{\partial}{\partial x_j} \\ &= \sum_{k=1}^{r} b^{ik} \left(\sum_{j=1}^{N} b^{jk} \frac{\partial}{\partial x_j} \right) \\ &= \sum_{k=1}^{r} b^{ik} Y_k. \end{aligned}$$

Hence, using Theorem 7.2.13, we find that the subunit trajectories can be approximated uniformly by piecewise differentiable curves, of which each differentiable arc is Hill's diffusion trajectory. In fact, it suffices to note that if $\dot{\beta}(t) = Y_k(\beta(t))$ on $[t_1, t_2]$ and if $\dot{\beta}(t_0) = 0$ for some $t_0 \in [t_1, t_2]$, then, by the uniqueness property, we have $\beta(t) \equiv \beta(t_0)$ on this interval; so that the trace of integral curves of Y_k is unchanged when this arc is dropped. This implies that the condition $\dot{\beta}(t) \neq 0$ on $[t_1, t_2]$ may be assumed.

Moreover, by virtue of Theorem 7.2.13, it follows from formula (2) that the drift trajectories (integral curves of X_0) can be approximated uniformly by piecewise differentiable curves, of which each differentiable arc is Hill's diffusion or drift trajectory.

Therefore we conclude that any finite number of subunit and drift trajectories can be approximated uniformly by a finite number of Hill's diffusion and drift trajectories.

Theorem 7.2.4 is proved. ■

Notes

Section 7.1: The maximum principles in this section are adapted from Oleĭnik-Radkevič [1] and Gilbarg-Trudinger [1]. The boundary point lemma, Lemma 7.1.7, was proved independently by E. Hopf [2] and Oleĭnik

[1]. For a general study of maximum principles, the reader might refer to Protter-Weinberger [1].

Section 7.2: Theorem 7.2.1 is inspired by the work of Feffermann-Phong [1]. Our proof of Theorem 7.2.1 follows Bony [1] and Amano [1]; see also Redheffer [1]. As mentioned in the text, the virtue of this theorem is that the notion of a subunit trajectory is coordinate-free.

It seems quite likely that there is an intimate connection between propagation of maximums and propagation of singularities for degenerate elliptic differential operators of second order. See Taira [8].

8 Elliptic Boundary Value Problems

This chapter is devoted to general boundary value problems for second-order elliptic differential operators. We begin in Section 8.1 with a summary of the basic facts about existence, uniqueness and regularity of solutions of the Dirichlet problem in the framework of Hölder spaces. In Section 8.2, using the calculus of pseudo-differential operators, we prove existence, uniqueness and regularity theorems for the Dirichlet problem in the framework of Sobolev spaces. In Section 8.3 we formulate general boundary value problems, and show that these problems can be reduced to the study of pseudo-differential operators on the boundary. The virtue of this reduction is that there is no difficulty in taking adjoints after restricting the attention to the boundary, whereas boundary value problems in general do not have adjoints. This allows us to discuss the existence theory more easily. In Section 8.4 we study the basic questions of existence and uniqueness of solutions of general boundary value problems with spectral parameter, which will play a fundamental role in constructing Markov processes in Chapter 10.

8.1. The Dirichlet Problem —(1)—

In this section we shall state the classical existence, uniqueness and regularity theorems for the Dirichlet problem in the framework of *Hölder spaces.*

Let Ω be a bounded domain in $\mathbf{R}^n$ with boundary $\partial\Omega$. We let

$$A = \sum_{i,j=1}^{n} a^{ij}(x) \frac{\partial^2}{\partial x_i \, \partial x_j} + \sum_{i=1}^{n} b^i(x) \frac{\partial}{\partial x_i} + c(x)$$

be a second-order *elliptic* differential operator with real coefficients such that:

1. $a^{ij} \in C^\theta(\Omega)$ with $0 < \theta < 1$, $a^{ij} = a^{ji}$ and there exists a constant $a_0 > 0$ such that

$$\sum_{i,j=1}^{n} a^{ij}(x)\xi_i\xi_j \geq a_0|\xi|^2, \qquad x \in \Omega,\ \xi \in \mathbf{R}^n.$$

2. $b^i \in C^\theta(\Omega)$.
3. $c \in C^\theta(\Omega)$ and $c \leq 0$ in Ω.

We are interested in the following Dirichlet problem: given functions f and φ defined in Ω and on $\partial\Omega$, respectively, find a function u in Ω such that

$$\begin{cases} Au = f & \text{in } \Omega, \\ u|_{\partial\Omega} = \varphi & \text{on } \partial\Omega. \end{cases} \tag{D}$$

The next theorem summarizes the basic facts about the Dirichlet problem in the framework of Hölder spaces.

8.1.1 Theorem. (i) (Existence and Uniqueness) *Suppose that the domain Ω is of class C^2. If $f \in C^\theta(\Omega)$ and $\varphi \in C(\partial\Omega)$, then problem (D) has a unique solution u in $C(\bar{\Omega}) \cap C^{2+\theta}(\Omega)$.*

(ii) (Interior Regularity) *Suppose that the functions a^{ij}, b^i and c belong to $C^{k+\theta}(\Omega)$ for some non-negative integer k. If $u \in C^2(\Omega)$ and $Au = f \in C^{k+\theta}(\Omega)$, then we have $u \in C^{k+2+\theta}(\Omega)$.*

(iii) (Global Regularity) *Suppose that the domain Ω is of class $C^{k+2+\theta}$ and that the functions a^{ij}, b^i and c belong to $C^{k+\theta}(\bar{\Omega})$ for some non-negative integer k. If $f \in C^{k+\theta}(\bar{\Omega})$ and $\varphi \in C^{k+2+\theta}(\partial\Omega)$, then a solution $u \in C(\bar{\Omega}) \cap C^2(\Omega)$ of problem (D) belongs to $C^{k+2+\theta}(\bar{\Omega})$.*

8.2. The Dirichlet Problem —(2)—

In this section, by using the theory of pseudo-differential operators, we shall consider the Dirichlet problem in the framework of *Sobolev spaces*. This is a generalization of the classical potential approach to the Dirichlet problem.

Let Ω be a bounded domain in $\mathbf{R}^n$ with C^∞ boundary $\partial\Omega$. Its closure $\bar{\Omega} = \Omega \cup \partial\Omega$ is an n-dimensional, compact C^∞ manifold with boundary. By virtue of Theorems 2.13.2 and 2.13.3, we may suppose that (cf. Figure 5-1):

(a) The domain Ω is a relatively compact open subset of an n-dimensional, compact C^∞ manifold M without boundary.

(b) In a neighborhood W of $\partial\Omega$ in M, a normal coordinate t is chosen so that the points of W are represented as (x', t), $x' \in \partial\Omega$, $-1 < t < 1$; $t > 0$ in Ω, $t < 0$ in $M \setminus \bar{\Omega}$ and $t = 0$ only on $\partial\Omega$.

(c) The manifold M is equipped with a strictly positive density μ which, on W, is the product of a strictly positive density ω on $\partial\Omega$ and the Lebesgue measure dt on $(-1, 1)$.

We let

$$A = \sum_{i,j=1}^{n} a^{ij}(x) \frac{\partial^2}{\partial x_i \, \partial x_j} + \sum_{i=1}^{n} b^i(x) \frac{\partial}{\partial x_i} + c(x)$$

be a second-order *elliptic* differential operator with real coefficients such that:

1. $a^{ij} \in C^\infty(M)$, $a^{ij} = a^{ji}$ and there exists a constant $a_0 > 0$ such that

$$\sum_{i,j=1}^{n} a^{ij}(x)\xi_i\xi_j \geq a_0|\xi|^2 \qquad \text{on } T^*(M).$$

   Here $T^*(M)$ is the cotangent bundle of M.

2. $b^i \in C^\infty(M)$.
3. $c \in C^\infty(M)$ and $c \leq 0$ in M.

Further, for simplicity, suppose that:

$$\text{The function } c \text{ does not vanish identically on } M. \tag{1}$$

First we construct a volume potential for A, which plays the same role for A as the Newtonian potential plays for the Laplacian.

8.2.1 Theorem.

(i) *The operator* $A\colon C^\infty(M) \to C^\infty(M)$ *is bijective, and its inverse* Q *is an elliptic operator in* $L_{\mathrm{cl}}^{-2}(M)$.

(ii) *The operators* A *and* Q *extend respectively to isomorphisms*

$$A\colon H^s(M) \to H^{s-2}(M),$$

$$Q\colon H^{s-2}(M) \to H^s(M),$$

for each $s \in \mathbf{R}$, *which are still inverses of each other.*

Proof. We apply Theorem 6.7.14 to the operator A.

Since A is elliptic on M, applying Theorem 7.2.1′ to our situation, we obtain that

$$N(A) = \{u \in C^\infty(M);\ Au = 0 \text{ in } M\}$$
$$= \{\text{constant functions}\}.$$

In view of hypothesis (1), this implies that

$$N(A) = \{0\}.$$

On the other hand, since the principal symbol $-\sum_{i,j} a^{ij}(x)\xi_i\xi_j$ of A is real, it follows from Corollary 6.7.12 that

$$\text{ind } A = 0.$$

Therefore, Theorem 8.2.1 follows from an application of Theorem 6.7.14. ■

Next we construct a surface potential for A, which is a generalization of the classical Poisson kernel for the Laplacian.

We let

$$Kv = \gamma_0(Q(v \otimes \delta)), \qquad v \in C^\infty(\partial\Omega),$$

where $v \otimes \delta$ is a distribution on M defined by

$$\langle v \otimes \delta, \varphi \cdot \mu \rangle = \langle v, \varphi(\cdot, 0) \cdot \omega \rangle, \qquad \varphi \in C^\infty(M).$$

In view of part (ii) of Theorem 6.8.1, it follows that the operator K is in $L_{\text{cl}}^{-1}(\partial\Omega)$ and maps $C^\infty(\partial\Omega)$ continuously into itself.

Further we have:

8.2.2 Theorem.

(i) *The operator K is an elliptic operator in $L_{\text{cl}}^{-1}(\partial\Omega)$.*

(ii) *The operator $K: C^\infty(\partial\Omega) \to C^\infty(\partial\Omega)$ is bijective, and its inverse L is an elliptic operator in $L_{\text{cl}}^{1}(\partial\Omega)$.*

Furthermore, the operators K and L extend respectively to isomorphisms

$$K: H^s(\partial\Omega) \to H^{s+1}(\partial\Omega),$$

$$L: H^{s+1}(\partial\Omega) \to H^s(\partial\Omega),$$

for each $s \in \mathbf{R}$, which are still inverses of each other.

Proof. (i) We calculate the homogeneous principal symbol of $K \in L^{-1}_{\mathrm{cl}}(\partial\Omega)$. In a neighborhood W of $\partial\Omega$ in M, we can write the operator $A = A(x, D)$ uniquely in the form

$$A(x, D) = A_2(x)\, D_t^2 + A_1(x, D_{x'})\, D_t + A_0(x, D_{x'}), \qquad x = (x', t), \tag{2}$$

where $A_j(x, D_{x'})$ $(j = 0, 1, 2)$ is a differential operator of order $2 - j$ acting along the surfaces parallel to $\partial\Omega$. We denote by $a_1(x, \xi')$ and $a_0(x, \xi')$ the principal symbols of $A_1(x, D_{x'})$ and $A_0(x, D_{x'})$, respectively. Since A is elliptic on M, it follows that:

1. $A_2(x) < 0,\ x \in W$;
2. $a_1(x, \xi')^2 - 4A_2(x)a_0(x, \xi') < 0,\ x = (x', t) \in W,\ \xi' \in T^*_{x'}(\partial\Omega) \setminus \{0\}$.

Hence the principal symbol of A can be decomposed as follows:

$$A_2(x)\xi_n^2 + a_1(x, \xi')\xi_n + a_0(x, \xi') = A_2(x)(\xi_n - \xi_n^+(x, \xi'))(\xi_n - \xi_n^-(x, \xi'))$$

where:

$$\xi_n^{\pm}(x, \xi') = -\frac{a_1(x, \xi') \pm \sqrt{-1}\,(4A_2(x)a_0(x, \xi') - a_1(x, \xi')^2)^{1/2}}{2A_2(x)}. \tag{3}$$

Since the principal symbol of Q $(= A^{-1})$ is

$$\frac{1}{A_2(x)(\xi_n - \xi_n^+(x, \xi'))(\xi_n - \xi_n^-(x, \xi'))},$$

applying formula (6.8.2) to our situation, we obtain that the homogeneous principal symbol $k(x', \xi')$ of K is given by the following:

$$\begin{aligned} k(x', \xi') &= \frac{1}{2\pi}\int_\Gamma \frac{d\xi_n}{A_2(x', 0)(\xi_n - \xi_n^+(x', 0, \xi'))(\xi_n - \xi_n^-(x', 0, \xi'))} \\ &= -\frac{1}{(4A_2(x', 0)a_0(x', 0, \xi') - a_1(x', 0, \xi')^2)^{1/2}}. \end{aligned}$$

This proves that $K \in L^{-1}_{\mathrm{cl}}(\partial\Omega)$ is elliptic.

(ii) We apply Theorem 6.7.14 to the operator K.

First, since the homogeneous principal symbol $k(x', \xi')$ of K is real, we obtain from Corollary 6.7.12 that

$$\operatorname{ind} K = 0.$$

Now we show that

$$N(K) = \{v \in C^\infty(\partial\Omega);\ Kv = 0\} = \{0\};$$

then part (ii) of Theorem 8.2.2 follows from an application of Theorem 6.7.14.

Suppose that $v \in C^\infty(\partial\Omega)$ and $Kv = 0$. Then, applying part (i) of Theorem 6.8.1 to our situation, we obtain that

$$\begin{cases} Q(v \otimes \delta)|_\Omega \in C^\infty(\bar{\Omega}), & (4) \\ Q(v \otimes \delta)|_{M \setminus \bar{\Omega}} \in C^\infty(M \setminus \Omega), & (5) \end{cases}$$

and also

$$Q(v \otimes \delta)|_{\partial\Omega} = Kv = 0. \tag{6}$$

But we have

$$A(Q(v \otimes \delta)|_\Omega) = AQ(v \otimes \delta)|_\Omega = v \otimes \delta|_\Omega = 0 \qquad \text{in } \Omega, \tag{7}$$

since A is a differential (hence local) operator. Therefore, in view of (4), (7) and (6), we can apply the maximum principle (Corollary 7.1.5) to obtain that

$$Q(v \otimes \delta) = 0 \qquad \text{on } \bar{\Omega}. \tag{8}$$

This gives that

$$Q(v \otimes \delta) = (Q(v \otimes \delta)|_{M \setminus \bar{\Omega}})^0.$$

Thus it follows from an application of the jump formula (5.6.3) that

$$\begin{aligned} v \otimes \delta &= AQ(v \otimes \delta) \\ &= A(Q(v \otimes \delta)|_{M \setminus \bar{\Omega}})^0 \\ &= (AQ(v \otimes \delta)|_{M \setminus \bar{\Omega}})^0 + \frac{1}{i}\{A_2(x)(D_t Q(v \otimes \delta)|_{\partial'\Omega}) \otimes \delta \\ &\quad + A_2(x)(Q(v \otimes \delta)|_{\partial'\Omega}) \otimes D_t\delta + A_1(x, D_{x'})(Q(v \otimes \delta)|_{\partial'\Omega}) \otimes \delta\} \\ &= \frac{1}{i}\{A_2(x)(D_t Q(v \otimes \delta)|_{\partial'\Omega}) + A_1(x, D_{x'})(Q(v \otimes \delta)|_{\partial'\Omega})\} \otimes \delta \\ &\quad + \frac{1}{i} A_2(x)(Q(v \otimes \delta)|_{\partial'\Omega}) \otimes D_t\delta \qquad (i = \sqrt{-1}), \end{aligned} \tag{9}$$

where

$$u|_{\partial'\Omega} = \text{the trace of } u \text{ on } \partial\Omega \text{ from } M \setminus \bar{\Omega}.$$

In order that formula (9) hold, the last term on the right-hand side must vanish; hence we have

$$Q(v \otimes \delta)|_{\partial'\Omega} = 0, \tag{10}$$

since $A_2(x) < 0$ in W. But we have

$$A(Q(v \otimes \delta)|_{M\setminus\bar{\Omega}}) = AQ(v \otimes \delta)|_{M\setminus\bar{\Omega}} = v \otimes \delta|_{M\setminus\bar{\Omega}} = 0. \tag{11}$$

Therefore, in view of (5), (11) and (10), we can apply the maximum principle to obtain that

$$Q(v \otimes \delta) = 0 \qquad \text{on } M \setminus \Omega. \tag{12}$$

Consequently it follows from (8) and (12) that

$$Q(v \otimes \delta) = 0 \qquad \text{on } M.$$

Since the operator Q is invertible, this implies that

$$v \otimes \delta = 0 \qquad \text{on } M,$$

so that

$$v = 0 \qquad \text{on } \partial\Omega.$$

The proof of Theorem 8.2.2 is complete. ■

The next uniqueness theorem for the Dirichlet problem will play a fundamental role in the sequel.

8.2.3 Theorem. *If $u \in H^s(\bar{\Omega})$ $(s \in \mathbf{R})$ satisfies*

$$\begin{cases} Au = 0 & \text{in } \Omega, \\ \gamma_0 u = 0 & \text{on } \partial\Omega, \end{cases} \tag{13}$$

then $u = 0$ in Ω.

Proof. Since $u \in H^s(\bar{\Omega})$ and $Au = 0$ in Ω, applying Theorem 5.6.5, we find that the distribution u has sectional traces $\gamma_j u$ of any order $j = 0, 1, 2, \ldots$, and

$$\gamma_j u \in H^{s-j-1/2}(\partial\Omega).$$

In a neighborhood W of $\partial\Omega$ in M, we can write the operator $A = A(x, D)$ uniquely in the form (2):

$$A(x, D) = A_2(x)\, D_t^2 + A_1(x, D_{x'})\, D_t + A_0(x, D_{x'}).$$

Then it follows from an application of the jump formula (5.6.3) that

$$A(u^0) = (Au)^0 + \frac{1}{i} A_2(\gamma_1 u \otimes \delta) + \frac{1}{i} A_2(\gamma_0 u \otimes D_t\delta) + \frac{1}{i} A_1\gamma_0 u \otimes \delta$$

$$= \frac{1}{i} A_2(\gamma_1 u \otimes \delta),$$

since $Au = 0$ in Ω and $\gamma_0 u = 0$. By Theorem 8.2.1, this gives that

$$u^0 = \frac{1}{i} Q(A_2(\gamma_1 u \otimes \delta)),$$

so that

$$u = \frac{1}{i} Q(A_2(\gamma_1 u \otimes \delta))|_{\Omega}. \tag{14}$$

In other words, every solution $u \in H^s(\bar{\Omega})$ of problem (13) can be expressed in the form (14). Thus we have

$$0 = \gamma_0 u = \frac{1}{i} K((A_2|_{\partial\Omega}) \cdot \gamma_1 u),$$

and hence

$$\gamma_1 u = 0,$$

since the operator K is invertible and $A_2 < 0$ on $\partial\Omega$. Therefore it follows from formula (14) that

$$u = 0 \qquad \text{in } \Omega.$$

This completes the proof. ■

We let

$$P\varphi = Q(L\varphi \otimes \delta)|_{\Omega}, \qquad \varphi \in C^\infty(\partial\Omega). \tag{15}$$

In view of Theorems 8.2.2 and 6.8.1, it follows that P maps $C^\infty(\partial\Omega)$ continuously into $C^\infty(\bar{\Omega})$, and it extends to a continuous linear operator

$$P\colon H^{s-1/2}(\partial\Omega) \to H^s(\bar{\Omega}) \tag{16}$$

for all $s \in \mathbf{R}$. Further we have, for all $\varphi \in H^{s-1/2}(\partial\Omega)$,

$$\begin{cases} AP\varphi = AQ(L\varphi \otimes \delta)|_{\Omega} = (L\varphi \otimes \delta)|_{\Omega} = 0 & \text{in } \Omega, \\ \gamma_0 P\varphi = KL\varphi = \varphi & \text{on } \partial\Omega. \end{cases} \tag{17}$$

The operator P is called the *Poisson operator*.

We let

$$N(A, s) = \{u \in H^s(\bar{\Omega});\ Au = 0 \text{ in } \Omega\}.$$

Since the injection $H^s(\bar{\Omega}) \to \mathscr{D}'(\Omega)$ is continuous, it follows that $N(A, s)$ is a closed subspace of $H^s(\bar{\Omega})$; hence it is a Hilbert space.

Then we have:

8.2.4 Theorem. *The Poisson operator P maps $H^{s-1/2}(\partial\Omega)$ isomorphically onto $N(A, s)$ for all $s \in \mathbf{R}$. Its inverse is the trace operator γ_0.*

Proof. In view of (16) and (17), it suffices to prove the surjectivity of P.

Let w be an arbitrary element of $N(A, s)$, and let

$$u = P\gamma_0 w.$$

Then Theorem 5.6.5 tells us that $\gamma_0 w \in H^{s-1/2}(\partial\Omega)$; hence by (16) we have $u \in H^s(\bar{\Omega})$. Further, in view of (17), it follows that

$$\begin{cases} A(w - u) = 0 & \text{in } \Omega, \\ \gamma_0(w - u) = 0 & \text{on } \partial\Omega. \end{cases}$$

Therefore, applying Theorem 8.2.3, we obtain that

$$w = u = P\gamma_0 w.$$

This proves the surjectivity of P, and also $P^{-1} = \gamma_0$. ∎

Combining Theorem 8.2.1 and Theorem 8.2.4, we can obtain:

8.2.5 Theorem. *Let $s \geq 2$. The Dirichlet problem*

$$\begin{cases} Au = f & \text{in } \Omega, \\ \gamma_0 u = \varphi & \text{on } \partial\Omega, \end{cases} \tag{D}$$

has a unique solution u in $H^s(\bar{\Omega})$ for any $f \in H^{s-2}(\bar{\Omega})$ and $\varphi \in H^{s-1/2}(\partial\Omega)$.

Proof. It suffices to note that the unique solution u of problem (D) is given by the following:

$$u = QEf + P(\varphi - \gamma_0(QEf)).$$

Here $E: H^{s-2}(\bar{\Omega}) \to H^{s-2}(M)$ is the Seeley extension operator. ∎

8.3. General Boundary Value Problems

In this section, by using the Dirichlet problem, we shall consider general boundary value problems for elliptic differential operators in the framework of *Sobolev spaces.*

Formulation of Boundary Value Problems

Let Ω be a bounded domain in $\mathbf{R}^n$ with C^∞ boundary $\partial\Omega$, and let

$$A = \sum_{i,j=1}^{n} a^{ij}(x) \frac{\partial^2}{\partial x_i \, \partial x_j} + \sum_{i=1}^{n} b^i(x) \frac{\partial}{\partial x_i} + c(x)$$

be a second-order *elliptic* differential operator with real coefficients as in Section 8.2:

1. $a^{ij} \in C^\infty(M)$, $a^{ij} = a^{ji}$ and there exists a constant $a_0 > 0$ such that

$$\sum_{i,j=1}^{n} a^{ij}(x)\xi_i\xi_j \geq a_0|\xi|^2 \qquad \text{on } T^*(M).$$

2. $b^i \in C^\infty(M)$.
3. $c \in C^\infty(M)$ and $c \leq 0$ on M.

Further we suppose that condition (8.2.1) is satisfied:

The function c does not vanish identically on M.

If $\sigma \leq \tau + 2$, we let

$$H_A^{\sigma,\tau} = \{u \in H^\sigma(\bar{\Omega});\, Au \in H^\tau(\bar{\Omega})\}.$$

We equip the space $H_A^{\sigma,\tau}$ with the inner product

$$(u, v)_{H_A^{\sigma,\tau}} = (u, v)_{H^\sigma(\bar{\Omega})} + (Au, Av)_{H^\tau(\bar{\Omega})},$$

and with the associated norm

$$\|u\|_{H_A^{\sigma,\tau}} = (\|u\|^2_{H^\sigma(\bar{\Omega})} + \|Au\|^2_{H^\tau(\bar{\Omega})})^{1/2}.$$

Then it is easy to see that $H_A^{\sigma,\tau}$ is a Hilbert space.

Every element $u \in H_A^{\sigma,\tau}$ can be decomposed as follows:

$$u = v + w \tag{1}$$

where

$$\begin{cases} v = QE(Au)|_{\Omega} \in H^{\tau+2}(\bar{\Omega}), \\ w = u - v \in N(A, \sigma). \end{cases}$$

Since the operators E: $H^s(\bar{\Omega}) \to H^s(M)$ and Q: $H^s(M) \to H^{s+2}(M)$ are continuous, it follows that the decomposition (1) is continuous; more precisely we have

$$\begin{cases} \|v\|_{H^{\tau+2}(\bar{\Omega})} \leq C\|Au\|_{H^{\tau}(\bar{\Omega})}; & (2) \\ \|w\|_{H^{\sigma}(\bar{\Omega})} \leq C\|u\|_{H_A^{\sigma,\tau}}. & (3) \end{cases}$$

Here and in the following the letter C denotes a generic positive constant.

Now we take

$$\tau \geq 0.$$

Then it follows from Theorem 5.6.3 that the trace maps γ_i: $H^{\tau+2}(\bar{\Omega}) \to H^{\tau-i+3/2}(\partial\Omega)$, $i = 0, 1$, are continuous:

$$|\gamma_i v|_{H^{\tau-i+3/2}(\partial\Omega)} \leq C\|v\|_{H^{\tau+2}(\bar{\Omega})}, \qquad v \in H^{\tau+2}(\bar{\Omega}). \tag{4}$$

On the other hand, applying Theorem 5.6.5, we obtain that the trace maps γ_j: $N(A, \sigma) \to H^{\sigma-j-1/2}(\partial\Omega)$, $j = 0, 1, 2, \ldots$, are continuous for all $\sigma \in \mathbf{R}$ (cf. inequality (5.6.1)):

$$|\gamma_i w|_{H^{\sigma-j-1/2}(\partial\Omega)} \leq C\|w\|_{H^{\sigma}(\bar{\Omega})}, \qquad w \in N(A, \sigma). \tag{5}$$

Therefore, if $u \in H_A^{\sigma,\tau}$, we can define its traces $\gamma_i u$, $i = 0, 1$, by the formulas

$$\gamma_i u = \gamma_i v + \gamma_i w, \qquad i = 0, 1, \tag{6}$$

and let

$$\gamma u = \{\gamma_0 u, \gamma_1 u\}.$$

Then we have:

8.3.1 Proposition. *If $\sigma \leq \tau + 2$ and $\tau \geq 0$, then the mapping*

$$\gamma: H_A^{\sigma,\tau} \to H^{\sigma-1/2}(\partial\Omega) \times H^{\sigma-3/2}(\partial\Omega)$$

is continuous.

Proof. It follows from inequalities (4) and (2) that

$$
\begin{aligned}
|\gamma_i v|_{H^{\sigma-i-1/2}(\partial\Omega)} &\leq |\gamma_i v|_{H^{\tau-i+3/2}(\partial\Omega)} \\
&\leq C\|v\|_{H^{\tau+2}(\Omega)} \\
&\leq C\|Au\|_{H^{\tau}(\Omega)}, \qquad (i = 0, 1).
\end{aligned} \tag{7}
$$

Further it follows from inequalities (5) and (3) that

$$
\begin{aligned}
|\gamma_i w|_{H^{\sigma-i-1/2}(\partial\Omega)} &\leq C\|w\|_{H^{\sigma}(\Omega)} \\
&\leq C\|u\|_{H_A^{\sigma,\tau}}, \qquad (i = 0, 1).
\end{aligned} \tag{8}
$$

In view of formulas (6), the continuity of γ follows from inequalities (7) and (8). ■

Let B_j ($j = 0, 1$) be a classical pseudo-differential operator of order m_j on $\partial\Omega$, and define

$$B\gamma u = B_0\gamma_0 u + B_1\gamma_1 u, \qquad u \in H_A^{\sigma,\tau}. \tag{9}$$

Then we have:

8.3.2 Proposition. *If $\sigma \leq \tau + 2$ and $\tau \geq 0$, then the mapping*

$$B\gamma\colon H_A^{\sigma,\tau} \to H^{\sigma-m-1/2}(\partial\Omega)$$

is continuous. Here $m = \max(m_0, m_1 + 1)$.

Proposition 8.3.2 follows immediately from Proposition 8.3.1, since the operators $B_0\colon H^{\sigma-1/2}(\partial\Omega) \to H^{\sigma-m_0-1/2}(\partial\Omega)$ and $B_1\colon H^{\sigma-3/2}(\partial\Omega) \to H^{\sigma-m_1-3/2}(\partial\Omega)$ are continuous.

Now we can formulate our boundary value problem for (A, B) as follows: given functions $f \in H^{\tau}(\bar{\Omega})$ and $\varphi \in H^{\tau-m+3/2}(\partial\Omega)$ ($\tau \geq 0$), find a function $u \in H^{\sigma}(\bar{\Omega})$ ($\sigma \leq \tau \leq 2$) such that

$$
\begin{cases}
Au = f & \text{in } \Omega, \\
B\gamma u = \varphi & \text{on } \partial\Omega.
\end{cases} \tag{+}
$$

Problem (+) is said to be *elliptic* (or *coercive*) if $\sigma = \tau + 2$, while it is said to be *subelliptic* if $\tau + 1 < \sigma < \tau + 2$.

Reduction to the Boundary

In this subsection we shall show that problem (+) can be reduced to the study of a pseudo-differential operator on the boundary.

Suppose that $u \in H_A^{\sigma,\tau}$ ($\sigma \leq \tau + 2$, $\tau \geq 0$) is a solution of problem (+):

$$\begin{cases} Au = f & \text{in } \Omega, \\ B\gamma u = \varphi & \text{on } \partial\Omega. \end{cases} \tag{+}$$

Then, by virtue of the decomposition (1) of u, this is equivalent to saying that $w \in H^\sigma(\bar{\Omega})$ is a solution of the problem

$$\begin{cases} Aw = 0 & \text{in } \Omega, \\ B\gamma w = \varphi - B\gamma v & \text{on } \partial\Omega. \end{cases} \tag{+'}$$

Here $v = QEf|_\Omega \in H^{\tau+2}(\bar{\Omega})$ and $w = u - v$. But, Theorem 8.2.4 tells us that the spaces $N(A, \sigma)$ and $H^{\sigma-1/2}(\partial\Omega)$ are isomorphic in such a way that

$$N(A, \sigma) \underset{P}{\overset{\gamma_0}{\rightleftarrows}} H^{\sigma-1/2}(\partial\Omega).$$

Therefore we find that $w \in H^\sigma(\bar{\Omega})$ is a solution of problem (+′) if and only if $\psi \in H^{\sigma-1/2}(\partial\Omega)$ is a solution of the equation

$$B\gamma P\psi = \varphi - B\gamma v \qquad \text{on } \partial\Omega. \tag{++}$$

Here $\psi = \gamma_0 w$, or equivalently, $w = P\psi$.

Summing up, we have:

8.3.3 Proposition. *Let $\sigma \leq \tau + 2$ and $\tau \geq 0$. For functions $f \in H^\tau(\bar{\Omega})$ and $\varphi \in H^{\tau-m+3/2}(\partial\Omega)$, there exists a solution $u \in H_A^{\sigma,\tau}$ of problem (+) if and only if there exists a solution $\psi \in H^{\sigma-1/2}(\partial\Omega)$ of problem (++). Furthermore, the solutions u and ψ are related as follows:*

$$u = QEf|_\Omega + P\psi.$$

We remark that equation (++) is a generalization of the classical *Fredholm integral equation.*

We let

$$\begin{aligned} T\colon C^\infty(\partial\Omega) &\to C^\infty(\partial\Omega) \\ \varphi &\mapsto B\gamma P\varphi. \end{aligned} \tag{10}$$

Then we have, by formula (9),

$$T = B_0 + B_1\Pi \tag{11}$$

where

$$\Pi = \gamma_1 P. \tag{12}$$

But, applying part (ii) of Theorem 6.8.1, we find that Π is a classical pseudo-differential operator of first order on $\partial\Omega$. Hence the operator T is a classical pseudo-differential operator of order m on $\partial\Omega$.

Consequently, Proposition 8.3.3 asserts that problem $(+)$ can be reduced to the study of the pseudo-differential operator T on the boundary $\partial\Omega$. We shall formulate this fact more precisely in terms of functional analysis.

First we remark that the operator $T: C^\infty(\partial\Omega) \to C^\infty(\partial\Omega)$ extends to a continuous linear operator $T: H^s(\partial\Omega) \to H^{s-m}(\partial\Omega)$ for all $s \in \mathbf{R}$. Then we have, by definition (10),

$$T\varphi = B\gamma P\varphi \qquad \text{for } \varphi \in H^{\sigma-1/2}(\partial\Omega), \tag{10'}$$

since the operators $P: H^{\sigma-1/2}(\partial\Omega) \to N(A, \sigma)$ and $B\gamma: H_A^{\sigma,\tau} \to H^{\sigma-m-1/2}(\partial\Omega)$ are both continuous.

We associate with problem $(+)$ a linear operator

$$\mathfrak{A}: H^\sigma(\bar{\Omega}) \to H^\tau(\bar{\Omega}) \times H^{\tau-m+3/2}(\partial\Omega)$$

as follows:

(a) The domain $D(\mathfrak{A})$ of $\mathfrak{A}$ is the space

$$D(\mathfrak{A}) = \{u \in H_A^{\sigma,\tau};\ B\gamma u \in H^{\tau-m+3/2}(\partial\Omega)\},$$

(b) $\mathfrak{A}u = \{Au, B\gamma u\}$, $u \in D(\mathfrak{A})$.

Since the operators $A: H_A^{\sigma,\tau} \to H^{\sigma-2}(\bar{\Omega})$ and $B\gamma: H_A^{\sigma,\tau} \to H^{\sigma-m-1/2}(\partial\Omega)$ are both continuous, it follows that $\mathfrak{A}$ is a *closed* operator. Further the operator $\mathfrak{A}$ is densely defined, since the domain $D(\mathfrak{A})$ contains $C^\infty(\bar{\Omega})$ and so it is dense in $H^\sigma(\bar{\Omega})$.

Similarly, we associate with equation $(++)$ a linear operator

$$\mathscr{T}: H^{\sigma-1/2}(\partial\Omega) \to H^{\tau-m+3/2}(\partial\Omega)$$

as follows:

(α) The domain $D(\mathscr{T})$ of $\mathscr{T}$ is the space

$$D(\mathscr{T}) = \{\varphi \in H^{\sigma-1/2}(\partial\Omega);\ T\varphi \in H^{\tau-m+3/2}(\partial\Omega)\}.$$

(β) $\mathscr{T}\varphi = T\varphi$, $\varphi \in D(\mathscr{T})$.

Then the operator $\mathscr{T}$ is a densely defined, *closed* operator, since the operator $T: H^{\sigma-1/2}(\partial\Omega) \to H^{\sigma-m-1/2}(\partial\Omega)$ is continuous, and since the domain $D(\mathscr{T})$ contains $C^\infty(\partial\Omega)$.

In what follows, we shall prove:

(I) The null space $N(\mathfrak{A})$ of $\mathfrak{A}$ has finite dimension if and only if the null space $N(\mathscr{T})$ of $\mathscr{T}$ has finite dimension, and we have

$$\dim N(\mathfrak{A}) = \dim N(\mathscr{T}).$$

(II) The range $R(\mathfrak{A})$ of $\mathfrak{A}$ is closed if and only if the range $R(\mathscr{T})$ of $\mathscr{T}$ is closed; and $R(\mathfrak{A})$ has finite codimension if and only if $R(\mathscr{T})$ has finite codimension, and we have

$$\operatorname{codim} R(\mathfrak{A}) = \operatorname{codim} R(\mathscr{T}).$$

(III) The operator $\mathfrak{A}$ is a Fredholm operator if and only if the operator $\mathscr{T}$ is a Fredholm operator, and we have

$$\operatorname{ind} \mathfrak{A} = \operatorname{ind} \mathscr{T}.$$

First we prove:

8.3.4 Theorem (null spaces). *The null spaces $N(\mathfrak{A})$ and $N(\mathscr{T})$ are isomorphic; hence we have*

$$\dim N(\mathfrak{A}) = \dim N(\mathscr{T}).$$

Proof. In view of assertion (10′), it follows from Theorem 8.2.4 that the spaces $N(\mathfrak{A})$ and $N(\mathscr{T})$ are isomorphic in such a way that

$$N(\mathfrak{A}) \underset{P}{\overset{\gamma_0}{\rightleftarrows}} N(\mathscr{T}).$$

This proves the theorem. ■

For the ranges $R(\mathfrak{A})$ and $R(\mathscr{T})$, we have:

8.3.5 Theorem (ranges). *The following two conditions are equivalent:*

(i) *The range $R(\mathfrak{A})$ is closed in $H^{\tau}(\bar{\Omega}) \times H^{\tau-m+3/2}(\partial\Omega)$.*
(ii) *The range $R(\mathscr{T})$ is closed in $H^{\tau-m+3/2}(\partial\Omega)$.*

Proof. (i) $\Rightarrow$ (ii): Let ψ be an arbitrary element of the closure of the range $R(\mathscr{T})$, and let $\{\varphi_j\}$ be a sequence in $D(\mathscr{T}) \subset H^{\sigma-1/2}(\partial\Omega)$ such that $\mathscr{T}\varphi_j \to \psi$ in $H^{\tau-m+3/2}(\partial\Omega)$. Then, letting $w_j = P\varphi_j$, we obtain that

$$\mathfrak{A}w_j = \{Aw_j, B\gamma w_j\} = \{0, \mathscr{T}\varphi_j\} \to \{0, \psi\} \qquad \text{in } H^{\tau}(\bar{\Omega}) \times H^{\tau-m+3/2}(\partial\Omega).$$

Thus it follows from condition (i) that there exists an element $w \in D(\mathfrak{A}) \subset H_A^{\sigma,\tau}$ such that $\mathfrak{A}w = \{0, \psi\}$, that is,

$$\begin{cases} Aw = 0 & \text{in } \Omega, \\ B\gamma w = \psi & \text{on } \partial\Omega. \end{cases}$$

But Theorem 8.2.4 tells us that the distribution w can be written as

$$w = P\varphi, \qquad \varphi = \gamma_0 w \in H^{\sigma - 1/2}(\partial\Omega).$$

Hence we have

$$T\varphi = B\gamma P\varphi = B\gamma w = \psi \in H^{\tau - m + 3/2}(\partial\Omega).$$

This proves that $\varphi \in D(\mathscr{T})$ and so

$$\psi \in R(\mathscr{T}).$$

(ii) ⇒ (i): Let $\{f, \varphi\}$ be an arbitrary element of the closure of the range $R(\mathfrak{A})$, and let $\{u_j\}$ be a sequence in $D(\mathfrak{A}) \subset H_A^{\sigma,\tau}$ such that

$$\mathfrak{A}u_j = \{Au_j, B\gamma u_j\} \to \{f, \varphi\} \qquad \text{in } H^\tau(\bar{\Omega}) \times H^{\tau - m + 3/2}(\partial\Omega).$$

We decompose the u_j as in formula (1):

$$u_j = v_j + w_j$$

where

$$\begin{cases} v_j = QE(Au_j)|_\Omega \in H^{\tau+2}(\bar{\Omega}), \\ w_j = u_j - v_j \in N(A, \sigma). \end{cases}$$

Since the operators E: $H^\tau(\bar{\Omega}) \to H^\tau(M)$ and Q: $H^\tau(M) \to H^{\tau+2}(M)$ are continuous, it follows that

$$v_j = QE(Au_j)|_\Omega \to QEf|_\Omega \qquad \text{in } H^{\tau+2}(\bar{\Omega}).$$

Thus, letting

$$v = QEf|_\Omega,$$

we obtain from Proposition 8.3.2, with $\sigma = \tau + 2$, that

$$B\gamma v_j \to B\gamma v \qquad \text{in } H^{\tau - m + 3/2}(\partial\Omega),$$

so that

$$B\gamma w_j = B\gamma u_j - B\gamma v_j \to \varphi - B\gamma v \qquad \text{in } H^{\tau - m + 3/2}(\partial\Omega).$$

But we have, by Theorem 8.2.4,

$$w_j = P\varphi_j, \qquad \varphi_j = \gamma_0 w_j \in H^{\sigma - 1/2}(\partial\Omega),$$

and hence

$$B\gamma w_j = B\gamma P\varphi_j = T\varphi_j \in R(\mathcal{T}).$$

Therefore it follows from condition (ii) that there exists an element $\psi \in D(\mathcal{T}) \subset H^{\sigma - 1/2}(\partial\Omega)$ such that

$$\mathcal{T}\psi = \varphi - B\gamma v.$$

We let

$$u = v + P\psi.$$

Then we have $u \in H^\sigma(\bar{\Omega})$ and

$$\begin{cases} Au = Av = AQEf|_\Omega = Ef|_\Omega = f \in H^\tau(\bar{\Omega}), \\ B\gamma u = B\gamma v + B\gamma P\psi = B\gamma v + \mathcal{T}\psi = \varphi \in H^{\tau - m + 3/2}(\partial\Omega). \end{cases}$$

This proves that $u \in D(\mathfrak{A})$ and so

$$\{f, \varphi\} \in R(\mathfrak{A}).$$

The proof of Theorem 8.3.5 is complete. ■

To study the relation between codim $R(\mathfrak{A})$ and codim $R(\mathcal{T})$, we consider the transposes $\mathfrak{A}'$ and $\mathcal{T}'$. Here the transpose $\mathfrak{A}'$ of $\mathfrak{A}$ is a closed linear operator from $H_{\bar{\Omega}}^{-\tau}(M) \times H^{-\tau + m - 3/2}(\partial\Omega)$ into $H_{\bar{\Omega}}^{-\sigma}(M)$ such that

$$\langle \mathfrak{A}u, \{v, \psi\}\rangle = \langle u, \mathfrak{A}'\{v, \psi\}\rangle, \qquad u \in D(\mathfrak{A}),\ \{v, \psi\} \in D(\mathfrak{A}'),$$

and the transpose $\mathcal{T}'$ of $\mathcal{T}$ is a closed linear operator from $H^{-\tau + m - 3/2}(\partial\Omega)$ into $H^{-\sigma + 1/2}(\partial\Omega)$ such that

$$\langle \mathcal{T}\varphi, \psi\rangle = \langle \varphi, \mathcal{T}'\psi\rangle, \qquad \varphi \in D(\mathcal{T}),\ \psi \in D(\mathcal{T}').$$

Then we have:

8.3.6 Theorem. *Suppose that the ranges $R(\mathfrak{A})$ and $R(\mathcal{T})$ are closed. Then the following two conditions are equivalent:*

(i) *The null space $N(\mathfrak{A}')$ has finite dimension.*
(ii) *The null space $N(\mathcal{T}')$ has finite dimension.*

Moreover, in this case, we have

$$\dim N(\mathfrak{A}') = \dim N(\mathcal{T}'). \tag{13}$$

Proof. (i) $\Rightarrow$ (ii): Suppose that the null space $N(\mathfrak{A}')$ has dimension ℓ, and let $\{\{v_j, \psi_j\}\}_{j=1}^{\ell} \subset H_{\bar{\Omega}}^{-\tau}(M) \times H^{-\tau+m-3/2}(\partial\Omega)$ be a basis of $N(\mathfrak{A}')$. We show that the family $\{\psi_j\}_{j=1}^{\ell}$ is a basis of the null space $N(\mathscr{T}')$.

To do so, in view of the closed range theorem (Theorem 3.4.6), it suffices to prove that an arbitrary element ψ of $H^{\tau-m+3/2}(\partial\Omega)$ belongs to the range $R(\mathscr{T})$ if and only if we have

$$\langle \psi, \psi_j \rangle = 0, \qquad j = 1, \ldots, \ell.$$

The "only if" part follows immediately. In fact, if $\psi = \mathscr{T}\varphi$ with $\varphi \in D(\mathscr{T})$, then, letting $w = P\varphi$, we obtain that

$$\langle \psi, \psi_j \rangle = \langle \mathscr{T}\varphi, \psi_j \rangle = \langle \mathfrak{A}w, \{v_j, \psi_j\} \rangle = \langle w, \mathfrak{A}'\{v_j, \psi_j\} \rangle = 0,$$

since $B\gamma w = \mathscr{T}\varphi$ and $\{v_j, \psi_j\} \in N(\mathfrak{A}')$.

To prove the "if" part, suppose that an element $\psi \in H^{\tau-m+3/2}(\partial\Omega)$ satisfies

$$\langle \psi, \psi_j \rangle = 0, \qquad 1 \leq j \leq \ell.$$

Then it follows that

$$\langle \{0, \psi\}, \{v_j, \psi_j\} \rangle = \langle \psi, \psi_j \rangle = 0, \qquad 1 \leq j \leq \ell.$$

Since the family $\{\{v_j, \psi_j\}\}_{j=1}^{\ell}$ is a basis of $N(\mathfrak{A}')$, applying the closed range theorem, we obtain that the element $\{0, \psi\}$ belongs to the space ${}^0N(\mathfrak{A}') = R(\mathfrak{A})$, that is, there exists an element $w \in D(\mathfrak{A}) \subset H_A^{\sigma,\tau}$ such that

$$\begin{cases} Aw = 0 & \text{in } \Omega, \\ B\gamma w = \psi & \text{on } \partial\Omega. \end{cases}$$

In view of Theorem 8.2.4, this implies that

$$\psi = B\gamma P\varphi = \mathscr{T}\varphi \in R(\mathscr{T}),$$

with $\varphi = \gamma_0 w \in D(\mathscr{T})$.

(ii) $\Rightarrow$ (i): Suppose that the null space $N(\mathscr{T}')$ has dimension ℓ, and let $\{\psi_j\}_{j=1}^{\ell} \subset H^{-\tau+m-3/2}(\partial\Omega)$ be a basis of $N(\mathscr{T}')$. We let

$$v_j = -E'Q'(B\gamma)'\psi_j \tag{14}$$

where the operators

$$E' : H^{-\tau}(M) \to H_{\bar{\Omega}}^{-\tau}(M),$$

$$Q' : H^{-\tau-2}(M) \to H^{-\tau}(M),$$

$$(B\gamma)' : H^{-\tau+m-3/2}(\partial\Omega) \to H^{-\tau-2}(M),$$

are the transposes of E, Q and $B\gamma$, respectively. We show that the family $\{\{v_j, \psi_j\}\}_{j=1}^{\ell}$ is a basis of the null space $N(\mathfrak{A}')$.

To do so, in view of the closed range theorem (Theorem 3.4.6), it suffices to prove that an arbitrary element $\{f, \varphi\}$ of $H^\tau(\bar{\Omega}) \times H^{\tau-m+3/2}(\partial\Omega)$ belongs to the range $R(\mathfrak{A})$ if and only if we have

$$\langle\{f, \varphi\}, \{v_j, \psi_j\}\rangle = 0, \qquad 1 \leq j \leq \ell. \tag{15}$$

In view of formula (14), it follows that

$$\begin{aligned}\langle\{f, \varphi\}, \{v_j, \psi_j\}\rangle &= \langle\varphi, \psi_j\rangle - \langle B\gamma QEf, \psi_j\rangle \\ &= \langle\varphi - B\gamma QEf, \psi_j\rangle.\end{aligned}$$

Since the family $\{\psi_j\}_{j=1}^{\ell}$ is a basis of $N(\mathscr{T}')$, applying the closed range theorem, we obtain that condition (15) holds if and only if we have

$$\varphi - B\gamma QEf \in {}^0N(\mathscr{T}') = R(\mathscr{T}).$$

But, in view of Proposition 8.3.3, this is equivalent to saying that

$$\{f, \varphi\} \in R(\mathfrak{A}).$$

Finally we remark that formula (13) is clear from the above proof. ■

8.3.7 Corollary. *Suppose that the ranges $R(\mathfrak{A})$ and $R(\mathscr{T})$ are closed. Then the following two conditions are equivalent:*

(i) *The range $R(\mathfrak{A})$ has finite codimension.*
(ii) *The range $R(\mathscr{T})$ has finite codimension.*

Moreover, in this case, we have

$$\text{codim } R(\mathfrak{A}) = \text{codim } R(\mathscr{T}).$$

Corollary 8.3.7 is an immediate consequence of the closed range theorem (Theorem 3.4.6) and Theorem 8.3.6.

Combining Theorems 8.3.4–8.3.6 and Corollary 8.3.7, we obtain:

8.3.8 Theorem (indices). *The following two conditions are equivalent:*

(i) *The operator $\mathfrak{A}$ is a Fredholm operator.*
(ii) *The operator $\mathscr{T}$ is a Fredholm operator.*

Moreover, in this case, we have

$$\text{ind } \mathfrak{A} = \text{ind } \mathscr{T}.$$

The next theorem states that $\mathfrak{A}$ has regularity property if and only if $\mathscr{T}$ has.

8.3.9 Theorem (regularity). *Let* $\sigma \leq \tau + 2$, $\tau \geq 0$ *and* $t < \sigma$. *Then the following two conditions are equivalent*:

(i) *If* $u \in H^t(\bar{\Omega})$, $Au \in H^\tau(\bar{\Omega})$ *and* $B\gamma u \in H^{\tau - m + 3/2}(\partial\Omega)$, *then we have* $u \in H^\sigma(\bar{\Omega})$.

(ii) *If* $\varphi \in H^{t - 1/2}(\partial\Omega)$ *and* $T\varphi \in H^{\tau - m + 3/2}(\partial\Omega)$, *then we have* $\varphi \in H^{\sigma - 1/2}(\partial\Omega)$.

Proof. (i) $\Rightarrow$ (ii): Suppose that $\varphi \in H^{t - 1/2}(\partial\Omega)$ and $T\varphi \in H^{\tau - m + 3/2}(\partial\Omega)$. Then, letting $u = P\varphi$, we obtain that

$$u \in H^t(\bar{\Omega}), \qquad Au = 0, \qquad B\gamma u = T\varphi \in H^{\tau - m + 3/2}(\partial\Omega).$$

Hence it follows from condition (i) that

$$u \in H^\sigma(\bar{\Omega}).$$

In view of Theorem 8.2.4, this implies that

$$\varphi = \gamma_0 u \in H^{\sigma - 1/2}(\partial\Omega).$$

(ii) $\Rightarrow$ (i): Suppose that $u \in H^t(\bar{\Omega})$, $Au \in H^\tau(\bar{\Omega})$ and $B\gamma u \in H^{\tau - m + 3/2}(\partial\Omega)$. Then the distribution u can be decomposed as in formula (1):

$$u = v + w$$

where

$$\begin{cases} v = QE(Au)|_\Omega \in H^{\tau + 2}(\bar{\Omega}), \\ w = u - v \in N(A, t). \end{cases}$$

Theorem 8.2.4 tells us that the distribution w can be written as

$$w = P\varphi, \qquad \varphi = \gamma_0 w \in H^{t - 1/2}(\partial\Omega).$$

Hence we have

$$T\varphi = B\gamma P\varphi = B\gamma w = B\gamma u - B\gamma v \in H^{\tau - m + 3/2}(\partial\Omega).$$

Therefore it follows from condition (ii) that

$$\varphi \in H^{\sigma - 1/2}(\partial\Omega),$$

so that

$$w = P\varphi \in H^\sigma(\bar{\Omega}).$$

This proves that

$$u = v + w \in H^\sigma(\bar{\Omega}). \qquad \blacksquare$$

8.3.10 Corollary. *The following two conditions are equivalent:*

(i) $N(\mathfrak{A}) \subset C^{\infty}(\bar{\Omega})$.
(ii) $N(\mathscr{T}) \subset C^{\infty}(\partial\Omega)$.

For the null spaces $N(\mathfrak{A}')$ and $N(\mathscr{T}')$, we have:

8.3.11 Theorem. *Suppose that the null spaces $N(\mathfrak{A}')$ and $N(\mathscr{T}')$ have finite dimension. Then the following two conditions are equivalent:*

(i) $N(\mathfrak{A}') \subset C^{\infty}(\bar{\Omega}) \times C^{\infty}(\partial\Omega)$.
(ii) $N(\mathscr{T}') \subset C^{\infty}(\partial\Omega)$.

Proof. (i) $\Rightarrow$ (ii): This is clear from the proof of the implication (i) $\Rightarrow$ (ii) of Theorem 8.3.6.

(ii) $\Rightarrow$ (i): We know from the proof of the implication (ii) $\Rightarrow$ (i) of Theorem 8.3.6 that:

If the family $\{\psi_j\}_{j=1}^{\ell}$ is a basis of the null space $N(\mathscr{T}')$ ($\ell = \dim N(\mathscr{T}')$), then the family

$$\{\{-E'Q'(B\gamma)'\psi_j, \psi_j\}\}_{j=1}^{\ell} \tag{16}$$

is a basis of the null space $N(\mathfrak{A}')$.

But we have, by formula (9),

$$(B\gamma)'\psi_j = \gamma_0'(B_0'\psi_j) + \gamma_1'(B_1'\psi_j).$$

We remark that $\gamma_0'(B_0'\psi_j)$ and $\gamma_1'(B_1'\psi_j)$ are distributions on M with support in $\partial\Omega$. If condition (ii) is satisfied, that is, if $\{\psi_j\}_{j=1}^{\ell} \subset C^{\infty}(\partial\Omega)$, applying part (i) of Theorem 6.8.1 to our situation, we obtain that

$$Q'(\gamma_0'(B_0'\psi_j))|_{\Omega},\ Q'(\gamma_1'(B_1'\psi_j))|_{\Omega} \in C^{\infty}(\bar{\Omega}),$$

and also

$$Q'(\gamma_0'(B_0'\psi_j))|_{M\setminus\bar{\Omega}},\ Q'(\gamma_1'(B_1'\psi_j))|_{M\setminus\bar{\Omega}} \in C^{\infty}(M\setminus\Omega).$$

Hence it follows from an application of Proposition 5.5.2 that

$$E'Q'(B\gamma)'\psi_j = E'Q'\gamma_0'(B_0'\psi_j) + E'Q'\gamma_1'(B_1'\psi_j) \in C^{\infty}(\bar{\Omega}).$$

In view of assertion (16), this proves condition (i). ■

The next theorem states that *a priori* estimates for $\mathfrak{A}$ are entirely equivalent to corresponding *a priori* estimates for $\mathcal{T}$.

8.3.12 Theorem (estimates). *Let $\sigma \le \tau + 2$, $\tau \ge 0$ and $t < \sigma$. Then the following two estimates are equivalent:*

$$\text{(i)}\quad \|u\|_{H^\sigma(\bar\Omega)} \le C(\|Au\|_{H^\tau(\bar\Omega)} + |B\gamma u|_{H^{\tau-m+3/2}(\partial\Omega)} + \|u\|_{H^t(\bar\Omega)}), \quad u \in D(\mathfrak{A}). \tag{17}$$

$$\text{(ii)}\quad |\varphi|_{H^{\sigma-1/2}(\partial\Omega)} \le C(|\mathcal{T}\varphi|_{H^{\tau-m+3/2}(\partial\Omega)} + |\varphi|_{H^{t-1/2}(\partial\Omega)}), \quad \varphi \in D(\mathcal{T}). \tag{18}$$

Here and in the following the letter C denotes a generic positive constant.

Proof. (i) $\Rightarrow$ (ii): Taking $u = P\varphi$ with $\varphi \in D(\mathcal{T})$ in estimate (17), we obtain that

$$\|P\varphi\|_{H^\sigma(\bar\Omega)} \le C(|\mathcal{T}\varphi|_{H^{\tau-m+3/2}(\partial\Omega)} + \|P\varphi\|_{H^t(\bar\Omega)}). \tag{19}$$

But Theorem 8.2.4 tells us that the Poisson operator P maps $H^{s-1/2}(\partial\Omega)$ isomorphically onto $N(A, s)$ for all $s \in \mathbf{R}$. Thus estimate (18) follows from estimate (19).

(ii) $\Rightarrow$ (i): Every element $u \in D(\mathfrak{A})$ can be decomposed as in formula (1):

$$u = v + w$$

where

$$\begin{cases} v = QE(Au)|_\Omega \in H^{\tau+2}(\bar\Omega), \\ w = u - v \in N(A, \sigma). \end{cases}$$

Then we have, by estimate (2),

$$\|v\|_{H^\sigma(\bar\Omega)} \le \|v\|_{H^{\tau+2}(\bar\Omega)} \le C\|Au\|_{H^\tau(\bar\Omega)}. \tag{20}$$

Further, applying estimate (18) to the distribution $\gamma_0 w$, we obtain that

$$\begin{aligned} &|\gamma_0 w|_{H^{\sigma-1/2}(\partial\Omega)} \\ &\quad \le C(|\mathcal{T}(\gamma_0 w)|_{H^{\tau-m+3/2}(\partial\Omega)} + |\gamma_0 w|_{H^{t-1/2}(\partial\Omega)}) \\ &\quad = C(|B\gamma w|_{H^{\tau-m+3/2}(\partial\Omega)} + |\gamma_0 w|_{H^{t-1/2}(\partial\Omega)}) \\ &\quad \le C(|B\gamma u|_{H^{\tau-m+3/2}(\partial\Omega)} + |B\gamma v|_{H^{\tau-m+3/2}(\partial\Omega)} + |\gamma_0 w|_{H^{t-1/2}(\partial\Omega)}). \end{aligned}$$

In view of Theorem 8.2.4, this gives that

$$\begin{aligned}\|w\|_{H^\sigma(\bar{\Omega})} &\leq C(|B\gamma u|_{H^{\tau-m+3/2}(\partial\Omega)} + |B\gamma v|_{H^{\tau-m+3/2}(\partial\Omega)} + \|w\|_{H^t(\bar{\Omega})}) \\ &\leq C(|B\gamma u|_{H^{\tau-m+3/2}(\partial\Omega)} + |B\gamma v|_{H^{\tau-m+3/2}(\partial\Omega)} \\ &\quad + \|u\|_{H^t(\bar{\Omega})} + \|v\|_{H^t(\bar{\Omega})}). \end{aligned} \tag{21}$$

But it follows from Proposition 8.3.2 with $\sigma = \tau + 2$ that

$$|B\gamma v|_{H^{\tau-m+3/2}(\partial\Omega)} \leq C\|v\|_{H^{\tau+2}(\bar{\Omega})} \leq C\|Au\|_{H^\tau(\bar{\Omega})}. \tag{22}$$

Thus, carrying (20) and (22) into (21), we obtain that

$$\|w\|_{H^\sigma(\bar{\Omega})} \leq C(\|Au\|_{H^\tau(\bar{\Omega})} + |B\gamma u|_{H^{\tau-m+3/2}(\partial\Omega)} + \|u\|_{H^t(\bar{\Omega})}). \tag{23}$$

Estimate (17) follows from estimates (20) and (23). ■

8.4. Existence and Uniqueness Theorem for General Boundary Value Problems

Let Ω be a bounded domain in $\mathbf{R}^n$ with C^∞ boundary $\partial\Omega$, and let

$$A = \sum_{i,j=1}^{n} a^{ij}(x) \frac{\partial^2}{\partial x_i\, \partial x_j} + \sum_{i=1}^{n} b^i(x) \frac{\partial}{\partial x_i} + c(x)$$

be a second-order *elliptic* differential operator with real coefficients such that:

1. $a^{ij} \in C^\infty(\bar{\Omega})$, $a^{ij} = a^{ji}$ and there exists a constant $a_0 > 0$ such that

$$\sum_{i,j=1}^{n} a^{ij}(x)\xi_i\xi_j \geq a_0|\xi|^2, \qquad x \in \bar{\Omega},\ \xi \in \mathbf{R}^n.$$

2. $b^i \in C^\infty(\bar{\Omega})$.
3. $c \in C^\infty(\bar{\Omega})$ and $c \leq 0$ in Ω.

In this section we shall consider the following boundary value problem: given functions f and φ defined in Ω and on $\partial\Omega$, respectively, find a function u in Ω such that

$$\begin{cases} (A - \alpha)u = f & \text{in } \Omega, \\ \quad Bu \equiv B_0(u|_{\partial\Omega}) + B_1\left(\left.\dfrac{\partial u}{\partial \nu}\right|_{\partial\Omega}\right) = \varphi & \text{on } \partial\Omega. \end{cases} \tag{$*$}$$

Here:

1. α is a non-negative, spectral parameter.
2. $B_j\,(j = 0, 1)$ is a classical pseudo-differential operator of order m_j on $\partial\Omega$.
3. ν is the unit exterior normal to $\partial\Omega$.

We shall prove an existence and uniqueness theorem for problem $(*)$ in the framework of *Sobolev spaces* when $\alpha \to +\infty$. For this purpose, we make use of a method essentially due to Agmon and Nirenberg (cf. Agmon [1], Lions-Magenes [1]). This is a technique of treating a spectral parameter as a second-order elliptic differential operator of an extra variable and relating the old problem to a new one with the additional variable. The following presentation of this technique is due to Fujiwara [1].

We introduce an auxiliary variable y of the unit circle

$$S = \mathbf{R}/2\pi\mathbf{Z},$$

and replace the parameter α by the differential operator

$$-\frac{\partial^2}{\partial y^2}.$$

We consider instead of problem $(*)$ the following boundary value problem: given functions $\tilde{f}$ and $\tilde{\varphi}$ defined in $\Omega \times S$ and on $\partial\Omega \times S$, respectively, find a function $\tilde{u}$ in $\Omega \times S$ such that

$$\begin{cases} \Lambda\tilde{u} \equiv \left(A + \dfrac{\partial^2}{\partial y^2}\right)\tilde{u} = \tilde{f} & \text{in } \Omega \times S, \\ B\tilde{u} \equiv B_0(\tilde{u}|_{\partial\Omega \times S}) + B_1\left(\dfrac{\partial \tilde{u}}{\partial \nu}\bigg|_{\partial\Omega \times S}\right) = \tilde{\varphi} & \text{on } \partial\Omega \times S. \end{cases} \qquad (\tilde{*})$$

Then, roughly speaking, the most important relationship between problem $(*)$ and problem $(\tilde{*})$ is stated as follows:

If the index of problem $(\tilde{*})$ *is* finite, *then the index of problem* $(*)$ *is equal to* zero *for all* $\alpha \geq 0$. (1)

Statement of Results

We state assertion (1) more precisely. Let $s \geq \max(2, m + 1/2)$ where $m = \max(m_0, m_1 + 1)$, and $0 < \kappa \leq 2$. We associate with problem $(*)$ a densely defined, closed linear operator

$$\mathfrak{A}(\alpha)\colon H^{s-2+\kappa}(\bar{\Omega}) \to H^{s-2}(\bar{\Omega}) \times H^{s-m-1/2}(\partial\Omega)$$

as follows:

(a) The domain $D(\mathfrak{A}(\alpha))$ of $\mathfrak{A}(\alpha)$ is the space

$$D(\mathfrak{A}(\alpha)) = \{u \in H^{s-2+\kappa}(\bar{\Omega});\ (A - \alpha)u \in H^{s-2}(\bar{\Omega}),\ Bu \in H^{s-m-1/2}(\partial\Omega)\}.$$

(b) $\mathfrak{A}(\alpha)u = \{(A - \alpha)u,\ Bu\},\ u \in D(\mathfrak{A}(\alpha))$.

Similarly, we associate with problem $(\tilde{*})$ a densely defined, closed linear operator

$$\tilde{\mathfrak{A}}\colon H^{s-2+\kappa}(\bar{\Omega} \times S) \to H^{s-2}(\bar{\Omega} \times S) \times H^{s-m-1/2}(\partial\Omega \times S)$$

as follows:

($\tilde{\text{a}}$) The domain $D(\tilde{\mathfrak{A}})$ of $\tilde{\mathfrak{A}}$ is the space

$$D(\tilde{\mathfrak{A}}) = \{\tilde{u} \in H^{s-2+\kappa}(\bar{\Omega} \times S);\ \Lambda\tilde{u} \in H^{s-2}(\bar{\Omega} \times S),\ B\tilde{u} \in H^{s-m-1/2}(\partial\Omega \times S)\}.$$

($\tilde{\text{b}}$) $\tilde{\mathfrak{A}}\tilde{u} = \{\Lambda\tilde{u},\ B\tilde{u}\},\ \tilde{u} \in D(\tilde{\mathfrak{A}})$.

Now we can state our main result.

8.4.1 Theorem. *Let $s \geq \max(2, m + 1/2)$, $0 < \kappa \leq 2$ and $s - 5/2 + \kappa > 0$. Then the following two conditions are equivalent:*

(i) *The operator $\tilde{\mathfrak{A}}\colon H^{s-2+\kappa}(\bar{\Omega} \times S) \to H^{s-2}(\bar{\Omega} \times S) \times H^{s-m-1/2}(\partial\Omega \times S)$ is a Fredholm operator.*

(ii) *For all $\alpha \geq 0$, the operator $\mathfrak{A}(\alpha)\colon H^{s-2+\kappa}(\bar{\Omega}) \to H^{s-2}(\bar{\Omega}) \times H^{s-m-1/2}(\partial\Omega)$ is a Fredholm operator with index zero; and there exists a constant $R' > 0$ such that if $\alpha' = \ell^2$ with $\ell \in \mathbf{Z}$ and $\ell^2 \geq R'$, then the operator $\mathfrak{A}(\alpha')$ is bijective and we have, for all $u \in D(\mathfrak{A}(\alpha'))$,*

$$\begin{aligned} &\|u\|^2_{H^{s-2+\kappa}(\bar{\Omega})} + (\alpha')^{s-2+\kappa}\|u\|^2_{L^2(\Omega)} \\ &\quad \leq C'(\|(A - \alpha')u\|^2_{H^{s-2}(\bar{\Omega})} + (\alpha')^{s-2}\|(A - \alpha')u\|^2_{L^2(\Omega)} \\ &\qquad + |Bu|^2_{H^{s-m-1/2}(\partial\Omega)} + (\alpha')^{s-m-1/2}|Bu|^2_{L^2(\partial\Omega)}), \end{aligned} \tag{2}$$

with a constant $C' > 0$ independent of $\alpha' \geq R'$.

In the "subelliptic" case, that is, in the case $1 < \kappa \leq 2$, we can prove the following:

8.4.2 Corollary. *Let $s \geq \max(2, m + 1/2)$ and $1 < \kappa \leq 2$. Then the following two conditions are equivalent:*

(i) *The operator $\tilde{\mathfrak{A}}: H^{s-2+\kappa}(\bar{\Omega} \times S) \to H^{s-2}(\bar{\Omega} \times S) \times H^{s-m-1/2}(\partial\Omega \times S)$ is a Fredholm operator.*

(ii) *For all $\alpha \geq 0$, the operator $\mathfrak{A}(\alpha): H^{s-2+\kappa}(\bar{\Omega}) \to H^{s-2}(\bar{\Omega}) \times H^{s-m-1/2}(\partial\Omega)$ is a Fredholm operator with index zero; and there exists a constant $R > 0$ such that if $\alpha \geq R$ then the operator $\mathfrak{A}(\alpha)$ is bijective and we have, for all $u \in D(\mathfrak{A}(\alpha))$,*

$$\begin{aligned} &\|u\|^2_{H^{s-2+\kappa}(\bar{\Omega})} + \alpha^{s-2+\kappa}\|u\|^2_{L^2(\Omega)} \\ &\quad \leq C(\|(A-\alpha)u\|^2_{H^{s-2}(\bar{\Omega})} + \alpha^{s-2}\|(A-\alpha)u\|^2_{L^2(\Omega)} \\ &\quad\quad + |Bu|^2_{H^{s-m-1/2}(\partial\Omega)} + \alpha^{s-m-1/2}|Bu|^2_{L^2(\partial\Omega)}), \end{aligned} \tag{3}$$

with a constant $C > 0$ independent of $\alpha \geq R$.

8.4.3 Remark. Problem $(*)$ is elliptic (or coercive) if and only if $\kappa = 2$, and it is subelliptic if and only if $1 < \kappa < 2$. In the elliptic case, Corollary 8.4.2 is proved by Agranovich and Vishik [1] (cf. [1], Theorem 4.1 and Theorem 5.1).

Proof of Theorem 8.4.1

(1) First we reduce the study of problem $(*)$ to that of a pseudo-differential operator on the boundary.

Applying Theorem 8.2.4 to the operator $A - \alpha$ $(\alpha \geq 0)$, we obtain the following results:

(a) The Dirichlet problem

$$\begin{cases} (A-\alpha)w = 0 & \text{in } \Omega, \\ \gamma_0 w = \varphi & \text{on } \partial\Omega, \end{cases} \tag{D}$$

has a unique solution w in $H^t(\bar{\Omega})$ for any $\varphi \in H^{t-1/2}(\partial\Omega)$ $(t \in \mathbf{R})$.

(b) The mapping

$$P(\alpha): H^{t-1/2}(\partial\Omega) \to H^t(\bar{\Omega}),$$

defined by $w = P(\alpha)\varphi$, is an isomorphism of $H^{t-1/2}(\partial\Omega)$ onto the space $N(A - \alpha, t) = \{u \in H^t(\bar{\Omega}); (A - \alpha)u = 0 \text{ in } \Omega\}$ for all $t \in \mathbf{R}$; and its inverse is the trace operator γ_0 on $\partial\Omega$.

We let

$$T(\alpha): C^\infty(\partial\Omega) \to C^\infty(\partial\Omega)$$
$$\varphi \mapsto BP(\alpha)\varphi.$$

Then the operator $T(\alpha)$ can be written as

$$T(\alpha) = B_0 + B_1\Pi(\alpha)$$

where

$$\Pi(\alpha): C^\infty(\partial\Omega) \to C^\infty(\partial\Omega)$$
$$\varphi \mapsto \frac{\partial}{\partial\nu}(P(\alpha)\varphi)|_{\partial\Omega}.$$

Applying part (ii) of Theorem 6.8.1, we find that $\Pi(\alpha)$ is a classical pseudo-differential operator of first order on $\partial\Omega$. Hence the operator $T(\alpha)$ is a classical pseudo-differential operator of order m on $\partial\Omega$, and it extends to a continuous linear operator $T(\alpha): H^t(\partial\Omega) \to H^{t-m}(\partial\Omega)$ for all $t \in \mathbf{R}$. Thus we can introduce a densely defined, closed linear operator

$$\mathscr{T}(\alpha): H^{s-5/2+\kappa}(\partial\Omega) \to H^{s-m-1/2}(\partial\Omega)$$

as follows:

(α) The domain $D(\mathscr{T}(\alpha))$ of $\mathscr{T}(\alpha)$ is the space

$$D(\mathscr{T}(\alpha)) = \{\varphi \in H^{2-5/2+\kappa}(\partial\Omega); T(\alpha)\varphi \in H^{s-m-1/2}(\partial\Omega)\}.$$

(β) $\mathscr{T}(\alpha)\varphi = T(\alpha)\varphi$, $\varphi \in D(\mathscr{T}(\alpha))$.

Then, arguing as in Section 8.3 ($\sigma = s - 2 + \kappa$, $\tau = s - 2$), we can prove the following:

(I) The null space $N(\mathfrak{A}(\alpha))$ of $\mathfrak{A}(\alpha)$ has finite dimension if and only if the null space $N(\mathscr{T}(\alpha))$ of $\mathscr{T}(\alpha)$ has finite dimension, and we have

$$\dim N(\mathfrak{A}(\alpha)) = \dim N(\mathscr{T}(\alpha)).$$

(II) The range $R(\mathfrak{A}(\alpha))$ of $\mathfrak{A}(\alpha)$ is closed if and only if the range $R(\mathscr{T}(\alpha))$ of $\mathscr{T}(\alpha)$ is closed; and $R(\mathfrak{A}(\alpha))$ has finite codimension if and only if $R(\mathscr{T}(\alpha))$ has finite codimension, and we have

$$\operatorname{codim} R(\mathfrak{A}(\alpha)) = \operatorname{codim} R(\mathscr{T}(\alpha)).$$

(III) The operator $\mathfrak{A}(\alpha)$ is a Fredholm operator if and only if the operator $\mathscr{T}(\alpha)$ is a Fredholm operator, and we have

$$\operatorname{ind} \mathfrak{A}(\alpha) = \operatorname{ind} \mathscr{T}(\alpha).$$

(2) Similarly, we reduce the study of problem ($\tilde{*}$) to that of a pseudo-differential operator on the boundary.

Applying Theorem 8.2.4 to the operator $\Lambda = A + \partial^2/\partial y^2$, we obtain the following results:

($\tilde{a}$) The Dirichlet problem

$$\begin{cases} \Lambda \tilde{w} = 0 & \text{in } \Omega \times S, \\ \gamma_0 \tilde{w} = \tilde{\varphi} & \text{on } \partial\Omega \times S, \end{cases} \tag{$\tilde{D}$}$$

has a unique solution $\tilde{w}$ in $H^t(\bar{\Omega} \times S)$ for any $\tilde{\varphi} \in H^{t-1/2}(\partial\Omega \times S)$ ($t \in \mathbf{R}$).

($\tilde{b}$) The mapping

$$\tilde{P}: H^{t-1/2}(\partial\Omega \times S) \to H^t(\bar{\Omega} \times S),$$

defined by $\tilde{w} = \tilde{P}\tilde{\varphi}$, is an isomorphism of $H^{t-1/2}(\partial\Omega \times S)$ onto the space $N(\Lambda, t) = \{u \in H^t(\bar{\Omega} \times S); \Lambda\tilde{u} = 0 \text{ in } \Omega \times S\}$ for all $t \in \mathbf{R}$; and its inverse is the trace operator γ_0 on $\partial\Omega \times S$.

We let

$$\begin{aligned} \tilde{T}: C^\infty(\partial\Omega \times S) &\to C^\infty(\partial\Omega \times S) \\ \tilde{\varphi} &\mapsto B\tilde{P}\tilde{\varphi}. \end{aligned}$$

Then the operator $\tilde{T}$ can be written as

$$\tilde{T} = B_0 + B_1\tilde{\Pi}$$

where

$$\begin{aligned} \tilde{\Pi}: C^\infty(\partial\Omega \times S) &\to C^\infty(\partial\Omega \times S) \\ \tilde{\varphi} &\mapsto \frac{\partial}{\partial \nu}(\tilde{P}\tilde{\varphi})|_{\partial\Omega \times S}. \end{aligned}$$

Applying part (ii) of Theorem 6.8.1, we find that $\tilde{\Pi}$ is a classical pseudo-differential operator of first order on $\partial\Omega \times S$. Hence the operator $\tilde{T}$ is a classical pseudo-differential operator of order m on $\partial\Omega \times S$, and it extends to a continuous linear operator $\tilde{T}: H^t(\partial\Omega \times S) \to H^{t-m}(\partial\Omega \times S)$ for all $t \in \mathbf{R}$. We can define a densely defined, closed linear operator

$$\tilde{\mathscr{T}}: H^{s-5/2+\kappa}(\partial\Omega \times S) \to H^{s-m-1/2}(\partial\Omega \times S)$$

as follows:

($\tilde{\alpha}$) The domain $D(\tilde{\mathscr{T}})$ of $\tilde{\mathscr{T}}$ is the space

$$D(\tilde{\mathscr{T}}) = \{\tilde{\varphi} \in H^{s-5/2+\kappa}(\partial\Omega \times S);\ \tilde{T}\tilde{\varphi} \in H^{s-m-1/2}(\partial\Omega \times S)\}.$$

($\tilde{\beta}$) $\tilde{\mathscr{T}}\tilde{\varphi} = \tilde{T}\tilde{\varphi},\ \tilde{\varphi} \in D(\tilde{\mathscr{T}})$.

Then we have the following results, analogous to results (I), (II) and (III):

($\tilde{\text{I}}$) The null space $N(\tilde{\mathfrak{A}})$ of $\tilde{\mathfrak{A}}$ has finite dimension if and only if the null space $N(\tilde{\mathscr{T}})$ of $\tilde{\mathscr{T}}$ has finite dimension, and we have

$$\dim N(\tilde{\mathfrak{A}}) = \dim N(\tilde{\mathscr{T}}).$$

($\tilde{\text{II}}$) The range $R(\tilde{\mathfrak{A}})$ of $\tilde{\mathfrak{A}}$ is closed if and only if the range $R(\tilde{\mathscr{T}})$ of $\tilde{\mathscr{T}}$ is closed; and $R(\tilde{\mathfrak{A}})$ has finite codimension if and only if $R(\tilde{\mathscr{T}})$ has finite codimension, and we have

$$\operatorname{codim} R(\tilde{\mathfrak{A}}) = \operatorname{codim} R(\tilde{\mathscr{T}}).$$

($\widetilde{\text{III}}$) The operator $\tilde{\mathfrak{A}}$ is a Fredholm operator if and only if the operator $\tilde{\mathscr{T}}$ is a Fredholm operator, and we have

$$\operatorname{ind} \tilde{\mathfrak{A}} = \operatorname{ind} \tilde{\mathscr{T}}.$$

(3) Now we study the null spaces $N(\tilde{\mathscr{T}})$ and $N(\mathscr{T}(\alpha'))$ when $\alpha' = \ell^2, \ell \in \mathbf{Z}$. In doing so, we need a lemma on the Fourier expansion:

8.4.4 Lemma. *Let $M = \partial\Omega$ or $M = \bar{\Omega}$. Then every $\tilde{\varphi} \in H^t(M \times S)$ $(t \in \mathbf{R})$ can be expanded as follows:*

$$\begin{cases} \tilde{\varphi} = \sum_{\ell \in \mathbf{Z}} \varphi_\ell \otimes e^{i\ell y} \qquad \text{in } H^t(M \times S), \\ \varphi_\ell \in H^t(M). \end{cases}$$

Furthermore, if $t \geq 0$, *we have,*

$$|\tilde{\varphi}|^2_{H^t(M \times S)} \approx \sum_{\ell \in \mathbf{Z}} (|\varphi_\ell|^2_{H^t(M)} + (1 + \ell^2)^t |\varphi_\ell|^2_{L^2(M)}). \tag{4}$$

Here the symbol $\approx$ denotes equivalent norms.

Proof. Considering the double $\hat{\Omega}$ of $\bar{\Omega}$ in the case $M = \bar{\Omega}$, we may suppose that M is a compact C^∞ manifold *without* boundary. In fact, it suffices to note that

$H^t(\bar{\Omega} \times S) =$ the space of restrictions to $\Omega \times S$ of elements of $H^t(\hat{\Omega} \times S)$;

$H^t(\bar{\Omega}) =$ the space of restrictions to Ω of elements of $H^t(\hat{\Omega})$.

Let $\{\chi_j\}$ be the eigenfunctions of the Laplace-Beltrami operator $-\Delta_M$ on M and $\{\lambda_j\}$ its corresponding eigenvalues:

$$-\Delta_M \chi_j = \lambda_j \chi_j, \qquad \lambda_j \geq 0.$$

Then we obtain from Theorem 6.7.16 that:

(a) $H^t(M) = \{\varphi \in \mathscr{D}'(M); \sum_j (1 + \lambda_j)^t |(\varphi, \chi_j)|^2 < +\infty\}$;
(b) Every $\varphi \in H^t(M)$ can be expanded as

$$\varphi = \sum_j (\varphi, \chi_j)\chi_j \qquad \text{in } H^t(M).$$

Similarly, applying Theorem 6.7.16 with $M = M \times S$, we obtain that:

(ã) $H^t(M \times S) = \{\tilde{\varphi} \in \mathscr{D}'(M \times S)$;

$$\sum_{j,\ell} (1 + \lambda_j + \ell^2)^t |(\tilde{\varphi}, \chi_j \otimes e^{i\ell y})|^2 < +\infty\};$$

($\tilde{\text{b}}$) Every $\tilde{\varphi} \in H^t(M \times S)$ can be expanded as

$$\tilde{\varphi} = \sum_{j,\ell} (\tilde{\varphi}, \chi_j \otimes e^{i\ell y})\chi_j \otimes e^{i\ell y} \qquad \text{in } H^t(M \times S).$$

Therefore we have the expansion

$$\tilde{\varphi} = \sum_\ell \varphi_\ell \otimes e^{i\ell y} \qquad \text{in } H^t(M \times S),$$

with

$$\varphi_\ell = \sum_j (\tilde{\varphi}, \chi_j \otimes e^{i\ell y})\chi_j \in H^t(M).$$

In fact, it suffices to note the following:

$$\sum_j (1 + \lambda_j)^t |(\varphi_\ell, \chi_j)|^2$$

$$= \sum_j (1 + \lambda_j)^t |(\tilde{\varphi}, \chi_j \otimes e^{i\ell y})|^2$$

$$\leq \begin{cases} \sum_j (1 + \lambda_j + \ell^2)^t |(\tilde{\varphi}, \chi_j \otimes e^{i\ell y})|^2 \\ \quad \leq |\tilde{\varphi}|^2_{H^t(M \times S)} & \text{if } t \geq 0, \\ (1 + \ell^2)^{-t} \sum_j (1 + \lambda_j + \ell^2)^t |(\tilde{\varphi}, \chi_j \otimes e^{i\ell y})|^2 \\ \quad \leq (1 + \ell^2)^{-t} |\tilde{\varphi}|^2_{H^t(M \times S)} & \text{if } t < 0. \end{cases}$$

Furthermore, if $t \geq 0$, we have the inequalities

$$\frac{1}{2}[(1 + \lambda_j)^t + (1 + \ell^2)^t] \leq (1 + \lambda_j + \ell^2)^t$$

$$\leq 2^t[(1 + \lambda_j)^t + (1 + \ell^2)^t].$$

Hence this gives that

$$|\tilde{\varphi}|^2_{H^t(M \times S)} = \sum_{j, \ell} (1 + \lambda_j + \ell^2)^t |(\tilde{\varphi}, \chi_j \otimes e^{i\ell y})|^2$$

$$\approx \sum_{j, \ell} (1 + \lambda_j)^t |(\tilde{\varphi}, \chi_j \otimes e^{i\ell y})|^2$$

$$+ \sum_{j, \ell} (1 + \ell^2)^t |(\tilde{\varphi}, \chi_j \otimes e^{i\ell y})|^2$$

$$= \sum_\ell |\varphi_\ell|^2_{H^t(M)} + \sum_\ell (1 + \ell^2)^t |\varphi_\ell|^2_{L^2(M)}.$$

Lemma 8.4.4 is proved. ■

The next lemma will play a fundamental role in the sequel.

8.4.5 Lemma. *We have, for all $\varphi \in \mathscr{D}'(\partial\Omega)$ and $\ell \in \mathbf{Z}$,*

$$\tilde{P}(\varphi \otimes e^{i\ell y}) = P(\ell^2)\varphi \otimes e^{i\ell y} \qquad \text{in } \mathscr{D}'(\Omega \times S). \tag{5}$$

$$\tilde{T}(\varphi \otimes e^{i\ell y}) = T(\ell^2)\varphi \otimes e^{i\ell y} \qquad \text{in } \mathscr{D}'(\partial\Omega \times S). \tag{6}$$

Proof. First we remark that

$$\mathscr{D}'(\partial\Omega) = \bigcup_{t \in \mathbf{R}} H^t(\partial\Omega),$$

since $\partial\Omega$ is compact. Thus one may suppose that $\varphi \in H^t(\partial\Omega)$ for some $t \in \mathbf{R}$.

We let

$$\tilde{w} = P(\ell^2)\varphi \otimes e^{i\ell y}.$$

Then we have

$$\tilde{w} \in H^{t+1/2}(\bar{\Omega} \times S),$$

since $P(\ell^2)\varphi \in H^{t+1/2}(\bar{\Omega})$ (cf. the proof of Lemma 8.4.4). Further the distribution $\tilde{w}$ satisfies

$$\begin{cases} \Lambda\tilde{w} = ((A - \ell^2)P(\ell^2)\varphi) \otimes e^{i\ell y} = 0 & \text{in } \Omega \times S, \\ \tilde{w}|_{\partial\Omega \times S} = \varphi \otimes e^{i\ell y} & \text{on } \partial\Omega \times S. \end{cases}$$

Thus, by the uniqueness of solutions of problem $(\tilde{D})$, we have

$$\tilde{P}(\varphi \otimes e^{i\ell y}) = \tilde{w} = P(\ell^2)\varphi \otimes e^{i\ell y} \qquad \text{in } H^{t+1/2}(\bar{\Omega} \times S),$$

and hence

$$\begin{aligned} \tilde{T}(\varphi \otimes e^{i\ell y}) &= B\tilde{P}(\varphi \otimes e^{i\ell y}) \\ &= BP(\ell^2)\varphi \otimes e^{i\ell y} \\ &= T(\ell^2)\varphi \otimes e^{i\ell y} \qquad \text{in } H^{t-m}(\partial\Omega \times S). \end{aligned}$$

This proves the lemma. ■

Now we can prove the most important relationship between the null spaces $N(\tilde{\mathscr{T}})$ and $N(\mathscr{T}(\alpha'))$ when $\alpha' = \ell^2$, $\ell \in \mathbf{Z}$.

8.4.6 Proposition. *The following two conditions are equivalent*:

(i) $\dim N(\tilde{\mathscr{T}}) < \infty$.

(ii) *There exists a finite subset I of* $\mathbf{Z}$ *such that*

$$\begin{cases} \dim N(\mathscr{T}(\ell^2)) < \infty & \text{if } \ell \in I, \\ \dim N(\mathscr{T}(\ell^2)) = 0 & \text{if } \ell \notin I. \end{cases}$$

Moreover, in this case, we have

$$\begin{cases} N(\tilde{\mathcal{T}}) = \bigoplus_{\ell \in I} N(\mathcal{T}(\ell^2)) \otimes e^{i\ell y}. \\ \dim N(\tilde{\mathcal{T}}) = \sum_{\ell \in I} \dim N(\mathcal{T}(\ell^2)). \end{cases}$$

Proof. Applying Lemma 8.4.4, we obtain that every $\tilde{\varphi} \in H^{s-5/2+\kappa}(\partial\Omega \times S)$ can be expanded as follows:

$$\begin{cases} \tilde{\varphi} = \sum_{\ell \in \mathbf{Z}} \varphi_\ell \otimes e^{i\ell y} \qquad \text{in } H^{s-5/2+\kappa}(\partial\Omega \times S), \\ \varphi_\ell \in H^{s-5/2+\kappa}(\partial\Omega). \end{cases}$$

Thus we have, by formula (6),

$$\tilde{T}\tilde{\varphi} = \sum_{\ell \in \mathbf{Z}} T(\ell^2)\varphi_\ell \otimes e^{i\ell y} \qquad \text{in } H^{s-5/2+\kappa-m}(\partial\Omega \times S).$$

By the uniqueness of the Fourier expansion, this gives that

$$N(\tilde{\mathcal{T}}) = \bigoplus_{\ell \in \mathbf{Z}} N(\mathcal{T}(\ell^2)) \otimes e^{i\ell y} \qquad \text{(formal sum)}.$$

Hence it is easy to see that conditions (i) and (ii) are equivalent, since the spaces $N(\mathcal{T}(\ell^2)) \otimes e^{i\ell y}$ are linearly independent. ■

(4) Next we study the ranges $R(\tilde{\mathcal{T}})$ and $R(\mathcal{T}(\alpha'))$ when $\alpha' = \ell^2$, $\ell \in \mathbf{Z}$. First we have:

8.4.7 Lemma. *If the range $R(\tilde{\mathcal{T}})$ is closed in $H^{s-m-1/2}(\partial\Omega \times S)$, then the range $R(\mathcal{T}(\ell^2))$ is closed in $H^{s-m-1/2}(\partial\Omega)$ for all $\ell \in \mathbf{Z}$.*

Proof. Let ψ be an arbitrary element of the closure of the range $R(\mathcal{T}(\ell^2))$ in $H^{s-m-1/2}(\partial\Omega)$, and let $\{\varphi^{(k)}\}$ be a sequence in $D(\mathcal{T}(\ell^2)) \subset H^{s-5/2+\kappa}(\partial\Omega)$ such that

$$\mathcal{T}(\ell^2)\varphi^{(k)} \to \psi \qquad \text{in } H^{s-m-1/2}(\partial\Omega).$$

We let

$$\tilde{\varphi}^{(k)} = \varphi^{(k)} \otimes e^{i\ell y}.$$

Then, using Lemmas 8.4.5 and 8.4.4, we find that

$$\begin{cases} \tilde{\varphi}^{(k)} \in D(\tilde{\mathscr{T}}), \\ \tilde{\mathscr{T}}\tilde{\varphi}^{(k)} = \mathscr{T}(\ell^2)\varphi^{(k)} \otimes e^{i\ell y} \to \psi \otimes e^{i\ell y} \qquad \text{in } H^{s-m-1/2}(\partial\Omega \times S). \end{cases}$$

Since the range $R(\tilde{\mathscr{T}})$ is closed, there exists an element $\tilde{\varphi} \in D(\tilde{\mathscr{T}}) \subset H^{s-5/2+\kappa}(\partial\Omega \times S)$ such that

$$\tilde{\mathscr{T}}\tilde{\varphi} = \psi \otimes e^{i\ell y}.$$

But Lemma 8.4.4 tells us that $\tilde{\varphi}$ can be expanded as follows:

$$\begin{cases} \tilde{\varphi} = \sum_{k\in\mathbf{Z}} \varphi_k \otimes e^{iky} \qquad \text{in } H^{s-5/2+\kappa}(\partial\Omega \times S), \\ \varphi_k \in H^{s-5/2+\kappa}(\partial\Omega). \end{cases}$$

Hence we have, by formula (6),

$$\psi \otimes e^{i\ell y} = \tilde{T}\tilde{\varphi} = \sum_{k\in\mathbf{Z}} T(k^2)\varphi_k \otimes e^{iky}.$$

Therefore, by the uniqueness of the Fourier expansion, it follows that

$$T(\ell^2)\varphi_\ell = \psi \in H^{s-m-1/2}(\partial\Omega).$$

This proves that $\varphi_\ell \in D(\mathscr{T}(\ell^2))$ and so $\psi \in R(\mathscr{T}(\ell^2))$.

Lemma 8.4.7 is proved. ■

To study relationships between codim $R(\tilde{\mathscr{T}})$ and codim $R(\mathscr{T}(\ell^2))$ ($\ell \in \mathbf{Z}$), we consider the adjoints $\tilde{\mathscr{T}}^*$ and $\mathscr{T}(\ell^2)^*$. The adjoint $\tilde{\mathscr{T}}^*$ of $\tilde{\mathscr{T}}$ is a closed linear operator from $H^{-s+m+1/2}(\partial\Omega \times S)$ into $H^{-s+5/2-\kappa}(\partial\Omega \times S)$ such that

$$(\tilde{\mathscr{T}}\tilde{\varphi}, \tilde{\psi}) = (\tilde{\varphi}, \tilde{\mathscr{T}}^*\tilde{\psi}), \qquad \tilde{\varphi} \in D(\tilde{\mathscr{T}}), \tilde{\psi} \in D(\tilde{\mathscr{T}}^*),$$

and the adjoint $\mathscr{T}(\ell^2)^*$ of $\mathscr{T}(\ell^2)$ is a closed linear operator from $H^{-s+m+1/2}(\partial\Omega)$ into $H^{-s+5/2-\kappa}(\partial\Omega)$ such that

$$(\mathscr{T}(\ell^2)\varphi, \psi) = (\varphi, \mathscr{T}(\ell^2)^*\psi), \qquad \varphi \in D(\mathscr{T}(\ell^2)), \psi \in D(\mathscr{T}(\ell^2)^*).$$

The next lemma allows us to give a characterization of the adjoints $\tilde{\mathscr{T}}^*$ and $\mathscr{T}(\ell^2)^*$ ($\ell \in \mathbf{Z}$) in terms of pseudo-differential operators.

8.4.8 Lemma. *Let M be a compact C^∞ manifold without boundary. If T is a classical pseudo-differential operator of order m on M, we define a densely defined, closed linear operator*

$$\mathscr{T}: H^{s-5/2+\kappa}(M) \to H^{s-m-1/2}(M) \qquad (s \in \mathbf{R})$$

as follows:

(a) *The domain* $D(\mathscr{T})$ *of* $\mathscr{T}$ *is the space*

$$D(\mathscr{T}) = \{\varphi \in H^{s-5/2+\kappa}(M);\ T\varphi \in H^{s-m-1/2}(M)\}.$$

(b) $\mathscr{T}\varphi = T\varphi$, $\varphi \in D(\mathscr{T})$.

Then the adjoint $\mathscr{T}^*$ *of* $\mathscr{T}$ *is characterized as follows*:

(c) *The domain* $D(\mathscr{T}^*)$ *of* $\mathscr{T}^*$ *is contained in the space* $\{\psi \in H^{-s+m+1/2}(M);\ T^*\psi \in H^{-s+5/2-\kappa}(M)\}$, *where* $T^* \in L^m_{\text{cl}}(M)$ *is the adjoint of* T.

(d) $\mathscr{T}^*\psi = T^*\psi$, $\psi \in D(\mathscr{T}^*)$.

Proof. Let ψ be an arbitrary element of $D(\mathscr{T}^*) \subset H^{-s+m+1/2}(M)$, and let $\{\psi_j\}$ be a sequence in $C^\infty(M)$ such that $\psi_j \to \psi$ in $H^{-s+m+1/2}(M)$. Then we have, for all $\varphi \in C^\infty(M) \subset D(\mathscr{T})$,

$$\begin{aligned}(\mathscr{T}^*\psi, \varphi) &= (\psi, \mathscr{T}\varphi)\\ &= (\psi, T\varphi)\\ &= \lim_j\, (\psi_j, T\varphi)\\ &= \lim_j\, (T^*\psi_j, \varphi)\\ &= (T^*\psi, \varphi),\end{aligned}$$

so that

$$T^*\psi = \mathscr{T}^*\psi \in H^{-s+5/2+\kappa}(M).$$

This proves the lemma. ■

Applying Lemma 8.4.8 to the pseudo-differential operators $\tilde{T}$ and $T(\alpha)$ $(\alpha \geq 0)$, we obtain:

8.4.9 Lemma. *The null spaces* $N(\tilde{\mathscr{T}}^*)$ *and* $N(\mathscr{T}(\alpha)^*)$ $(\alpha \geq 0)$ *are characterized respectively as follows*:

$$\begin{cases} N(\tilde{\mathscr{T}}^*) = \{\tilde{\psi} \in H^{-s+5/2-\kappa}(\partial\Omega \times S);\ \tilde{T}^*\tilde{\psi} = 0\}.\\ N(\mathscr{T}(\alpha)^*) = \{\psi \in H^{-s+5/2-\kappa}(\partial\Omega);\ T(\alpha)^*\psi = 0\}.\end{cases} \tag{7}$$

Furthermore we have:

8.4.10 Lemma. *The following two conditions are equivalent:*

(i) $\dim N(\tilde{\mathscr{T}}^*) < \infty$.
(ii) *There exists a finite subset J of $\mathbf{Z}$ such that*

$$\begin{cases} \dim N(\mathscr{T}(\ell^2)^*) < \infty & \text{if } \ell \in J, \\ \dim N(\mathscr{T}(\ell^2)^*) = 0 & \text{if } \ell \notin J. \end{cases}$$

Moreover, in this case, we have

$$\dim N(\tilde{\mathscr{T}}^*) = \sum_{\ell \in J} \dim N(\mathscr{T}(\ell^2)^*).$$

Proof. Passing to the adjoint in formula (6), we have, for all $\psi \in \mathscr{D}'(\partial\Omega)$ and $\ell \in \mathbf{Z}$,

$$T^*(\psi \otimes e^{i\ell y}) = T(\ell^2)^*\psi \otimes e^{i\ell y} \qquad \text{in } \mathscr{D}'(\partial\Omega \times S). \tag{8}$$

In fact, if $\varphi \in C^\infty(\partial\Omega)$ and $k \in \mathbf{Z}$, we have

$$\begin{aligned} (T^*(\psi \otimes e^{i\ell y}), \varphi \in e^{iky}) &= (\psi \otimes e^{i\ell y}, T(\varphi \otimes e^{iky})) \\ &= (\psi \otimes e^{i\ell y}, T(k^2)\varphi \otimes e^{iky})) \\ &= \begin{cases} 2\pi(\psi, T(\ell^2)\varphi) & \text{if } k = \ell, \\ 0 & \text{if } k \neq \ell, \end{cases} \end{aligned}$$

and also

$$\begin{aligned} &(T(\ell^2)^*\psi \otimes e^{i\ell y}, \varphi \otimes e^{iky}) \\ &\quad = \begin{cases} 2\pi(T(\ell^2)^*\psi, \varphi) = 2\pi(\psi, T(\ell^2)\varphi) & \text{if } k = \ell, \\ 0 & \text{if } k \neq \ell. \end{cases} \end{aligned}$$

This proves formula (8), since the set $\{\varphi \otimes e^{iky}; \varphi \in C^\infty(\partial\Omega), k \in \mathbf{Z}\}$ is dense in $C^\infty(\partial\Omega \times S)$.

By virtue of formulas (7) and (8), arguing as in the proof of Proposition 8.4.6, we can prove that conditions (i) and (ii) are equivalent. ■

The next proposition gives the most important relationship between codim $R(\tilde{\mathscr{T}})$ and codim $R(\mathscr{T}(\alpha'))$ when $\alpha' = \ell^2$, $\ell \in \mathbf{Z}$.

8.4.11 Proposition. *Suppose that the ranges $R(\tilde{\mathscr{T}})$ and $R(\mathscr{T}(\ell^2))$, $\ell \in \mathbf{Z}$, are closed. Then the following two conditions are equivalent:*

(i) codim $R(\tilde{\mathscr{T}}) < \infty$.

(ii) *There exists a finite subset J of $\mathbf{Z}$ such that*

$$\begin{cases} \operatorname{codim} R(\mathscr{T}(\ell^2)) < \infty & \text{if } \ell \in J, \\ \operatorname{codim} R(\mathscr{T}(\ell^2)) = 0 & \text{if } \ell \notin J. \end{cases}$$

Moreover, in this case, we have

$$\operatorname{codim} R(\tilde{\mathscr{T}}) = \sum_{\ell \in J} \operatorname{codim} R(\mathscr{T}(\ell^2)).$$

Proposition 8.4.11 is an immediate consequence of the closed range theorem (Theorem 3.4.7) and Lemma 8.4.10.

Proof of Theorem 8.4.1. (i)$\Rightarrow$(ii): 1) Suppose that the operator $\tilde{\mathfrak{A}}$: $H^{s-2+\kappa}(\bar{\Omega} \times S) \to H^{s-2}(\bar{\Omega} \times S) \times H^{s-m-1/2}(\partial\Omega \times S)$ is a Fredholm operator. Then it follows from result ($\widetilde{\text{III}}$) that the operator $\tilde{\mathscr{T}}$: $H^{s-5/2+\kappa}(\partial\Omega \times S) \to H^{s-m-1/2}(\partial\Omega \times S)$ is a Fredholm operator, and ind $\tilde{\mathfrak{A}}$ = ind $\tilde{\mathscr{T}}$. Therefore, applying Proposition 8.4.6, Lemma 8.4.7 and Proposition 8.4.11, we obtain the following results:

(a) There exists a finite subset I of $\mathbf{Z}$ such that

$$\begin{cases} \dim N(\mathscr{T}(\ell^2)) < \infty & \text{if } \ell \in I, \\ \dim N(\mathscr{T}(\ell^2)) = 0 & \text{if } \ell \notin I. \end{cases}$$

(b) The range $R(\mathscr{T}(\ell^2))$ is closed for all $\ell \in \mathbf{Z}$, and there exists a finite subset J of $\mathbf{Z}$ such that

$$\begin{cases} \operatorname{codim} R(\mathscr{T}(\ell^2)) < \infty & \text{if } \ell \in J, \\ \operatorname{codim} R(\mathscr{T}(\ell^2)) = 0 & \text{if } \ell \notin J. \end{cases}$$

In other words, the operator $\mathscr{T}(\ell^2)$: $H^{s-5/2+\kappa}(\partial\Omega) \to H^{s-m-1/2}(\partial\Omega)$ is a Fredholm operator for all $\ell \in \mathbf{Z}$, and it is bijective if $\ell \notin (I \cup J)$. Hence, in view of results (I), (II) and(III), it follows that the operator $\mathfrak{A}(\ell^2)$: $H^{s-2+\kappa}(\bar{\Omega}) \to H^{s-2}(\bar{\Omega}) \times H^{s-m-1/2}(\partial\Omega)$ is a Fredholm operator for all $\ell \in \mathbf{Z}$, and it is bijective if $\ell \notin (I \cup J)$.

2) Next we show that

$$\operatorname{ind} \mathfrak{A}(\alpha) = 0 \qquad \text{for all } \alpha \geq 0.$$

First observe that the domain $D(\mathfrak{A}(\alpha))$ does not depend on $\alpha \geq 0$. Take an integer ℓ such that $\ell \notin (I \cup J)$, and let

$$\alpha' = \ell^2.$$

Then we have, for all $u \in D(\mathfrak{A}(\alpha))$,

$$\begin{aligned}\mathfrak{A}(\alpha)u &= \{(A - \alpha)u, Bu\} \\ &= \{(A - \alpha')u, Bu\} + \{(\alpha' - \alpha)u, 0\} \\ &= \mathfrak{A}(\alpha')u + \{(\alpha' - \alpha)u, 0\},\end{aligned}$$

that is,

$$\mathfrak{A}(\alpha) = \mathfrak{A}(\alpha') + \{(\alpha' - \alpha)I, 0\}.$$

Since the operator $\mathfrak{A}(\alpha')$ is bijective, this gives that

$$\mathfrak{A}(\alpha)\mathfrak{A}(\alpha')^{-1} = \mathrm{I} + \{(\alpha - \alpha')I, 0\}\mathfrak{A}(\alpha')^{-1}.$$

But it follows from an application of the closed graph theorem (Theorem 3.4.3) that the inverse $\mathfrak{A}(\alpha')^{-1}\colon H^{s-2}(\overline{\Omega}) \times H^{s-m-1/2}(\partial\Omega) \to H^{s-2+\kappa}(\overline{\Omega})$ is continuous. Further Rellich's theorem tells us that the injection $H^{s-2+\kappa}(\overline{\Omega}) \to H^{s-2}(\overline{\Omega})$ is compact for $\kappa > 0$. Thus we find that the operator

$$\{(\alpha - \alpha')I, 0\}\mathfrak{A}(\alpha')^{-1}\colon H^{s-2}(\overline{\Omega}) \times H^{s-m-1/2}(\partial\Omega) \to H^{s-2}(\overline{\Omega}) \times H^{s-m-1/2}(\partial\Omega)$$

is compact. Therefore it follows from an application of Theorem 3.6.3 that

$$\operatorname{ind}(\mathfrak{A}(\alpha)\mathfrak{A}(\alpha')^{-1}) = 0.$$

Hence we have, by Corollary 3.7.3,

$$\begin{aligned}\operatorname{ind} \mathfrak{A}(\alpha) &= \operatorname{ind}((\mathfrak{A}(\alpha)\mathfrak{A}(\alpha')^{-1})\mathfrak{A}(\alpha')) \\ &= \operatorname{ind}(\mathfrak{A}(\alpha)\mathfrak{A}(\alpha')^{-1}) + \operatorname{ind} \mathfrak{A}(\alpha') \\ &= 0.\end{aligned}$$

3) Finally we show that:

There exists a constant $R' > 0$ such that if $\alpha' = \ell^2$ with $\ell \in \mathbf{Z}$ and $\ell^2 \geq R'$, then we have inequality (2) for all $u \in D(\mathfrak{A}(\alpha'))$. (9)

Applying Theorem 3.7.6 with

$$X = H^{s-5/2+\kappa}(\bar{\Omega} \times S),$$

$$Y = H^{s-2}(\bar{\Omega} \times S) \times H^{s-m-1/2}(\partial\Omega \times S),$$

$$Z = L^2(\Omega \times S),$$

$$T = \tilde{\mathfrak{A}},$$

we obtain that there exists a constant $\tilde{C} > 0$ such that

$$\|\tilde{u}\|^2_{H^{s-2+\kappa}(\bar{\Omega}\times S)} \leq \tilde{C}(\|\Lambda\tilde{u}\|^2_{H^{s-2}(\bar{\Omega}\times S)} + |B\tilde{u}|^2_{H^{s-m-1/2}(\partial\Omega\times S)} + \|\tilde{u}\|^2_{L^2(\Omega\times S)}), \qquad \tilde{u} \in D(\tilde{\mathfrak{A}}). \tag{10}$$

Now take

$$\tilde{u} = u \otimes e^{i\ell y}, \qquad u \in D(\mathfrak{A}(\ell^2)), \qquad \ell \in \mathbf{Z}.$$

Then we can apply inequality (4) to obtain the following:

(a) $\|\tilde{u}\|^2_{H^{s-2+\kappa}(\bar{\Omega}\times S)} \approx \|u\|^2_{H^{s-2+\kappa}(\bar{\Omega})} + (1 + \ell^2)^{s-2+\kappa}\|u\|^2_{L^2(\Omega)}$.

(b) $$\|\Lambda\tilde{u}\|^2_{H^{s-2}(\bar{\Omega}\times S)} = \|(A - \ell^2)u \otimes e^{i\ell y}\|^2_{H^{s-2}(\bar{\Omega}\times S)} \approx \|(A-\ell^2)u\|^2_{H^{s-2}(\bar{\Omega})} + (1+\ell^2)^{s-2}\|(A-\ell^2)u\|^2_{L^2(\Omega)}.$$

(c) $$|B\tilde{u}|^2_{H^{s-m-1/2}(\partial\Omega\times S)} = |Bu \otimes e^{i\ell y}|^2_{H^{s-m-1/2}(\partial\Omega\times S)} \approx |Bu|^2_{H^{s-m-1/2}(\partial\Omega)} + (1+\ell^2)^{s-m-1/2}|Bu|^2_{L^2(\partial\Omega)}.$$

Therefore, carrying these inequalities (a)–(c) into inequality (10), we have, with a constant $\tilde{C}' > 0$ independent of $\alpha' = \ell^2$,

$$\|u\|^2_{H^{s-2+\kappa}(\bar{\Omega})} + (\alpha')^{s-2+\kappa}\|u\|^2_{L^2(\Omega)} \leq \tilde{C}'(\|(A-\alpha')u\|^2_{H^{s-2}(\bar{\Omega})} + (\alpha')^{s-2}\|(A-\alpha')u\|^2_{L^2(\Omega)} + |Bu|^2_{H^{s-m-1/2}(\partial\Omega)} + (\alpha')^{s-m-1/2}|Bu|^2_{L^2(\partial\Omega)} + \|u\|_{L^2(\Omega)}).$$

But, since $s - 2 + \kappa > 0$, we can eliminate the last term $\|u\|_{L^2(\Omega)}$ on the right-hand side if α' is sufficiently large. This proves assertion (9).

(ii) $\Rightarrow$ (i): 1) Suppose that condition (ii) is satisfied. Then it follows from results (I), (II), (III) that:

The operator $\mathscr{T}(\alpha)$: $H^{s-5/2+\kappa}(\partial\Omega) \to H^{s-m-1/2}(\partial\Omega)$ is a Fredholm operator with index zero for all $\alpha \geq 0$, and it is bijective if $\alpha = \ell^2$, $\ell \in \mathbf{Z}$ and $\ell^2 \geq R'$.

Thus Proposition 8.4.6 tells us that

$$\dim N(\tilde{\mathscr{T}}) < \infty.$$

2) We show that:

The range $R(\tilde{\mathscr{T}})$ is closed and has finite codimension: codim $R(\tilde{\mathscr{T}}) < \infty$.

Then condition (i) follows from result ($\widetilde{\text{III}}$).

In view of Proposition 8.4.11, it suffices to prove the closedness of $R(\tilde{\mathscr{T}})$. To do so, we show that there exists a constant $\tilde{C} > 0$ such that

$$|\tilde{\varphi}|^2_{H^{s-5/2+\kappa}(\partial\Omega\times S)} \leq \tilde{C}(|\tilde{T}\tilde{\varphi}|^2_{H^{s-m-1/2}(\partial\Omega\times S)} + |\tilde{\varphi}|^2_{L^2(\partial\Omega\times S)}), \qquad \tilde{\varphi} \in D(\tilde{\mathscr{T}}). \quad (11)$$

Then the closedness of $R(\tilde{\mathscr{T}})$ follows from an application of Theorem 3.7.6.

We let

$$I = \{\ell \in \mathbf{Z};\, \ell^2 < R'\}.$$

Since the operators $\mathscr{T}(\ell^2)$, $\ell \in I$, are Fredholm operators and I is a finite set, applying Theorem 3.7.6 with

$$X = H^{s-5/2+\kappa}(\partial\Omega),$$

$$Y = H^{s-m-1/2}(\partial\Omega),$$

$$Z = L^2(\partial\Omega),$$

$$T = \mathscr{T}(\ell^2),$$

we obtain that there exists a constant $C_I > 0$ independent of $\ell \in I$ such that

$$|\varphi|^2_{H^{s-5/2+\kappa}(\partial\Omega)} \leq C_I(|T(\ell^2)\varphi|^2_{H^{s-m-1/2}(\partial\Omega)} + |\varphi|^2_{L^2(\partial\Omega)}), \qquad \varphi \in D(\mathscr{T}(\ell^2)).$$

Thus, using inequality (4) with $M = \partial\Omega$, we have

$$\begin{aligned}|\varphi \otimes e^{i\ell y}|^2_{H^{s-5/2+\kappa}(\partial\Omega\times S)} \\ \leq \tilde{C}_I(|T(\ell^2)\varphi \otimes e^{i\ell y}|^2_{H^{s-m-1/2}(\partial\Omega\times S)} + |\varphi \otimes e^{i\ell y}|^2_{L^2(\partial\Omega\times S)}), \\ \varphi \in D(\mathscr{T}(\ell)^2). \quad (12)\end{aligned}$$

Here $\tilde{C}_I > 0$ is a constant independent of $\ell \in I$.

On the other hand, using inequality (4), we find that inequality (2) is equivalent to the following:

$$\begin{aligned}\|u \otimes e^{i\ell y}\|^2_{H^{s-2+\kappa}(\bar{\Omega}\times S)} \\ \leq \tilde{C}'(\|\Lambda(u \otimes e^{i\ell y})\|^2_{H^{s-2}(\bar{\Omega}\times S)} + |Bu \otimes e^{i\ell y}|^2_{H^{s-m-1/2}(\partial\Omega\times S)}), \\ u \in D(\mathfrak{A}(\ell^2)).\end{aligned}$$

Here $\ell \in \mathbf{Z} \setminus I$ and $\tilde{C}' > 0$ is a constant independent of $\ell \in \mathbf{Z} \setminus I$. We take

$$u = P(\ell^2)\varphi, \qquad \varphi \in D(\mathscr{T}(\ell^2)), \qquad \ell \in \mathbf{Z} \setminus I.$$

Then, in view of formula (5), we have

$$|\varphi \otimes e^{i\ell y}|^2_{H^{s-5/2+\kappa}(\partial\Omega \times S)} \leq \tilde{C}_{II} |T(\ell^2)\varphi \otimes e^{i\ell y}|^2_{H^{s-m-1/2}(\partial\Omega \times S)},$$
$$\varphi \in D(\mathscr{T}(\ell^2)), \tag{13}$$

where $\tilde{C}_{II} > 0$ is a constant independent of $\ell \in \mathbf{Z} \setminus I$. In fact, it suffices to note that the Poisson operator $\tilde{P}$ is an isomorphism of $H^{s-5/2+\kappa}(\partial\Omega \times S)$ onto the space $N(\Lambda, s-2+\kappa) = \{\tilde{u} \in H^{s-2+\kappa}(\bar{\Omega} \times S); \Lambda\tilde{u} = 0 \text{ in } \Omega \times S\}$.

Now let $\tilde{\varphi}$ be an arbitrary element of $D(\tilde{\mathscr{T}})$. Then, using Lemmas 8.4.4 and 8.4.5, we find that $\tilde{\varphi}$ can be expanded as follows:

$$\begin{cases} \tilde{\varphi} = \sum_{\ell \in \mathbf{Z}} \varphi_\ell \otimes e^{i\ell y} \qquad \text{in } H^{s-5/2+\kappa}(\partial\Omega \times S), \\ \varphi_\ell \in D(\mathscr{T}(\ell^2)). \end{cases}$$

Therefore, combining inequalities (12) and (13), we obtain that

$$\begin{aligned}
&|\tilde{T}\tilde{\varphi}|^2_{H^{s-m-1/2}(\partial\Omega \times S)} \\
&\quad = \sum_{\ell \in \mathbf{Z}} |T(\ell^2)\varphi_\ell \otimes e^{i\ell y}|^2_{H^{s-m-1/2}(\partial\Omega \times S)} \\
&\quad = \sum_{\ell \in I} |T(\ell^2)\varphi_\ell \otimes e^{i\ell y}|^2_{H^{s-m-1/2}(\partial\Omega \times S)} \\
&\qquad + \sum_{\ell \in \mathbf{Z} \setminus I} |T(\ell^2)\varphi_\ell \otimes e^{i\ell y}|^2_{H^{s-m-1/2}(\partial\Omega \times S)} \\
&\quad \geq \frac{1}{\tilde{C}_{\mathrm{I}}} \sum_{\ell \in I} |\varphi_\ell \otimes e^{i\ell y}|^2_{H^{s-5/2+\kappa}(\partial\Omega \times S)} - \sum_{\ell \in I} |\varphi_\ell \otimes e^{i\ell y}|^2_{L^2(\partial\Omega \times S)} \\
&\qquad + \frac{1}{\tilde{C}_{\mathrm{II}}} \sum_{\ell \in \mathbf{Z} \setminus I} |\varphi_\ell \otimes e^{i\ell y}|^2_{H^{s-5/2+\kappa}(\partial\Omega \times S)} \\
&\quad \geq \min\left(\frac{1}{\tilde{C}_{\mathrm{I}}}, \frac{1}{\tilde{C}_{\mathrm{II}}}\right) \sum_{\ell \in \mathbf{Z}} |\varphi_\ell \otimes e^{i\ell y}|^2_{H^{s-5/2+\kappa}(\partial\Omega \times S)} \\
&\qquad - \sum_{\ell \in \mathbf{Z}} |\varphi_\ell \otimes e^{i\ell y}|^2_{L^2(\partial\Omega \times S)} \\
&\quad = \min\left(\frac{1}{\tilde{C}_{\mathrm{I}}}, \frac{1}{\tilde{C}_{\mathrm{II}}}\right) |\tilde{\varphi}|^2_{H^{s-5/2+\kappa}(\partial\Omega \times S)} - |\tilde{\varphi}|^2_{L^2(\partial\Omega \times S)}.
\end{aligned}$$

This proves inequality (11).

The proof of Theorem 8.4.1 is now complete. ■

Proof of Corollary 8.4.2

In view of Theorem 8.4.1, it suffices to prove the following:

If the operator $\tilde{\mathfrak{A}}: H^{s-2+\kappa}(\bar{\Omega} \times S) \to H^{s-2}(\bar{\Omega} \times S) \times H^{s-m-1/2}(\partial\Omega \times S)$ *is a Fredholm operator and if* $1 < \kappa < 2$, *then there exists a constant* $R > 0$ *such that if* $\alpha \geq R$ *the operator* $\mathfrak{A}(\alpha): H^{s-2+\kappa}(\bar{\Omega}) \to H^{s-2}(\bar{\Omega}) \times H^{s-m-1/2}(\partial\Omega)$ *is bijective and we have inequality* (3) *for all* $u \in D(\mathfrak{A}(\alpha))$.

(1) First we show that:

There exists a constant $R > 0$ such that if $\alpha \geq R$ then we have, for all $w \in D(\mathfrak{A}(\alpha))$ satisfying $Bw = 0$ on $\partial\Omega$,

$$\begin{aligned} &\|w\|^2_{H^{s-2+\kappa}(\bar{\Omega})} + \alpha^{s-2+\kappa}\|w\|^2_{L^2(\Omega)} \\ &\quad \leq C''(\|(A-\alpha)w\|^2_{H^{s-2}(\bar{\Omega})} + \alpha^{s-2}\|(A-\alpha)w\|^2_{L^2(\Omega)}), \end{aligned} \tag{14}$$

with a constant $C'' > 0$ independent of $\alpha \geq R$.

Choose a function $\zeta \in C_0^\infty(\mathbf{R})$ such that $0 \leq \zeta \leq 1$ on $\mathbf{R}$ and $\operatorname{supp} \zeta \subset [\pi/3, 5\pi/3]$, and let

$$\tilde{w}(x, y) = w(x) \otimes \zeta(y)e^{i\sqrt{\alpha}y}, \qquad i = \sqrt{-1}, \qquad \alpha \geq 0.$$

Then we have

$$\begin{aligned} \Lambda\tilde{w} &= \left(A + \frac{\partial^2}{\partial y^2}\right)\tilde{w} \\ &= (A-\alpha)w \otimes \zeta e^{i\sqrt{\alpha}y} + 2(i\sqrt{\alpha})w \otimes \zeta' e^{i\sqrt{\alpha}y} + w \otimes \zeta'' e^{i\sqrt{\alpha}y} \qquad \text{in } \Omega \times S, \end{aligned}$$

and

$$B\tilde{w} = Bw \otimes \zeta e^{i\sqrt{\alpha}y} = 0 \qquad \text{on } \partial\Omega \times \mathrm{S}.$$

Hence, using inequality (4), we find that

$$\tilde{w} \in D(\tilde{\mathfrak{A}}).$$

Therefore, applying inequality (10) to the function $\tilde{w} = w \otimes \zeta e^{i\sqrt{\alpha}y}$, we obtain that

$$\begin{aligned} &\|w \otimes \zeta e^{i\sqrt{\alpha}y}\|^2_{H^{s-2+\kappa}(\bar{\Omega}\times S)} \\ &\quad \leq \tilde{C}(\|\Lambda(w \otimes \zeta e^{i\sqrt{\alpha}y})\|^2_{H^{s-2}(\bar{\Omega}\times S)} + \|w \otimes \zeta e^{i\sqrt{\alpha}y}\|^2_{L^2(\Omega\times S)}) \\ &\quad \leq \tilde{C}'(\|(A-\alpha)w \otimes \zeta e^{i\sqrt{\alpha}y}\|^2_{H^{s-2}(\bar{\Omega}\times S)} + \|w \otimes \zeta'' e^{i\sqrt{\alpha}y}\|^2_{H^{s-2}(\bar{\Omega}\times S)} \\ &\qquad + 4\alpha\|w \otimes \zeta' e^{i\sqrt{\alpha}y}\|^2_{H^{s-2}(\bar{\Omega}\times S)} + \|w \otimes \zeta e^{i\sqrt{\alpha}y}\|^2_{L^2(\Omega\times S)}). \end{aligned} \tag{15}$$

We can estimate each term of (15) as follows (cf. the proof of inequality (4)):

1) $$\|(A-\alpha)w \otimes \zeta e^{i\sqrt{\alpha}y}\|^2_{H^{s-2}(\bar{\Omega}\times S)}$$
$$\approx \|(A-\alpha)w\|^2_{H^{s-2}(\bar{\Omega})}\|\zeta e^{i\sqrt{\alpha}y}\|^2_{L^2(S)} + \|(A-\alpha)w\|^2_{L^2(\Omega)}\|\zeta e^{i\sqrt{\alpha}y}\|^2_{H^{s-2}(S)}$$
$$\leq c_1(\|(A-\alpha)w\|^2_{H^{s-2}(\bar{\Omega})} + \alpha^{s-2}\|(A-\alpha)w\|^2_{L^2(\Omega)}).$$

Here $c_1 > 0$ is a constant depending only on ζ and s.

In fact, it suffices to note the following:

$$\|\zeta e^{i\sqrt{\alpha}y}\|^2_{H^{s-2}(S)} = \left\|\left(1-\frac{d^2}{dy^2}\right)^{(s-2)/2}(\zeta e^{i\sqrt{\alpha}y})\right\|^2_{L^2(S)}$$
$$\approx \int_{\mathbf{R}} (1+\eta^2)^{s-2}|\hat{\zeta}(\eta-\sqrt{\alpha})|^2\, d\eta$$
$$= \int_{\mathbf{R}} (1+(\eta+\sqrt{\alpha})^2)^{s-2}|\hat{\zeta}(\eta)|^2\, d\eta$$
$$\leq 4^{s-2}\left(\int_{\mathbf{R}} (1+\eta^2)^{s-2}|\hat{\zeta}(\eta)|^2\, d\eta + \alpha^{s-2}\int_{\mathbf{R}} |\hat{\zeta}(\eta)|^2\, d\eta\right)$$
$$= 4^{s-2}(\|\zeta\|^2_{H^{s-2}(S)} + \alpha^{s-2}\|\zeta\|^2_{L^2(S)}).$$

2) $$\|w \otimes \zeta'' e^{i\sqrt{\alpha}y}\|^2_{H^{s-2}(\bar{\Omega}\times S)}$$
$$\approx \|w\|^2_{H^{s-2}(\bar{\Omega})}\|\zeta'' e^{i\sqrt{\alpha}y}\|^2_{L^2(S)} + \|w\|^2_{L^2(\Omega)}\|\zeta'' e^{i\sqrt{\alpha}y}\|^2_{H^{s-2}(S)}$$
$$\leq c_2(\|w\|^2_{H^{s-2}(\bar{\Omega})} + \alpha^{s-2}\|w\|^2_{L^2(\Omega)}).$$

Here $c_2 > 0$ is a constant depending only on ζ and s.

3) $$\|w \otimes \zeta' e^{i\sqrt{\alpha}y}\|^2_{H^{s-2}(\bar{\Omega}\times S)} \leq c_3(\|w\|^2_{H^{s-2}(\bar{\Omega})} + \alpha^{s-2}\|w\|^2_{L^2(\Omega)}).$$

Here $c_3 > 0$ is a constant depending only on ζ and s.

4) $$\|w \otimes \zeta e^{i\sqrt{\alpha}y}\|^2_{L^2(\Omega\times S)} \approx \|w\|^2_{L^2(\Omega)}.$$

5) $\|w \otimes \zeta e^{i\sqrt{\alpha}y}\|^2_{H^{s-2+\kappa}(\bar{\Omega}\times S)}$
$$\approx \|w\|^2_{H^{s-2+\kappa}(\bar{\Omega})}\|\zeta e^{i\sqrt{\alpha}y}\|^2_{L^2(S)} + \|w\|^2_{L^2(\Omega)}\|\zeta e^{i\sqrt{\alpha}y}\|^2_{H^{s-2+\kappa}(S)}$$
$$\geq c_4\|w\|^2_{H^{s-2+\kappa}(\bar{\Omega})} + c_5\alpha^{s-2+\kappa}\|w\|^2_{L^2(\Omega)} - c_6\|w\|^2_{L^2(\Omega)}.$$

Here c_4, c_5, c_6 are positive constants depending only on ζ and s.

In fact, we have

$$\|\zeta e^{i\sqrt{\alpha}y}\|^2_{H^{s-2+\kappa}(S)} \approx \int_{\mathbf{R}} (1 + (\eta + \sqrt{\alpha})^2)^{s-2+\kappa} |\hat{\zeta}(\eta)|^2 \, d\eta$$

$$\geq 4^{-(s-2+\kappa)} \alpha^{s-2+\kappa} \int_{\mathbf{R}} |\hat{\zeta}(\eta)|^2 \, d\eta$$

$$- \int_{\mathbf{R}} (1 + \eta^2)^{s-2+\kappa} |\hat{\zeta}(\eta)|^2 \, d\eta$$

$$= 4^{-(s-2+\kappa)} \alpha^{s-2+\kappa} \|\zeta\|^2_{L^2(S)} - \|\zeta\|^2_{H^{s-2+\kappa}(S)}.$$

Therefore, carrying these inequalities 1)–5) into inequality (15), we have, with a constant $C_1 > 0$ independent of $\alpha \geq 1$,

$$\|w\|^2_{H^{s-2+\kappa}(\bar{\Omega})} + \alpha^{s-2+\kappa} \|w\|^2_{L^2(\Omega)}$$

$$\leq C_1(\|(A - \alpha)w\|^2_{H^{s-2}(\bar{\Omega})} + \alpha^{s-2} \|(A - \alpha)w\|^2_{L^2(\Omega)}$$

$$+ \alpha \|w\|^2_{H^{s-2}(\bar{\Omega})} + \alpha^{s-1} \|w\|^2_{L^2(\Omega)}). \tag{16}$$

To eliminate the term $\alpha\|w\|^2_{H^{s-2}(\bar{\Omega})}$ on the right-hand side of (16), we need the following interpolation inequalities:

(a) For every $\varepsilon > 0$, there exists a constant $C_\varepsilon > 0$ such that

$$\|u\|^2_{H^{s-1}(\bar{\Omega})} \leq \varepsilon \|u\|^2_{H^{s-2+\kappa}(\bar{\Omega})} + C_\varepsilon \|u\|^2_{L^2(\Omega)}, \qquad u \in H^{s-2+\kappa}(\bar{\Omega}). \tag{17}$$

(b) There exists a constant $C_2 > 0$ independent of $\alpha \geq 0$ such that

$$\alpha \|v\|^2_{H^{s-2}(\bar{\Omega})} \leq C_2(\|v\|^2_{H^{s-1}(\bar{\Omega})} + \alpha^{s-1} \|v\|^2_{L^2(\Omega)}), \qquad v \in H^{s-1}(\bar{\Omega}). \tag{18}$$

Inequality (18) is an immediate consequence of the following inequality:

$$\alpha(1 + |\xi|^2)^{s-2} \leq (1 + |\xi|^2)^{s-1} + \alpha^{s-1}, \qquad \alpha \geq 0, \, \xi \in \mathbf{R}^n.$$

Applying inequalities (17) and (18) to the function w, we have

$$\alpha \|w\|^2_{H^{s-2}(\bar{\Omega})} \leq \varepsilon C_2 \|w\|^2_{H^{s-2+\kappa}(\bar{\Omega})} + C_2(C_\varepsilon + \alpha^{s-1}) \|w\|^2_{L^2(\Omega)},$$

and hence (taking $\varepsilon = 1/2C_1C_2$)

$$\alpha C_1 \|w\|^2_{H^{s-2}(\bar{\Omega})} \leq \frac{1}{2} \|w\|^2_{H^{s-2+\kappa}(\bar{\Omega})} + C_3 \alpha^{s-1} \|w\|^2_{L^2(\Omega)},$$

with a constant $C_3 > 0$ independent of $\alpha \geq 1$. Therefore, carrying this into inequality (16), we have, with another constant $C_4 > 0$,

$$\|w\|^2_{H^{s-2+\kappa}(\bar{\Omega})} + \alpha^{s-2+\kappa}\|w\|^2_{L^2(\Omega)}$$
$$\leq C_4(\|(A-\alpha)w\|^2_{H^{s-2}(\bar{\Omega})} + \alpha^{s-2}\|(A-\alpha)w\|^2_{L^2(\Omega)} + \alpha^{s-1}\|w\|^2_{L^2(\Omega)}).$$

Since $s-2+\kappa > s-1$, we can eliminate the last term $\alpha^{s-1}\|w\|^2_{L^2(\Omega)}$ on the right-hand side if α is sufficiently large. This proves inequality (14).

(2) Inequality (14) tells us that the operator $\mathfrak{A}(\alpha)$ is injective for all $\alpha \geq R$; hence it is bijective for all $\alpha \geq R$, since ind $\mathfrak{A}(\alpha) = 0$ for all $\alpha \geq 0$.

(3) Finally we show that inequality (3) holds for all $u \in D(\mathfrak{A}(\alpha))$.

We may suppose that:

1. $R' = \ell_0^2$ for some positive integer ℓ_0,
2. $R \geq R'$,

where R' is the constant in condition (ii) of Theorem 8.4.1 and R is the constant in step (1). Thus, for any $\alpha \geq R$, we can choose a positive integer $\ell \geq \ell_0$ such that

$$\ell^2 \leq \alpha \leq (\ell+1)^2.$$

We let

$$\alpha' = \ell^2.$$

Then we have

$$\begin{cases} \alpha' \leq \alpha \leq (\ell+1)^2 \leq 4\alpha', \\ \alpha - \alpha' \leq 2\ell + 1 \leq 3\sqrt{\alpha'}. \end{cases} \tag{19}$$

Now let u be an arbitrary element of $D(\mathfrak{A}(\alpha))$. Since $\alpha' = \ell^2 \geq \ell_0^2 = R'$, it follows from Theorem 8.4.1 that there exists a unique solution $v \in H^{s-2+\kappa}(\bar{\Omega})$ of the problem

$$\begin{cases} (A-\alpha')v = 0 & \text{in } \Omega, \\ Bv = Bu & \text{on } \partial\Omega, \end{cases} \tag{20}$$

and that

$$\|v\|^2_{H^{s-2+\kappa}(\bar{\Omega})} + (\alpha')^{s-2+\kappa}\|v\|^2_{L^2(\Omega)}$$
$$\leq C'(|Bu|^2_{H^{s-m-1/2}(\partial\Omega)} + (\alpha')^{s-m-1/2}|Bu|^2_{L^2(\partial\Omega)}).$$

By inequalities (19), this gives that

$$\|v\|^2_{H^{s-2+\kappa}(\bar{\Omega})} + \alpha^{s-2+\kappa}\|v\|^2_{L^2(\Omega)} \leq C_5(|Bu|^2_{H^{s-m-1/2}(\partial\Omega)} + \alpha^{s-m-1/2}|Bu|^2_{L^2(\partial\Omega)}). \tag{21}$$

Here $C_5 > 0$ is a constant independent of $\alpha \geq R$.

We let

$$w = u - v.$$

Then, in view of (20), it follows that

$$\begin{cases} (A - \alpha)w = (A - \alpha)u - (\alpha' - \alpha)v & \text{in } \Omega, \\ Bw = 0 & \text{on } \partial\Omega. \end{cases}$$

Thus we can apply inequality (14) to obtain that

$$\begin{aligned} &\|w\|^2_{H^{s-2+\kappa}(\bar{\Omega})} + \alpha^{s-2+\kappa}\|w\|^2_{L^2(\Omega)} \\ &\leq C''(\|(A - \alpha)u - (\alpha' - \alpha)v\|^2_{H^{s-2}(\bar{\Omega})} \\ &\quad + \alpha^{s-2}\|(A - \alpha)u - (\alpha' - \alpha)v\|^2_{L^2(\Omega)}) \\ &\leq 2C''(\|(A - \alpha)u\|^2_{H^{s-2}(\bar{\Omega})} + (\alpha' - \alpha)^2\|v\|^2_{H^{s-2}(\bar{\Omega})} \\ &\quad + \alpha^{s-2}\|(A - \alpha)u\|^2_{L^2(\Omega)} + \alpha^{s-2}(\alpha' - \alpha)^2\|v\|^2_{L^2(\Omega)}). \end{aligned}$$

Further, in view of inequalities (19) and (18), this gives that

$$\begin{aligned} &\|w\|^2_{H^{s-2+\kappa}(\bar{\Omega})} + \alpha^{s-2+\kappa}\|w\|^2_{L^2(\Omega)} \\ &\leq C_6(\|(A - \alpha)u\|^2_{H^{s-2}(\bar{\Omega})} + \alpha^{s-2}\|(A - \alpha)u\|^2_{L^2(\Omega)} \\ &\quad + \|v\|^2_{H^{s-1}(\bar{\Omega})} + \alpha^{s-1}\|v\|_{L^2(\Omega)}). \end{aligned} \tag{22}$$

Here $C_6 > 0$ is a constant independent of $\alpha \geq R$.

Since $s - 2 + \kappa > s - 1$, combining inequalities (21) and (22), we obtain inequality (3).

The proof of Corollary 8.4.2 is complete. ■

Notes

Section 8.1: Theorem 8.1.1 is adapted from the book of Gilbarg-Trudinger [1], where a thorough treatment of quasilinear elliptic equations is given.

Section 8.2: The proof of Theorem 8.2.3, based on the jump formula, may conceivably be new. Theorem 8.2.4 is an expression of the fact that every solution u of the equation $Au = 0$ can be expressed by means of a single layer potential.

Section 8.3: The main idea of the proof of Theorems 8.3.4–8.3.12 is due to Hörmander [2] and Seeley [3], and details were carried out by Taira [1].

Section 8.4: Theorem 8.4.1 and Corollary 8.4.2 are adapted from Taira [5]. It is worth pointing out here that the key lemma in the proof of Theorem 8.4.1 is Lemma 8.4.5 which follows from the unique solvability of the Dirichlet problem. Hence the methods and results in this section can be extended to treat general boundary value problems for *degenerate* elliptic differential operators of second order which enjoy an existence and uniqueness theory for the Dirichlet problem in the framework of appropriate functions spaces. For detailed study of the Dirichlet problem for such operators, the reader might refer to Oleĭnik-Radkevič [1] and Stroock-Varadhan [2], which are based on the work of Fichera [1].

There are many topics on elliptic boundary value problems which we have not touched on. The reader is referred especially to Agmon [1], Lions-Magenes [1] and Rempel-Schulze [1] for more material.

9 Markov Processes, Semigroups and Boundary Value Problems

This chapter is devoted to the functional analytic approach to the study of Markov processes. In Section 9.1, we summarize the basic definitions and results about Markov processes, and formulate Markov processes in terms of transition functions. From the viewpoint of functional analysis, the transition function is something more convenient than the Markov process itself. In fact, we can associate with each transition function in a natural way a family of bounded linear operators acting on the space of continuous functions on the state space, and the so-called Markov property implies that this family forms a semigroup. Transition functions and their associated semigroups are studied in Section 9.2. These semigroups are called Feller semigroups. In Section 9.3, using the Hille–Yosida theory of semigroups, we characterize Feller semigroups in terms of their infinitesimal generators. In Sections 9.4 and 9.5, we describe analytically the infinitesimal generator of a Feller semigroup when the state space is the closure of a bounded domain in Euclidean space. The infinitesimal generator of a Feller semigroup is described by an integro-differential operator and a boundary condition. Hence we are reduced to the study of boundary value problems in the theory of partial differential equations. In Section 9.6, we consider, conversely, under which conditions on an integro-differential operator and a boundary condition one can construct a Feller semigroup. We prove general existence theorems for Feller semigroups in terms of boundary value problems in the case when an integro-differential operator and a boundary condition are differential operators. In other words, we prove general existence theorems for Markov processes with continuous paths (diffusion processes). The construction of Feller semigroups will be carried out in Chapter 10.

9.1. Markov Processes and Transition Functions

Definition of a Markov Process

Let K be a locally compact, separable metric space and $\mathscr{B}$ the σ-algebra of all Borel sets in K, that is, the smallest σ-algebra containing all open sets in K. Let $(\Omega, \mathscr{F}, P)$ be a probability space. A function X defined on Ω taking values in K is called a *random variable* if it satisfies

$$\{X \in E\} = X^{-1}(E) \in \mathscr{F} \qquad \text{for all } E \in \mathscr{B}.$$

We express this by saying that X is $\mathscr{F}/\mathscr{B}$-measurable. A family $\{x_t\}_{t \geq 0}$ of random variables is called a *stochastic process*, and may be thought of as the motion in time of a physical particle. The space K is called the *state space* and Ω the *sample space*. For a fixed $\omega \in \Omega$, the function $x_t(\omega)$, $t \geq 0$, defines in the state space K a *trajectory* or *path* of the process corresponding to the sample point ω.

In this generality the notion of a stochastic process is of course not so interesting. The most important class of stochastic processes is the class of Markov processes which is characterized by the Markov property. Intuitively, the *Markov property* is that the prediction of subsequent motion of a particle, knowing its position at time t, does not depend on what has been observed during the time interval $[0, t]$; that is, the "future" is independent of the "past" for a known "present".

This vague idea can be made precise and effective in several ways.

If $\{Z_\lambda\}_{\lambda \in \Lambda}$ is a family of random variables, we let

$\sigma(Z_\lambda; \lambda \in \Lambda) =$ the smallest σ-algebra, contained in $\mathscr{F}$, with respect to which all Z_λ are measurable.

If $\{x_t\}_{t \geq 0}$ is a stochastic process, we introduce three sub-σ-algebras of $\mathscr{F}$ as follows:

$$\begin{cases} \mathscr{F}_{\leq t} = \sigma(x_s; 0 \leq s \leq t), \\ \mathscr{F}_{=t} = \sigma(x_t), \\ \mathscr{F}_{\geq t} = \sigma(x_s; t \leq s < \infty). \end{cases}$$

Intuitively, an event in $\mathscr{F}_{\le t}$ is determined by the behavior of the process $\{x_s\}$ up to time t and an event in $\mathscr{F}_{\ge t}$ by its behavior after time t. Thus they represent respectively the "past" and "future" relative to the "present" moment.

Let $\mathscr{G}$ be a sub-σ-algebra of $\mathscr{F}$. The *conditional probability* $P(B|\mathscr{G})$ of $B \in \mathscr{F}$ for given $\mathscr{G}$ is a $\mathscr{G}$-measurable function on Ω which satisfies

$$P(A \cap B) = \int_A P(B|\mathscr{G})(\omega)\, dP(\omega), \qquad A \in \mathscr{G}.$$

We remark that the function $P(B|\mathscr{G})$ is determined up to a set in $\mathscr{G}$ of P-measure zero, that is, it is an equivalence class of $\mathscr{G}$-measurable functions on Ω with respect to the measure P.

9.1.1 Definition. A stochastic process $\{x_t\}$ is called a *Markov process* if one of the following equivalent conditions is satisfied:

(i) $P(A \cap B|\mathscr{F}_{=t}) = P(A|\mathscr{F}_{=t})P(B|\mathscr{F}_{=t})$, $A \in \mathscr{F}_{\le t}$, $B \in \mathscr{F}_{\ge t}$.
(ii) $P(B|\mathscr{F}_{\le t}) = P(B|\mathscr{F}_{=t})$, $B \in \mathscr{F}_{\ge t}$.

Intuitively, condition (i) means that, given the "present" of the process, the "past" event A and the "future" event B are conditionally independent. Condition (ii) means that the conditional probability of a "future" event B given the "present" is the same as the probability of B given the "present" and the "past".

An observer may record not only the trajectories of the process, but also some other occurrences, only indirectly related or entirely unrelated to the process. Thus we obtain a broader and more flexible formulation of the Markov property if we enlarge the "past" as follows:

Let $\{\mathscr{F}_t\}_{t \ge 0}$ be a family of sub-σ-algebras of $\mathscr{F}$ such that:

(a) If $s < t$, then $\mathscr{F}_s \subset \mathscr{F}_t$;
(b) For each $t \ge 0$, the function x_t is $\mathscr{F}_t/\mathscr{B}$-measurable, that is,

$$\{x_t \in E\} \in \mathscr{F}_t \qquad \text{for all } E \in \mathscr{B}.$$

We express property (a) by saying that the family $\{\mathscr{F}_t\}$ is increasing, and property (b) by saying that the process $\{x_t\}$ is adapted to $\{\mathscr{F}_t\}$. Note that the family $\{\mathscr{F}_{\le t}\}_{t \ge 0}$ satisfies both conditions and is the minimal possible one.

9.1.2 Definition. Let $\{x_t\}_{t \ge 0}$ be a stochastic process and $\{\mathscr{F}_t\}_{t \ge 0}$ an increasing family of sub-σ-algebras of $\mathscr{F}$. We say that $\{x_t\}$ is a *Markov process* with respect to $\{\mathscr{F}_t\}$ if it satisfies

(i) $\{x_t\}$ is adapted to $\{\mathscr{F}_t\}$;
(ii) $P(B|\mathscr{F}_t) = P(B|\mathscr{F}_{=t})$, $B \in \mathscr{F}_{\geq t}$.

This definition reduces to Definition 9.1.1 if $\mathscr{F}_t = \mathscr{F}_{\leq t}$. Clearly, choosing the family $\{\mathscr{F}_t\}$ as the "past" has the effect of making it harder for the Markov property to hold, while the property becomes more powerful.

Now we introduce a class of Markov processes which we will deal with in this book.

9.1.3 Definition. Suppose that we are given the following:

1) A locally compact, separable metric space K and the σ-algebra $\mathscr{B}$ of all Borel sets in K. A point ∂ is adjoined to K as the point at infinity if K is not compact, and as an isolated point if K is compact. We let

$$K_\partial = K \cup \{\partial\},$$

$$\mathscr{B}_\partial = \text{the } \sigma\text{-algebra in } K_\partial \text{ generated by } \mathscr{B}.$$

2) The space Ω of all mappings ω: $[0, \infty] \to K_\partial$ such that $\omega(\infty) = \partial$ and that if $\omega(t) = \partial$ then $\omega(s) = \partial$ for all $s \geq t$. We let ω_∂ be the constant map $\omega_\partial(t) = \partial$ for all $t \in [0, \infty]$.

3) For each $t \in [0, \infty]$, the coordinate map x_t defined by $x_t(\omega) = \omega(t)$, $\omega \in \Omega$.

4) For each $t \in [0, \infty]$, a mapping φ_t: $\Omega \to \Omega$ defined by $\varphi_t\omega(s) = \omega(t + s)$, $\omega \in \Omega$. Note that $\varphi_\infty\omega = \omega_\partial$ and $x_t \circ \varphi_s = x_{t+s}$ for all $t, s \in [0, \infty]$.

5) A σ-algebra $\mathscr{F}$ in Ω and an increasing family $\{\mathscr{F}_t\}_{0 \leq t \leq \infty}$ of sub-σ-algebras of $\mathscr{F}$.

6) For each $x \in K_\partial$, a probability measure P_x on $(\Omega, \mathscr{F})$.

We say that these elements define a (temporally homogeneous) *Markov process* $\mathscr{X} = (x_t, \mathscr{F}, \mathscr{F}_t, P_x)$ if the following four conditions are satisfied:

(i) For each $0 \leq t < \infty$, the function x_t is $\mathscr{F}_t/\mathscr{B}_\partial$-measurable, that is,

$$\{x_t \in E\} \in \mathscr{F}_t \qquad \text{for all } E \in \mathscr{B}_\partial.$$

(ii) For each $0 \leq t < \infty$ and $E \in \mathscr{B}$, the function

$$p_t(x, E) = P_x\{x_t \in E\} \tag{1}$$

is a Borel measurable function of $x \in K$.

(iii) $P_x\{\omega \in \Omega;\ x_0(\omega) = x\} = 1$ for each $x \in K_\partial$.

(iv) For all t, $h \in [0, \infty]$, $x \in K_\partial$ and $E \in \mathscr{B}_\partial$, we have

$$P_x\{x_{t+h} \in E \mid \mathscr{F}_t\} = p_h(x_t, E),$$

or equivalently

$$P_x(A \cap \{x_{t+h} \in E\}) = \int_A p_h(x_t(\omega), E)\, dP_x(\omega), \qquad A \in \mathscr{F}_t.$$

In this definition, the term "Markov process" means a family of Markov processes over $(\Omega, \mathscr{F}, P_x)$ with respect to $\{\mathscr{F}_t\}$, one Markov process for each of the measures P_x corresponding to all possible initial positions $x \in K_\partial$.

Here is an intuitive way of thinking about the above definition of a Markov process. The value $P_x(A)$, $A \in \mathscr{F}$, may be interpreted as the probability of the event A under the condition that a particle starts at position x; hence the value $p_t(x, E)$ expresses the transition probability that a particle starting at position x will be found in the set E at time t. The function p_t is called the *transition function* of the process $\mathscr{X}$. The transition function p_t specifies the probability structure of the process. The intuitive meaning of the crucial condition (iv) is that the future behavior of a particle, knowing its history up to time t, is the same as the behavior of a particle starting at $x_t(\omega)$, that is, a particle starts afresh. A particle moves in the space K until it "dies" at which time it reaches the point ∂; hence the point ∂ is called the *terminal point*.

With this interpretation in mind, we let

$$\zeta(\omega) = \inf\{t \in [0, \infty];\ x_t(\omega) = \partial\}.$$

The random variable ζ is called the *lifetime* of the process $\mathscr{X}$.

Transition Functions

From the viewpoint of analysis, the transition function is something more convenient than the Markov process itself. In fact, we can associate with each transition function in a natural way a family of bounded linear operators acting on the space of continuous functions on the state space, and the Markov property implies that this family forms a semigroup, as will be shown later on.

Our first job is thus to give the precise definition of a transition function adapted to the theory of semigroups:

9.1.4 Definition. Let K be a locally compact, separable metric space, and $\mathscr{B}$ the σ-algebra of all Borel sets in K. A function $p_t(x, E)$, defined for all $t \geq 0$,

$x \in K$ and $E \in \mathscr{B}$, is called a (temporally homogeneous) *Markov transition function* on K if it satisfies the following four conditions:

(a) $p_t(x, \cdot)$ is a non-negative measure on $\mathscr{B}$ and $p_t(x, K) \le 1$ for each $t \ge 0$ and $x \in K$.

(b) $p_t(\cdot, E)$ is a Borel measurable function for each $t \ge 0$ and $E \in \mathscr{B}$.

(c) $p_0(x, \{x\}) = 1$ for each $x \in K$.

(d) (The Chapman-Kolmogorov equation) For any $t, s \ge 0$, $x \in K$ and $E \in \mathscr{B}$, we have

$$p_{t+s}(x, E) = \int_K p_t(x, dy) p_s(y, E).$$

9.1.5 Remark. It is just condition (d) which reflects the Markov property that a particle starts afresh (cf. Figure 0-1).

In view of conditions (a) and (d), it follows that

$$\begin{aligned} p_{t+s}(x, K) &= \int_K p_t(x, dy) p_s(y, K) \\ &\le \int_K p_t(x, dy) \\ &= p_t(x, K). \end{aligned}$$

This implies that $p_t(x, K)$ is a non-increasing function of t. Hence the limit $p_{+0}(x, K) = \lim_{t \downarrow 0} p_t(x, K)$ exists.

A Markov transition function p_t is said to be *normal* if it satisfies

$$p_{+0}(x, K) = 1 \qquad \text{for all } x \in K.$$

The next theorem justifies our definition of a transition function, and hence it will be fundamental for our further study of Markov processes:

9.1.6 Theorem. *For every Markov process, the function p_t, defined by formula* (1), *is a Markov transition function. Conversely, every normal Markov transition function corresponds to some Markov process.*

Feller Transition Functions

Let (K, ρ) be a locally compact, separable metric space, and $B(K)$ the space of real-valued, bounded Borel measurable functions on K; $B(K)$ is a Banach space with the supremum norm

$$\|f\| = \sup_{x \in K} |f(x)|.$$

If p_t is a transition function on K, we let

$$T_t f(x) = \int_K p_t(x, dy) f(y), \qquad f \in B(K).$$

Then, applying Theorem 1.17.2 with $\mathscr{F} = \mathscr{B}$ and $\mathscr{H} = \{f \in B(K);\ T_t f$ is Borel measurable$\}$, we obtain that $\mathscr{H} = B(K)$, that is, the function $T_t f$ is Borel measurable whenever $f \in B(K)$. In fact, it suffices to note the following:

1. Condition (b) of Definition 9.1.4 implies condition (i) of Theorem 1.17.2.
2. An application of the monotone convergence theorem (Theorem 1.19.1) gives that condition (ii) of Theorem 1.17.2 is satisfied.

In view of condition (a) of Definition 9.1.4, it follows that, for each $t \geq 0$, the operator T_t is a non-negative, contraction linear operator on $B(K)$ into itself:

$$f \in B(K),\ 0 \leq f \leq 1 \text{ on } K \qquad \Rightarrow \qquad 0 \leq T_t f \leq 1 \text{ on } K.$$

Furthermore, we have, by condition (d) of Definition 9.1.4,

$$\begin{aligned} T_{t+s} f(x) &= \int_K p_{t+s}(x, dy) f(y) \\ &= \int_K \int_K p_t(x, dz) p_s(z, dy) f(y) \\ &= \int_K p_t(x, dz) T_s f(z) \\ &= T_t(T_s f)(x), \end{aligned}$$

so that the operators T_t form a *semigroup*:

$$T_{t+s} = T_t \cdot T_s, \qquad t, s \geq 0.$$

We also have, by condition (c) of Definition 9.1.4,

$$T_0 = I = \text{the identity operator}.$$

The Hille-Yosida theory of semigroups requires the strong continuity of $\{T_t\}_{t \geq 0}$:

$$\lim_{t \downarrow 0} \|T_t f - f\| = 0, \qquad f \in B(K), \tag{2}$$

that is,

$$\lim_{t\downarrow 0} \sup_{x\in K} \left| \int_K p_t(x, dy) f(y) - f(x) \right| = 0, \qquad f \in B(K). \tag{2'}$$

Now, taking $f = \chi_{\{x\}} \in B(K)$ in formula (2′), we obtain that

$$\lim_{t\downarrow 0} p_t(x, \{x\}) = 1, \qquad x \in K. \tag{3}$$

But, the Brownian motion transition function, the most important and interesting example, does not satisfy condition (3). Thus we shift out attention to continuous functions, instead of measurable functions.

Let $C(K)$ be the space of real-valued, bounded continuous functions on K; $C(K)$ is a Banach space with the supremum (maximum) norm

$$\|f\| = \sup_{x\in K} |f(x)|.$$

We say that a function $f \in C(K)$ converges to $a \in \mathbf{R}$ as $x \to \partial$ if, for each $\varepsilon > 0$, there exists a compact subset E of K such that

$$|f(x) - a| < \varepsilon \qquad \text{for all } x \in K \setminus E,$$

and write $\lim_{x\to\partial} f(x) = a$. Let $C_0(K)$ be the subspace of $C(K)$ which consists of all functions satisfying $\lim_{x\to\partial} f(x) = 0$; $C_0(K)$ is a closed subspace of $C(K)$. Note that $C_0(K)$ may be identified with $C(K)$ if K is compact.

Now we introduce a useful convention:

Any real-valued function f on K is extended to $K_\partial = K \cup \{\partial\}$ by setting $f(\partial) = 0$.

From this point of view, the space $C_0(K)$ is identified with the subspace of $C(K_\partial)$ which consists of all functions f satisfying $f(\partial) = 0$, and also

$$C(K_\partial) = \{\text{constant functions}\} + C_0(K).$$

Further, we extend a transition function p_t on K to a transition function p'_t on K_∂ as follows:

$$\begin{cases} p'_t(x, E) = p_t(x, E), & x \in K, E \in \mathscr{B}; \\ p'_t(x, \{\partial\}) = 1 - p_t(x, K), & x \in K; \\ p'_t(\partial, K) = 0, \quad p'_t(\partial, \{\partial\}) = 1. \end{cases}$$

We remark that our convention is consistent, since $T_t f(\partial) = f(\partial) = 0$.

Now we introduce some conditions on the measures $p_t(x, \cdot)$ related to continuity in $x \in K$, for every fixed $t \geq 0$.

9.1.7 Definition. A Markov transition function p_t is called a *Feller function* if the function

$$T_t f(x) = \int_K p_t(x, dy) f(y)$$

is a continuous function of $x \in K$ whenever f is bounded and continuous on K. That is, the Feller property is equivalent to saying that the space $C(K)$ is an invariant subspace of $B(K)$ for the operators T_t. We say that p_t is a C_0*-function* if the space $C_0(K)$ is an invariant subspace of $C(K)$ for the operators T_t.

9.1.8 Remark. The Feller property is equivalent to saying that the measures $p_t(x, \cdot)$ depend continuously on $x \in K$ in the usual weak topology, for every fixed $t \geq 0$ (cf. Section 3.3).

Path Functions of Markov Processes

It is naturally interesting and important to ask the following question:

Given a Markov transition function p_t, under which conditions on p_t does there exist a Markov process with transition function p_t whose paths are almost surely *continuous*?

A Markov process $\mathscr{X} = (x_t, \mathscr{F}, \mathscr{F}_t, P_x)$ is said to be *right-continuous* provided that for each $x \in K$

$$P_x\{\omega \in \Omega; \text{the mapping } t \to x_t(\omega) \text{ is a right-continuous function from } [0, \infty) \text{ into } K_\partial\} = 1.$$

Further we say that X is *continuous* provided that for each $x \in K$

$$P_x\{\omega \in \Omega; \text{the mapping } t \to x_t(\omega) \text{ is a continuous function from } [0, \zeta) \text{ into } K\} = 1.$$

Here ζ is the lifetime of the process $\mathscr{X}$.

Now we give some useful criteria for path-continuity in terms of transition functions (cf. Dynkin [1], Kinney [1], Seregin [1]):

9.1.9 Theorem. *Let K be a locally compact, separable metric space, and p_t a normal Markov transition function on K.*

(i) *Suppose that the following two conditions are satisfied:*

(*L*) *For each $s > 0$ and each compact $E \subset K$, we have*

$$\lim_{x \to \partial} \sup_{0 \leq t \leq s} p_t(x, E) = 0.$$

(*M*) *For each $\varepsilon > 0$ and each compact $E \subset K$, we have*

$$\lim_{t \downarrow 0} \sup_{x \in E} p_t(x, K \setminus U_\varepsilon(x)) = 0,$$

where $U_\varepsilon(x) = \{y \in K;\ \rho(y, x) < \varepsilon\}$ is an ε-neighborhood of x.

Then there exists a Markov process $\mathscr{X}$ with transition function p_t whose paths are right-continuous *on $[0, \infty)$ and have left-hand limits on $[0, \zeta)$ almost surely.*

(ii) *Suppose that condition (L) and the following condition (replacing condition (M)) are satisfied:*

(*N*) *For each $\varepsilon > 0$ and each compact $E \subset K$, we have*

$$\lim_{t \downarrow 0} \frac{1}{t} \sup_{x \in E} p_t(x, K \setminus U_\varepsilon(x)) = 0.$$

Then there exists a Markov process $\mathscr{X}$ with transition function p_t whose paths are almost surely continuous *on $[0, \zeta)$.*

9.1.10 Remarks. 1. Condition (*L*) is trivially satisfied if the state space K is compact.

2. It is known (cf. Dynkin [1], Lemma 6.2) that if the paths of a Markov process are right-continuous, then the transition function p_t satisfies

$$\lim_{t \downarrow 0} p_t(x, U_\varepsilon(x)) = 1, \qquad x \in K.$$

Strong Markov Processes

A Markov process is called a strong Markov process if the "starting afresh" property holds not only for every fixed moment but also for suitable random times.

Now we formulate precisely this "strong" Markov property. Let $\mathscr{X} = (x_t, \mathscr{F}, \mathscr{F}_t, P_x)$ be a Markov process. A mapping $\tau: \Omega \to [0, \infty]$ is called a *stopping time* or *Markov time* with respect to $\{\mathscr{F}_t\}$ if it satisfies

$$\{\tau \leq t\} \in \mathscr{F}_t \qquad \text{for all } t \in [0, \infty).$$

Intuitively, this means that the events $\{\tau \leq t\}$ depend on the process only up to time t, but not on the "future" after time t. We remark that any non-negative constant mapping is a stopping time.

If τ is a stopping time with respect to $\{\mathscr{F}_t\}$, we let

$$\mathscr{F}_\tau = \{A \in \mathscr{F}; A \cap \{\tau \leq t\} \in \mathscr{F}_t \text{ for all } t \in [0, \infty)\}.$$

Intuitively, one may think of $\mathscr{F}_\tau$ as the "past" up to random time τ. It is easy to verify that $\mathscr{F}_\tau$ is a σ-algebra. If $\tau \equiv t_0$ for some constant $t_0 \geq 0$, then $\mathscr{F}_\tau$ reduces to $\mathscr{F}_{t_0}$.

For each $t \in [0, \infty]$, we define a mapping

$$\Phi_t: [0, t] \times \Omega \to K_\partial$$

by

$$\Phi_t(s, \omega) = x_s(\omega).$$

We say that $\mathscr{X} = (x_t, \mathscr{F}, \mathscr{F}_t, P_x)$ is *progressively measurable* with respect to $\{\mathscr{F}_t\}$ if the mapping Φ_t is $\mathscr{B}_{[0,t]} \times \mathscr{F}_t/\mathscr{B}_\partial$-measurable for each $t \in [0, \infty]$, that is, if we have

$$\{\Phi_t \in E\} = \Phi_t^{-1}(E) \in \mathscr{B}_{[0,t]} \times \mathscr{F}_t \qquad \text{for all } E \in \mathscr{B}_\partial.$$

Here $\mathscr{B}_{[0,t]}$ is the σ-algebra of all Borel sets in the interval $[0, t]$. We remark that if $\mathscr{X}$ is progressively measurable and τ is a stopping time, then the mapping $x_\tau: \omega \to x_{\tau(\omega)}(\omega)$ is $\mathscr{F}_\tau/\mathscr{B}_\partial$-measurable.

9.1.11 Definition. A progressively measurable Markov process $\mathscr{X} = (x_t, \mathscr{F}, \mathscr{F}_t, P_x)$ is said to have the *strong Markov property* with respect to $\{\mathscr{F}_t\}$ if the following condition is satisfied:

For all $h \geq 0$, $x \in K_\partial$, $E \in \mathscr{B}_\partial$ and all stopping times τ, we have

$$\mathrm{P}_x\{\mathrm{x}_{\tau+h} \in E \mid \mathscr{F}_\tau\} = p_h(x_\tau, E),$$

or equivalently

$$P_x(A \cap \{x_{\tau+h} \in E\}) = \int_A p_h(x_{\tau(\omega)}(\omega), E)\, dP_x(\omega), \qquad A \in \mathscr{F}_\tau.$$

This expresses the idea of "starting afresh" at random times.

The next result gives a useful criterion for the strong Markov property:

9.1.12 Theorem. *Every right-continuous Markov process whose transition function has the C_0-property is a strong Markov process.*

We state a simple criterion for the strong Markov property in terms of transition functions. To do so, we introduce the following:

9.1.13 Definition. A Markov transition function p_t on K is said to be *uniformly stochastically continuous* on K if the following condition is satisfied:

For each $\varepsilon > 0$ and each compact $E \subset K$, we have

$$\lim_{t \downarrow 0} \sup_{x \in E} [1 - p_t(x, U_\varepsilon(x))] = 0, \tag{4}$$

where $U_\varepsilon(x) = \{y \in K; \rho(y, x) < \varepsilon\}$ is an ε-neighborhood of x.

We remark that every uniformly stochastically continuous transition function is normal and satisfies condition (M).

Combining part (i) of Theorem 9.1.9 and Theorem 9.1.12, we have the following:

9.1.14 Theorem. *Every uniformly stochastically continuous C_0-transition function which satisfies condition (L) is the transition function of some strong Markov process.*

A continuous strong Markov process is called a *diffusion process.*

The next result states a sufficient condition for the existence of a diffusion process with a prescribed Markov transition function:

9.1.15 Theorem. *Every uniformly stochastically continuous C_0-transition function which satisfies conditions (L) and (N) is the transition function of some diffusion process.*

This is an immediate consequence of part (ii) of Theorem 9.1.9 and Theorem 9.1.12.

9.2. Transition Functions and Feller Semigroups

In this section we study the semigroups associated with Feller transition functions.

Let (K, ρ) be a locally compact, separable metric space and $K_\partial = K \cup \{\partial\}$ its one-point compactification.

In Section 9.1, we have proved:

9.2.1 Theorem. *If p_t is a Feller transition function on K, then the associated operators $\{T_t\}_{t \geq 0}$, defined by*

$$T_t f(x) = \int_K p_t(x, dy) f(y), \qquad f \in C(K), \tag{1}$$

form a non-negative and contraction semigroup on $C(K)$:

(i) $f \in C(K)$, $0 \leq f \leq 1$ *on* $K \Rightarrow 0 \leq T_t f \leq 1$ *on* K.
(ii) $T_{t+s} = T_t \cdot T_s$, $t, s \geq 0$ (semigroup property); $T_0 = I$.

The purpose of this section is to prove a converse:

9.2.2 Theorem. *If $\{T_t\}_{t \geq 0}$ is a non-negative and contraction semigroup on $C_0(K)$, then there exists a unique C_0-transition function p_t on K such that formula* (1) *holds.*

Proof. We fix $t \geq 0$ and $x \in K$, and define

$$F(f) = T_t(f - f(\partial))(x) + f(\partial), \qquad f \in C(K_\partial).$$

Then it follows that F is a linear functional on $C(K_\partial)$, and it is non-negative and bounded with norm $\|F\| \leq 1$, since T_t is a non-negative and contraction operator on $C_0(K)$. Therefore, applying the Riesz representation theorem (Theorem 3.3.3) to the functional F, we obtain that there exists a unique finite, non-negative Borel measure $p_t(x, \cdot)$ on K_∂ such that

$$T_t(f - f(\partial))(x) + f(\partial) = F(f) = \int_{K_\partial} p_t(x, dy) f(y), \qquad f \in C(K_\partial). \tag{2}$$

Note that this formula reduces to formula (1) if $f \in C_0(K)$, that is, if $f(\partial) = 0$.

We show that the measures p_t satisfy conditions (a) through (d) of Definition 9.1.4.

(a) Since F is contractive, formula (2) with $f = 1$ gives that

$$p_t(x, K) \leq p_t(x, K_\partial) = \int_{K_\partial} p_t(x, dy) = F(1) \leq 1, \qquad x \in K.$$

(c) Since $T_0 = I$, it follows that $p_0(x, \{x\}) = 1$ for each $x \in K$.

(b) We prove that the function $p_t(\cdot, E)$ is Borel measurable for each $E \in \mathscr{B}$. To do so, it suffices to show that the collection

$$\mathscr{A}_\partial = \{E \in \mathscr{B}_\partial; p_t(\cdot, E) \text{ is } \mathscr{B}_\partial\text{-measurable}\}$$

coincides with the σ-algebra $\mathscr{B}_\partial$.

1) The collection $\mathscr{A}_\partial$ contains the collection $\mathscr{O}_\partial$ of all open subsets of K_∂:

$$\mathscr{A}_\partial \supset \mathscr{O}_\partial. \tag{3}$$

In fact, if $G \in \mathscr{O}_\partial$, we let (cf. Figure 9-1 below)

$$f_n(x) = \min\{n\rho(x, K_\partial \setminus G), 1\}, \qquad n = 1, 2, \ldots.$$

Then f_n is a function $C(K_\partial)$, and satisfies

$$\lim_{n\to\infty} f_n(x) = \begin{cases} 1 & \text{if } x \in G, \\ 0 & \text{if } x \in K_\partial \setminus G. \end{cases}$$

Thus, by virtue of the dominated convergence theorem (Theorem 1.19.3), we obtain from formula (2) with $f = f_n$ that

$$\lim_{n\to\infty} (T_t(f_n - f_n(\partial))(x) + f_n(\partial)) = \lim_{n\to\infty} \int_{K_\partial} p_t(x, dy) f_n(y) = p_t(x, G).$$

Since the functions $T_t(f_n - f_n(\partial))$ are continuous, this proves that the limit function $p_t(\cdot, G)$ is $\mathscr{B}_\partial$-measurable, and so $G \in \mathscr{A}_\partial$.

2) We have, by assertion (3),

$$d(\mathscr{O}_\partial) \subset d(\mathscr{A}_\partial). \tag{4}$$

3) The collection $\mathscr{A}_\partial$ is a d-system:

$$d(\mathscr{A}_\partial) = \mathscr{A}_\partial. \tag{5}$$

In fact:

(i) $K_\partial \in \mathscr{A}_\partial$, since $p_t(\cdot, K_\partial) = 1$.

(ii) If $A, B \in \mathscr{A}_\partial$ and $A \subset B$, then it follows that the function

$$p_t(\cdot, B \setminus A) = p_t(\cdot, B) - p_t(\cdot, A)$$

is $\mathscr{B}_\partial$-measurable. This proves that $B \setminus A \in \mathscr{A}_\partial$.

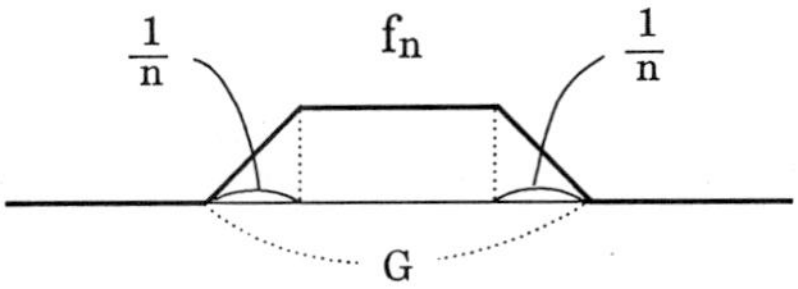

Figure 9-1

(iii) If $\{A_n\}_{n=1}^{\infty}$ is an increasing sequence of elements of $\mathscr{A}_\partial$, then it follows that the function

$$p_t\left(\cdot, \bigcup_{n=1}^{\infty} A_n\right) = \lim_{n\to\infty} p_t(\cdot, A_n)$$

is $\mathscr{B}_\partial$-measurable. This proves that $\bigcup_{n=1}^{\infty} A_n \in \mathscr{A}_\partial$.

4) Since $\mathscr{O}_\partial$ is a π-system, it follows from an application of the monotone class theorem (Theorem 1.16.1) that

$$d(\mathscr{O}_\partial) = \sigma(\mathscr{O}_\partial) = \mathscr{B}_\partial. \tag{6}$$

5) Combining assertions (6), (4) and (5), we obtain that

$$\mathscr{B}_\partial = d(\mathscr{O}_\partial) \subset d(\mathscr{A}_\partial) = \mathscr{A}_\partial \subset \mathscr{B}_\partial,$$

so that

$$\mathscr{A}_\partial = \mathscr{B}_\partial.$$

(d) In view of the semigroup property and Fubini's theorem (Theorem 1.19.4), it follows from formula (2) that for all $f \in C(K_\partial)$ we have

$$\begin{aligned}
\int_{K_\partial} p_{t+s}(x, dz) f(z) &= T_{t+s}(f - f(\partial)) + f(\partial) \\
&= T_t(T_s(f - f(\partial))) + f(\partial) \\
&= \int_{K_\partial} p_t(x, dy) \int_{K_\partial} p_s(y, dz)(f(z) - f(\partial)) + f(\partial) \\
&= \int_{K_\partial} \left(\int_{K_\partial} p_t(x, dy) p_s(y, dz) \right) f(z),
\end{aligned}$$

since $p_t(\cdot, K_\partial) = 1$. Hence the uniqueness part of the Riesz representation theorem gives that

$$p_{t+s}(x, E) = \int_{K_\partial} p_t(x, dy) p_s(y, E) = \int_K p_t(x, dy) p_s(y, E), \qquad E \in \mathscr{B},$$

since $p_s(\partial, \{\partial\}) = 1$ and so $p_s(\partial, K) = 0$.

Finally the C_0-property of p_t comes automatically, since $T_t\colon C_0(K) \to C_0(K)$.

The proof of Theorem 9.2.2 is now complete. ■

The C_0-property deals with continuity of a transition function $p_t(x, E)$ in x, and does not, by itself, have any concern with continuity in t. Now we give a necessary and sufficient condition on $p_t(x, E)$ in order that its associated semigroup $\{T_t\}_{t \geq 0}$ be strongly continuous in t on the space $C_0(K)$:

$$\lim_{s \to 0} \| T_{t+s} f - T_t f \| = 0, \qquad f \in C_0(K). \tag{7}$$

9.2.3 Theorem. *Let p_t be a C_0-transition function on K. Then the associated semigroup $\{T_t\}_{t \geq 0}$, defined by formula* (1), *is strongly continuous in t on $C_0(K)$ if and only if p_t is uniformly stochastically continuous on K and satisfies condition (L) of Theorem* 9.1.9.

9.2.4 Remark. Since the semigroup $\{T_t\}$ is a contraction semigroup, it follows from Remark 3.9.2 that the strong continuity (7) of $\{T_t\}$ in t for $t \geq 0$ is equivalent to the strong continuity at $t = 0$:

$$\lim_{t \downarrow 0} \| T_t f - f \| = 0, \qquad f \in C_0(K). \tag{7'}$$

Proof of Theorem 9.2.3. (i) The "if" part: Since continuous functions with compact support are dense in $C_0(K)$, it suffices to prove formula (7′) for all such functions f.

For any compact subset E of K containing supp f, we have

$$\begin{aligned} \| T_t f - f \| &\leq \sup_{x \in E} | T_t f(x) - f(x) | + \sup_{x \in K \setminus E} | T_t f(x) | \\ &\leq \sup_{x \in E} | T_t f(x) - f(x) | + \| f \| \sup_{x \in K \setminus E} p_t(x, \operatorname{supp} f). \end{aligned} \tag{8}$$

But, condition (L) implies that, for each $\varepsilon > 0$, one can find a compact subset E of K such that, for all sufficiently small $t > 0$,

$$\sup_{x \in K \setminus E} p_t(x, \operatorname{supp} f) < \varepsilon. \tag{9}$$

On the other hand, we have, for each $\delta > 0$,

$$T_t f(x) - f(x) = \int_{U_\delta(x)} p_t(x, dy)(f(y) - f(x))$$

$$+ \int_{K \setminus U_\delta(x)} p_t(x, dy)(f(y) - f(x)) - f(x)(1 - p_t(x, K)),$$

and hence

$$\sup_{x \in E} |T_t f(x) - f(x)|$$

$$\leq \sup_{\rho(x, y) < \delta} |f(y) - f(x)| + 3\|f\| \sup_{x \in E}[1 - p_t(x, U_\delta(x))].$$

Since f is uniformly continuous, we can choose a constant $\delta > 0$ such that

$$\sup_{\rho(x, y) < \delta} |f(y) - f(x)| < \varepsilon.$$

Further it follows from condition (9.1.4) with $\varepsilon = \delta$ (the uniform stochastic continuity of p_t) that, for all sufficiently small $t > 0$,

$$\sup_{x \in E}[1 - p_t(x, U_\delta(x))] < \varepsilon.$$

Hence we have, for all sufficiently small $t > 0$,

$$\sup_{x \in E} |T_t f(x) - f(x)| < \varepsilon(1 + 3\|f\|). \tag{10}$$

Therefore, carrying inequalities (9) and (10) into inequality (8), we obtain that, for all sufficiently small $t > 0$,

$$\|T_t f - f\| < \varepsilon(1 + 4\|f\|).$$

This proves formula (7′), that is, the strong continuity of $\{T_t\}$.

(ii) The "only if" part: For any $x \in K$ and $\varepsilon > 0$, we define (cf. Figure 9-2)

$$f_x(y) = \begin{cases} 1 - \dfrac{1}{\varepsilon}\rho(x, y) & \text{if } \rho(x, y) \leq \varepsilon, \\ 0 & \text{if } \rho(x, y) > \varepsilon. \end{cases} \tag{11}$$

Let E be an arbitrary compact subset of K. Then, for all sufficiently small $\varepsilon > 0$, the functions f_x, $x \in E$, are in $C_0(K)$ and satisfy

$$\|f_x - f_z\| \leq \frac{1}{\varepsilon}\rho(x, z), \qquad x, z \in E. \tag{12}$$

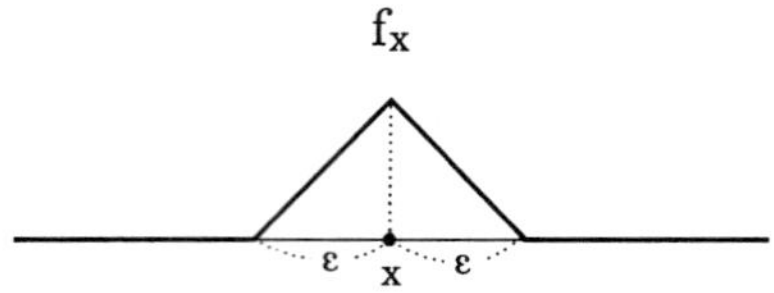

Figure 9-2

But, for any $\delta > 0$, by the compactness of E, we can find a finite number of points $x_1, \ldots, x_n$ of E such that

$$E = \bigcup_{k=1}^{n} U_{\delta\varepsilon/4}(x_k),$$

and hence

$$\min_{1 \le k \le n} \rho(x, x_k) \le \frac{\delta\varepsilon}{4} \qquad \text{for all } x \in E.$$

Thus, combining this with inequality (12), we obtain that

$$\min_{1 \le k \le n} \|f_x - f_{x_k}\| \le \frac{\delta}{4} \qquad \text{for all } x \in E. \tag{13}$$

Now we have, by formula (11),

$$\begin{aligned} 0 \le 1 - p_t(x, U_\varepsilon(x)) &\le 1 - \int_{K_\partial} p_t(x, dy) f_x(y) \\ &= f_x(x) - T_t f_x(x) \\ &\le \|f_x - T_t f_x\| \\ &\le \|f_x - f_{x_k}\| + \|f_{x_k} - T_t f_{x_k}\| \\ &\quad + \|T_t f_{x_k} - T_t f_x\| \\ &\le 2\|f_x - f_{x_k}\| + \|f_{x_k} - T_t f_{x_k}\|. \end{aligned}$$

In view of inequality (13), the first term on the last inequality is bounded by $\delta/2$ for the right choice of k. Further it follows from the strong continuity (7′) of $\{T_t\}$ that the second term tends to zero as $t \downarrow 0$ for each $k = 1, \ldots, n$.

Consequently, we have, for all sufficiently small $t > 0$,

$$\sup_{x \in E} [1 - p_t(x, U_\varepsilon(x))] \le \delta.$$

This proves condition (9.1.4), that is, the uniform stochastic continuity of p_t.

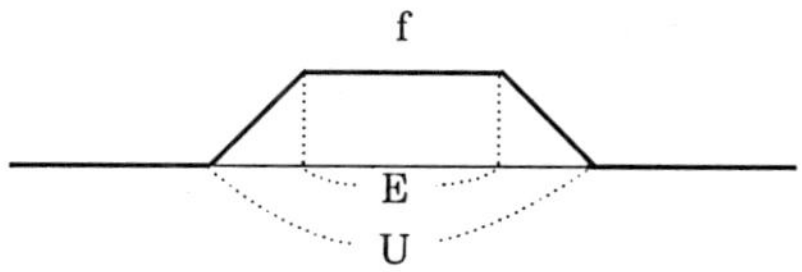

Figure 9-3

It remains to verify condition (L). We assume to the contrary that:

For some $s > 0$ and some compact $E \subset K$, there exist a constant $\varepsilon_0 > 0$, a sequence $\{t_k\}$, $t_k \downarrow t$ $(0 \leq t \leq s)$ and a sequence $\{x_k\}$, $x_k \to \partial$, such that

$$p_{t_k}(x_k, E) \geq \varepsilon_0. \tag{14}$$

Now take a relatively compact subset U of K containing E, and let (cf. Figure 9-3)

$$f(x) = \frac{\rho(x, K \setminus U)}{\rho(x, E) + \rho(x, K \setminus U)}.$$

Then the function f is in $C_0(K)$ and satisfies

$$T_t f(x) = \int_K p_t(x, dy) f(y) \geq p_t(x, E) \geq 0.$$

Therefore, combining this with inequality (14), we obtain that

$$T_{t_k} f(x_k) \geq p_{t_k}(x_k, E) \geq \varepsilon_0.$$

But we have

$$T_{t_k} f(x_k) \leq \| T_{t_k} f - T_t f \| + T_t f(x_k). \tag{15}$$

Since the semigroup $\{T_t\}$ is strongly continuous and $T_t f \in C_0(K)$, we can let $k \to \infty$ in inequality (15) to obtain that

$$\limsup_{k \to \infty} T_{t_k} f(x_k) = 0.$$

This is a contradiction.

The proof of Theorem 9.2.3 is complete. ■

9.2.5 Definition. A family $\{T_t\}_{t \geq 0}$ of bounded linear operators acting on $C_0(K)$ is called a *Feller semigroup* on K if it satisfies the following three conditions:

(i) $T_{t+s} = T_t \cdot T_s$, $t, s \geq 0$; $T_0 = I$.
(ii) $\{T_t\}$ is strongly continuous in t for $t \geq 0$:

$$\lim_{s \to 0} \| T_{t+s} f - T_t f \| = 0, \qquad f \in C_0(K).$$

(iii) $\{T_t\}$ is non-negative and contractive on $C_0(K)$:

$$f \in C_0(K),\ 0 \leq f \leq 1 \text{ on } K \quad \Rightarrow \quad 0 \leq T_t f \leq 1 \text{ on } K.$$

Combining Theorems 9.2.1, 9.2.2 and 9.2.3, we have the following:

9.2.6 Theorem. *If p_t is a uniformly stochastically continuous C_0-transition function on K which satisfies condition (L) of Theorem 9.1.9, then its associated operators $\{T_t\}_{t \geq 0}$, defined by formula (1), form a Feller semigroup on K. Conversely, if $\{T_t\}_{t \geq 0}$ is a Feller semigroup on K, then there exists a uniformly stochastically continuous C_0-transition function p_t on K, satisfying condition (L), such that formula (1) holds.*

9.3. Feller Semigroups and their Infinitesimal Generators

Let K be a *compact* metric space. We recall that the space $C_0(K)$ may be identified with $C(K)$.

If $\{T_t\}_{t \geq 0}$ is a Feller semigroup on K, we define the *infinitesimal generator* $\mathfrak{A}$ of $\{T_t\}$ by

$$\mathfrak{A}u = \lim_{t \downarrow 0} \frac{T_t u - u}{t}, \tag{1}$$

provided that the limit (1) exists in $C(K)$. More precisely, the generator $\mathfrak{A}$ is a linear operator from $C(K)$ into itself defined as follows:

1. The domain $D(\mathfrak{A})$ of $\mathfrak{A}$ is the set

$$D(\mathfrak{A}) = \{u \in C(K);\ \text{the limit (1) exists}\}.$$

2. $$\mathfrak{A}u = \lim_{t \downarrow 0} \frac{T_t u - u}{t}, \qquad u \in D(\mathfrak{A}).$$

The next theorem is a version of the Hille-Yosida theorem (Theorem 3.9.6) adapted to the present context.

9.3.1 Theorem. (i) *Let $\{T_t\}_{t \geq 0}$ be a Feller semigroup on a compact metric space K and let $\mathfrak{A}$ be its infinitesimal generator. Then we have:*

(a) *The domin $D(\mathfrak{A})$ is everywhere dense in $C(K)$.*

(b) *For each $\alpha > 0$, the equation $(\alpha I - \mathfrak{A})u = f$ has a unique solution u in $D(\mathfrak{A})$ for any $f \in C(K)$. Hence, for each $\alpha > 0$, the* Green operator *$(\alpha I - \mathfrak{A})^{-1}: C(K) \to C(K)$ can be defined by*

$$u = (\alpha I - \mathfrak{A})^{-1} f, \qquad f \in C(K).$$

(c) *For each $\alpha > 0$, the operator $(\alpha I - \mathfrak{A})^{-1}$ is non-negative on $C(K)$:*

$$f \in C(K),\ f \geq 0 \text{ on } K \quad \Rightarrow \quad (\alpha I - \mathfrak{A})^{-1} f \geq 0 \text{ on } K.$$

(d) *For each $\alpha > 0$, the operator $(\alpha I - \mathfrak{A})^{-1}$ is bounded on $C(K)$ with norm*

$$\|(\alpha I - \mathfrak{A})^{-1}\| \leq \frac{1}{\alpha}.$$

(ii) *Conversely, if $\mathfrak{A}$ is a linear operator from $C(K)$ into itself satisfying condition* (a) *and if there is a constant $\alpha_0 \geq 0$ such that for all $\alpha > \alpha_0$ conditions* (b) *through* (d) *are satisfied, then $\mathfrak{A}$ is the infinitesimal generator of some Feller semigroup $\{T_t\}_{t \geq 0}$ on K.*

Proof. In view of the Hille-Yosida theorem (Theorem 3.9.6), it suffices to show that the semigroup $\{T_t\}_{t \geq 0}$ is non-negative if and only if its resolvent $\{(\alpha I - \mathfrak{A})^{-1}\}_{\alpha > \alpha_0}$ is non-negative.

The "only if" part follows from expression (3.9.3) of $(\alpha I - \mathfrak{A})^{-1}$ in terms of $\{T_t\}$. The "if" part follows from definition (3.9.6) of T_t and expression (3.9.5) of $T_t(\alpha)$ in terms of $J_\alpha = \alpha(\alpha I - \mathfrak{A})^{-1}$. ■

9.3.2 Corollary. *Let $\mathfrak{A}$ be the infinitesimal generator of a Feller semigroup on K. Suppose that the constant function* 1 *belongs to the domain $D(\mathfrak{A})$ of $\mathfrak{A}$ and that for some constant c we have*

$$\mathfrak{A}1 \leq -c \qquad \text{on } K. \tag{2}$$

Then the operator $\mathfrak{A}' = \mathfrak{A} + cI$ is the infinitesimal generator of some Feller semigroup on K.

Proof. It follows from part (i) of Theorem 3.9.1 that, for all $\alpha > c$, the operator

$$(\alpha I - \mathfrak{A}')^{-1} = ((\alpha - c)I - \mathfrak{A})^{-1}$$

is defined and non-negative on the whole space $C(K)$. But, in view of inequality (2), it follows that

$$\alpha \leq \alpha - (\mathfrak{A}1 + c) = (\alpha I - \mathfrak{A}')1 \qquad \text{on } K,$$

so that

$$\begin{aligned} \alpha(\alpha I - \mathfrak{A}')^{-1}1 &\leq (\alpha I - \mathfrak{A}')^{-1}(\alpha I - \mathfrak{A}')1 \\ &= 1 \qquad \text{on } K. \end{aligned}$$

Hence we have, for all $\alpha > c$,

$$\begin{aligned} \|(\alpha I - \mathfrak{A}')^{-1}\| &= \|(\alpha I - \mathfrak{A}')^{-1}1\| \\ &\leq \frac{1}{\alpha}. \end{aligned}$$

Therefore, applying part (ii) of Theorem 9.3.1 to the operator $\mathfrak{A}'$, we obtain that $\mathfrak{A}'$ is the infinitesimal generator of some Feller semigroup on K. This completes the proof of Corollary 9.3.2. ■

Although Theorem 9.3.1 tells us precisely when a linear operator $\mathfrak{A}$ is the infinitesimal generator of some Feller semigroup, it is usually difficult to verify conditions (b) through (d). So we give useful criteria in terms of the maximum principle.

9.3.3 Theorem. (i) *Let B be a linear operator from $C(K)$ into itself, and suppose that:*

(α) *The domain $D(B)$ of B is everywhere dense in $C(K)$.*

(β) *There exists an open and dense subset K_0 of K such that if $u \in D(B)$ takes its positive maximum at a point x_0 of K_0, then we have*

$$Bu(x_0) \leq 0.$$

Then the operator B is closable in $C(K)$.

(ii) *Let B be as in part* (i), *and suppose that:*

(β') *If $u \in D(B)$ takes its positive maximum at a point x' of K, then we have*

$$Bu(x') \leq 0.$$

(γ) *For some $\alpha_0 \geq 0$, the range $R(\alpha_0 I - B)$ of $\alpha_0 I - B$ is everywhere dense in $C(K)$.*

Then the minimal closed extension $\bar{B}$ of B is the infinitesimal generator of some Feller semigroup on K.

Proof. (i) By Theorem 3.4.1, it suffices to show that:

$$\{u_n\} \subset D(B),\ u_n \to 0 \text{ and } Bu_n \to v \text{ in } C(K) \qquad \Rightarrow \qquad v = 0.$$

Replacing v by $-v$ if necessary, we assume to the contrary that:

The function v takes a positive value at some point of K.

Then, since K_0 is open and dense in K, we can find a point x_0 of K_0, a neighborhood U of x_0 contained in K_0 and a constant $\varepsilon > 0$ such that, for sufficiently large n,

$$Bu_n(x) > \varepsilon, \qquad x \in U. \tag{3}$$

On the other hand, by condition (α), there exists a function $h \in D(B)$ such that

$$\begin{cases} h(x_0) > 1, \\ h(x) < 0, \qquad x \in K \setminus U. \end{cases}$$

Therefore, since $u_n \to 0$ in $C(K)$, it follows that the function

$$u_n' = u_n + \frac{\varepsilon h}{1 + \|Bh\|}$$

satisfies

$$\begin{cases} u_n'(x_0) = u_n(x_0) + \dfrac{\varepsilon h(x_0)}{1 + \|Bh\|} > 0, \\[2ex] u_n'(x) = u_n(x) + \dfrac{\varepsilon h(x)}{1 + \|Bh\|} < 0, \qquad x \in K \setminus U, \end{cases}$$

if n is sufficiently large. This implies that the function $u_n' \in D(B)$ takes its positive maximum at a point x_n' of $U \subset K_0$. Hence we have, by condition (β),

$$Bu_n'(x_n') \leq 0.$$

But it follows from inequality (3) that

$$Bu_n'(x_n') = Bu_n(x_n') + \varepsilon \frac{Bh(x_n')}{1 + \|Bh\|} > Bu_n(x_n') - \varepsilon > 0.$$

This is a contradiction.

(ii) We apply part (ii) of Theorem 9.3.1 to the operator $\bar{B}$.

1) First we show that:

$$u \in D(B), (\alpha_0 I - B)u \geq 0 \text{ on } K \quad \Rightarrow \quad u \geq 0 \text{ on } K. \tag{4}$$

By condition (γ), we can find a function $v \in D(B)$ such that

$$(\alpha_0 I - B)v \geq 1 \quad \text{on } K. \tag{5}$$

Then we have, for any $\varepsilon > 0$,

$$\begin{cases} u + \varepsilon v \in D(B), \\ (\alpha_0 I - B)(u + \varepsilon v) \geq \varepsilon \quad \text{on } K. \end{cases}$$

In view of condition (β'), this implies that the function $-(u + \varepsilon v)$ does not take any positive maximum on K, so that

$$u + \varepsilon v \geq 0 \quad \text{on } K.$$

Thus, letting $\varepsilon \downarrow 0$, we obtain that

$$u \geq 0 \quad \text{on } K.$$

This proves assertion (4).

2) It follows from assertion (4) that the inverse $(\alpha_0 I - B)^{-1}$ of $\alpha_0 I - B$ is defined and non-negative on the range $R(\alpha_0 I - B)$. Further it is bounded with norm

$$\|(\alpha_0 I - B)^{-1}\| \leq \|v\|. \tag{6}$$

Here v is a function which satisfies condition (5).

In fact, since $g = (\alpha_0 I - B)v \geq 1$ on K, it follows that, for all $f \in C(K)$,

$$-\|f\|g \leq f \leq \|f\|g \quad \text{on } K.$$

Hence, by the non-negativity of $(\alpha_0 I - B)^{-1}$, we have, for all $f \in R(\alpha_0 I - B)$,

$$-\|f\|v \leq (\alpha_0 I - B)^{-1}f \leq \|f\|v \quad \text{on } K.$$

This proves inequality (6).

3) Next we show that

$$R(\alpha_0 I - \bar{B}) = C(K). \tag{7}$$

Let f be an arbitrary element of $C(K)$. By condition (γ), we can find a sequence $\{u_n\}$ in $D(B)$ such that $f_n = (\alpha_0 I - B)u_n \to f$ in $C(K)$. Since the inverse $(\alpha_0 I - B)^{-1}$ is bounded, it follows that $u_n = (\alpha_0 I - B)^{-1}f_n$ converges to some $u \in C(K)$, and hence $Bu_n = \alpha_0 u_n - f_n$ converges to $\alpha_0 u - f$ in $C(K)$. Thus we have

$$\begin{cases} u \in D(\bar{B}), \\ \bar{B}u = \alpha_0 u - f, \end{cases}$$

so that

$$(\alpha_0 I - \bar{B})u = f.$$

This proves assertion (7).

4) Further we show that:

$$u \in D(\bar{B}),\ (\alpha_0 I - \bar{B})u \geq 0 \text{ on } K \quad \Rightarrow \quad u \geq 0 \text{ on } K. \tag{4'}$$

Since $R(\alpha_0 I - \bar{B}) = C(K)$, in view of the proof of assertion (4), it suffices to show the following:

If $u \in D(\bar{B})$ takes its positive maximum at a point x' of K, then we have

$$\bar{B}u(x') \leq 0. \tag{8}$$

Assume to the contrary that

$$\bar{B}u(x') > 0.$$

Since there exists a sequence $\{u_n\}$ in $D(B)$ such that $u_n \to u$ and $Bu_n \to \bar{B}u$ in $C(K)$, we can find a neighborhood U of x' and a constant $\varepsilon > 0$ such that, for sufficiently large n,

$$Bu_n(x) > \varepsilon, \qquad x \in U. \tag{9}$$

Further, by condition (α), we can find a function $h \in D(B)$ such that

$$\begin{cases} h(x') > 1, \\ h(x) < 0, \qquad x \in K \setminus U. \end{cases}$$

Then the function

$$u_n' = u_n + \frac{\varepsilon h}{1 + \|Bh\|}$$

satisfies

$$\begin{cases} u_n'(x') > u(x') > 0, \\ u_n'(x) < u(x'), \qquad x \in K \setminus U, \end{cases}$$

if n is sufficiently large. This implies that the function $u_n' \in D(B)$ takes its positive maximum at a point x_n' of U. Hence we have, by condition (β'),

$$Bu_n'(x_n') \leq 0.$$

But it follows from inequality (9) that

$$Bu_n'(x_n') = Bu_n(x_n') + \varepsilon \frac{Bh(x_n')}{1 + \|Bh\|} > Bu_n(x_n') - \varepsilon > 0.$$

This is a contradiction.

5) In view of steps 3) and 4), we obtain that the inverse $(\alpha_0 I - \bar{B})^{-1}$ of $\alpha_0 I - \bar{B}$ is defined on the whole space $C(K)$, and is bounded with norm

$$\|(\alpha_0 I - \bar{B})^{-1}\| = \|(\alpha_0 I - \bar{B})^{-1} 1\|.$$

6) Finally we show that:

For *all* $\alpha > \alpha_0$ the inverse $(\alpha I - \bar{B})^{-1}$ of $\alpha I - \bar{B}$ is defined on the whole space $C(K)$, and is non-negative and bounded with norm (10)

$$\|(\alpha I - \bar{B})^{-1}\| \leq \frac{1}{\alpha}.$$

We let

$$G_{\alpha_0} = (\alpha_0 I - \bar{B})^{-1}.$$

First choose a constant $\alpha_1 > \alpha_0$ such that

$$(\alpha_1 - \alpha_0)\|G_{\alpha_0}\| < 1,$$

and let

$$\alpha_0 < \alpha \leq \alpha_1.$$

Then, for any $f \in C(K)$, the Neumann series

$$u = \left(I + \sum_{n=1}^{\infty} (\alpha_0 - \alpha)^n G_{\alpha_0}^n\right) G_{\alpha_0} f$$

converges in $C(K)$, and is a solution of the equation

$$u - (\alpha_0 - \alpha) G_{\alpha_0} u = G_{\alpha_0} f.$$

Hence we have

$$\begin{cases} u \in D(\bar{B}), \\ (\alpha I - \bar{B})u = f. \end{cases}$$

This proves that

$$R(\alpha I - \bar{B}) = C(K), \qquad \alpha_0 < \alpha \leq \alpha_1. \tag{11}$$

Thus, arguing as in the proof of step 1), we obtain that, for $\alpha_0 < \alpha \leq \alpha_1$,

$$u \in D(\bar{B}), (\alpha I - \bar{B})u \geq 0 \text{ on } K \qquad \Rightarrow \qquad u \geq 0 \text{ on } K. \tag{12}$$

Combining assertions (11) and (12), we find that for $\alpha_0 < \alpha \leq \alpha_1$ the inverse $(\alpha I - \bar{B})^{-1}$ is defined and non-negative on the whole space $C(K)$.

We let

$$G_\alpha = (\alpha I - \bar{B})^{-1}, \qquad \alpha_0 < \alpha \leq \alpha_1.$$

Then the operator G_α is bounded with norm

$$\|G_\alpha\| \leq \frac{1}{\alpha}. \tag{13}$$

In fact, in view of assertion (8), it follows that if $u \in D(\bar{B})$ takes its positive maximum at a point x' of K, then we have

$$\bar{B}u(x') \leq 0,$$

so that

$$\max_K u = u(x') \leq \frac{1}{\alpha}(\alpha I - \bar{B})u(x') \leq \frac{1}{\alpha}\|(\alpha I - \bar{B})u\|. \tag{14}$$

Similarly, if u takes its negative minimum at a point of K, then (replacing u by $-u$) we have

$$-\min_K u = \max_K(-u) \leq \frac{1}{\alpha}\|(\alpha I - \bar{B})u\|. \tag{15}$$

Inequality (13) follows from inequalities (14) and (15).

Summing up, we have proved assertion (10) for $\alpha_0 < \alpha \leq \alpha_1$.

Now suppose that assertion (10) is proved for $\alpha_0 < \alpha \leq \alpha_{n-1}$, $n = 2, 3, \ldots$. Then, taking

$$\alpha_n = 2\alpha_{n-1} - \frac{\alpha_1}{2},$$

or equivalently,

$$\alpha_n = \left(2^{n-2} + \frac{1}{2}\right)\alpha_1,$$

we have, for $\alpha_{n-1} < \alpha \leq \alpha_n$,

$$\begin{aligned}(\alpha - \alpha_{n-1})\|G_{\alpha_{n-1}}\| &\leq \frac{\alpha - \alpha_{n-1}}{\alpha_{n-1}} \\ &\leq \frac{\alpha_n - \alpha_{n-1}}{\alpha_{n-1}} \\ &= \frac{1}{1 + 2^{2-n}} \\ &< 1.\end{aligned}$$

Hence assertion (10) for $\alpha_{n-1} < \alpha \leq \alpha_n$ is proved just as in the proof of assertion (10) for $\alpha_0 < \alpha \leq \alpha_1$. This proves assertion (10).

Consequently, applying part (ii) of Theorem 9.3.1 to the operator $\bar{B}$, we obtain that $\bar{B}$ is the infinitesimal generator of some Feller semigroup on K.

The proof of Theorem 9.3.3 is now complete. ∎

9.3.4 Corollary. *Let $\mathfrak{A}$ be the infinitesimal generator of a Feller semigroup $\{T_t\}_{t \geq 0}$ on K and M a bounded linear operator on $C(K)$ into itself. Suppose that either M or $\mathfrak{A}' = \mathfrak{A} + M$ satisfies condition (β'). Then the operator $\mathfrak{A}'$ is the infinitesimal generator of some Feller semigroup on K.*

Proof. We apply part (ii) of Theorem 9.3.3 to the operator $\mathfrak{A}'$.

First note that $\mathfrak{A}' = \mathfrak{A} + M$ is a densely defined, closed linear operator from $C(K)$ into itself. Since the semigroup $\{T_t\}_{t \geq 0}$ is non-negative and contractive on $C(K)$, it follows that if $u \in D(\mathfrak{A})$ takes its positive maximum at a point x' of K, then we have

$$\mathfrak{A}u(x') = \lim_{t \downarrow 0} \frac{T_t u(x') - u(x')}{t} \leq 0.$$

This implies that if M satisfies condition (β'), so does $\mathfrak{A}' = \mathfrak{A} + M$.

We let

$$G_{\alpha_0} = (\alpha_0 I - \mathfrak{A})^{-1}, \qquad \alpha_0 > 0.$$

If α_0 is so large that

$$\|G_{\alpha_0} M\| \leq \|G_{\alpha_0}\| \cdot \|M\| \leq \frac{\|M\|}{\alpha_0} < 1,$$

then the Neumann series

$$u = \left(I + \sum_{n=1}^{\infty} (G_{\alpha_0} M)^n \right) G_{\alpha_0} f$$

converges in $C(K)$ for any $f \in C(K)$, and is a solution of the equation

$$u - G_{\alpha_0} Mu = G_{\alpha_0} f.$$

Hence we have

$$\begin{cases} u \in D(\mathfrak{A}) = D(\mathfrak{A}'), \\ (\alpha_0 I - \mathfrak{A}')u = f. \end{cases}$$

This proves that

$$R(\alpha_0 I - \mathfrak{A}') = C(K).$$

Therefore, applying part (ii) of Theorem 9.3.3 to the operator $\mathfrak{A}'$, we obtain that $\mathfrak{A}'$ is the infinitesimal generator of some Feller semigroup on K. ■

9.4. Infinitesimal Generators of Feller Semigroups —(1)—

Let K be a *compact* metric space, and let $C(K)$ be the Banach space of real-valued continuous functions on K with the supremum (maximum) norm.

Recall that a Feller semigroup $\{T_t\}_{t \geq 0}$ on K is a strongly continuous semigroup of bounded linear operators T_t acting on $C(K)$ such that:

$$f \in C(K),\ 0 \leq f \leq 1 \text{ on } K \quad \Rightarrow \quad 0 \leq T_t f \leq 1 \text{ on } K.$$

The infinitesimal generator $\mathfrak{A}$ of $\{T_t\}$ is defined by

$$\mathfrak{A}u = \lim_{t \downarrow 0} \frac{T_t u - u}{t}, \qquad u \in C(K), \tag{1}$$

provided that the limit exists in $C(K)$. That is, the generator $\mathfrak{A}$ is a linear operator from $C(K)$ into itself whose domain $D(\mathfrak{A})$ consists of all $u \in C(K)$ for which the limit (1) exists.

Theorem 9.3.1, a version of the Hille-Yosida theorem, tells us that a Feller semigroup is completely characterized by its infinitesimal generator. Therefore we are reduced to the study of the infinitesimal generators of Feller semigroups.

In this section and the next section, we shall describe analytically the infinitesimal generators of Feller semigroups in the case when K is the closure $\bar{D}$ of a *bounded domain* D in $\mathbf{R}^N$.

Our first job is to derive an explicit formula in the *interior* D of $\bar{D}$ for the infinitesimal generator $\mathfrak{A}$ of a Feller semigroup $\{T_t\}_{t\geq 0}$ on $\bar{D}$. The next result is adapted from Sato-Ueno [1], while the main idea of its proof is due to Wentzell (Ventcel') [1] (cf. the proof of Theorem 9.5.1 in Section 9.5).

9.4.1 Theorem. *Let D be a bounded domain in $\mathbf{R}^N$, and let $\{T_t\}_{t\geq 0}$ be a Feller semigroup on $\bar{D}$ and $\mathfrak{A}$ its infinitesimal generator. Suppose that for every point x^0 of D, there exist a local coordinate system $(x_1, \ldots, x_N)$ on a neighborhood of x^0 and continuous functions $\chi_1, \ldots, \chi_N$ on $\bar{D}$ such that $\chi_i = x_i$ in a neighborhood of x^0 and that the functions $1, \chi_1, \ldots, \chi_N$ and $\sum_{i=1}^N \chi_i^2$ belong to the domain $D(\mathfrak{A})$ of $\mathfrak{A}$.*

Then we have, for all $u \in D(\mathfrak{A}) \cap C^2(\bar{D})$,

$$\mathfrak{A}u(x^0) = \sum_{i,j=1}^N a^{ij}(x^0) \frac{\partial^2 u}{\partial x_i \, \partial x_j}(x^0) + \sum_{i=1}^N b^i(x^0) \frac{\partial u}{\partial x_i}(x^0) + c(x^0)u(x^0)$$
$$+ \int_{\bar{D}} e(x^0, dy)\left[u(y) - u(x^0) - \sum_{i=1}^N \frac{\partial u}{\partial x_i}(x^0)(\chi_i(y) - \chi_i(x^0))\right] \tag{2}$$

where:

1. *The matrix $(a^{ij}(x^0))$ is symmetric and positive semi-definite.*
2. $b^i(x^0) = \mathfrak{A}(\chi_i - \chi_i(x^0))(x^0)$.
3. $c(x^0) = \mathfrak{A}1(x^0)$.
4. *$e(x^0, \cdot)$ is a non-negative Borel measure on $\bar{D}$ such that* (3)

$$\begin{cases} e(x^0, \bar{D} \setminus U) < \infty, \\ \displaystyle\int_U e(x^0, dy)\left[\sum_{i=1}^N (\chi_i(y) - \chi_i(x^0))^2\right] < \infty, \end{cases}$$

for any neighborhood U of x^0.

9.4.2 Remark. Bony, Courrège and Priouret [1] give a more precise characterization of the infinitesimal generators of Feller semigroups in terms of the maximum principle (see [1], Théorèmes IX and XIV).

Proof of Theorem 9.4.1. By Theorem 9.2.6, there corresponds to a Feller semigroup $\{T_t\}_{t\geq 0}$ on $\bar{D}$ a unique Feller and uniformly stochastically continuous transition function p_t on $\bar{D}$ in the following manner:

$$T_t f(x) = \int_{\bar{D}} p_t(x, dy) f(y), \qquad f \in C(\bar{D}).$$

Since the functions 1, $\chi_1, \ldots, \chi_N$ and $\sum_{i=1}^N \chi_i^2$ belong to $D(\mathfrak{A})$, it follows that

$$\chi_1 - \chi_1(x^0), \ldots, \chi_N - \chi_N(x^0), \qquad \sum_{i=1}^N (\chi_i - \chi_i(x^0))^2 \in D(\mathfrak{A}).$$

Thus we have

$$\begin{aligned}
&\mathfrak{A}u(x^0)\\
&\quad = \lim_{t \downarrow 0} \frac{1}{t}(T_t u(x^0) - u(x^0))\\
&\quad = \lim_{t \downarrow 0} \frac{1}{t}\left(\int_{\bar{D}} p_t(x^0, dy)u(y) - u(x^0)\right)\\
&\quad = \lim_{t \downarrow 0} \left\{\frac{1}{t}(p_t(x^0, \bar{D}) - 1)u(x^0)\right.\\
&\qquad + \frac{1}{t}\sum_{i=1}^N \int_{\bar{D}} p_t(x^0, dy)(\chi_i(y) - \chi_i(x^0))\frac{\partial u}{\partial x_i}(x^0)\\
&\qquad \left. + \frac{1}{t}\int_{\bar{D}} p_t(x^0, dy)\left[u(y) - u(x^0) - \sum_{i=1}^N \frac{\partial u}{\partial x_i}(x^0)(\chi_i(y) - \chi_i(x^0))\right]\right\}\\
&\quad = c(x^0)u(x^0) + \sum_{i=1}^N b^i(x^0)\frac{\partial u}{\partial x_i}(x^0)\\
&\qquad + \lim_{t \downarrow 0} \frac{1}{t}\int_{\bar{D}\setminus\{x^0\}} p_t(x^0, dy)\tilde{u}(x^0, y)d(x^0, y),
\end{aligned} \tag{4}$$

where:

$$c(x^0) = \lim_{t \downarrow 0} \frac{T_t 1(x^0) - 1}{t} = \mathfrak{A}1(x^0),$$

$$b^i(x^0) = \lim_{t \downarrow 0} \frac{T_t(\chi_i - \chi_i(x^0))(x^0)}{t} = \mathfrak{A}(\chi_i - \chi_i(x^0))(x^0),$$

and

$$d(x^0, y) = \sum_{i=1}^N (\chi_i(y) - \chi_i(x^0))^2,$$

$$\tilde{u}(x^0, y) = \frac{u(y) - u(x^0) - \sum_{i=1}^N \frac{\partial u}{\partial x_i}(x^0)(\chi_i(y) - \chi_i(x^0))}{d(x^0, y)}, \qquad y \in \bar{D} \setminus \{x^0\}.$$

To rewrite the last term of formula (4), we define a non-negative measure $\tilde{p}_t(x^0, \cdot)$ on $\bar{D}$ by

$$\tilde{p}_t(x^0, E) = \frac{1}{t} \int_E p_t(x^0, dy) d(x^0, y), \qquad E \in \mathscr{B}_{\bar{D}}.$$

(Here and in the following $\mathscr{B}_K$ denotes the σ-algebra of all Borel sets in a metric space K.) Then we can rewrite formula (4) as follows:

$$\begin{aligned} \mathfrak{A}u(x^0) = c(x^0)u(x^0) &+ \sum_{i=1}^{N} b^i(x^0) \frac{\partial u}{\partial x_i}(x^0) \\ &+ \lim_{t \downarrow 0} \int_{\bar{D} \setminus \{x^0\}} \tilde{p}_t(x^0, dy) \tilde{u}(x^0, y). \end{aligned} \tag{4'}$$

We remark that, for all sufficiently small $t > 0$,

$$\begin{aligned} \tilde{p}_t(x^0, \bar{D}) &\leq \lim_{t \downarrow 0} \tilde{p}_t(x^0, \bar{D}) + 1 \\ &= \lim_{t \downarrow 0} \frac{1}{t} \int_{\bar{D}} p_t(x^0, dy) d(x^0, y) + 1 \\ &= \mathfrak{A}\left(\sum_{i=1}^{N} (\chi_i - \chi_i(x^0))^2 \right)(x^0) + 1. \end{aligned} \tag{5}$$

Now we introduce a compactification of $\bar{D} \setminus \{x^0\}$ to which the function $\tilde{u}(x^0, \cdot)$ may be continuously extended.

We let

$$z^{ij}(x^0, y) = \frac{(\chi_i(y) - \chi_i(x^0))(\chi_j(y) - \chi_j(x^0))}{d(x^0, y)}, \qquad y \in \bar{D} \setminus \{x^0\}.$$

Then the functions $z^{ij}(x^0, \cdot)$ satisfy

$$|z^{ij}(x^0, y)| \leq 1,$$

and the matrix $(z^{ij}(x^0, \cdot))$ is symmetric and positive semi-definite. We define a compact subspace M of symmetric, positive semi-definite matrices by

$$M = \{(z^{ij})_{1 \leq i, j \leq N};\ z^{ij} = z^{ji},\ (z^{ij}) \geq 0,\ |z^{ij}| \leq 1\},$$

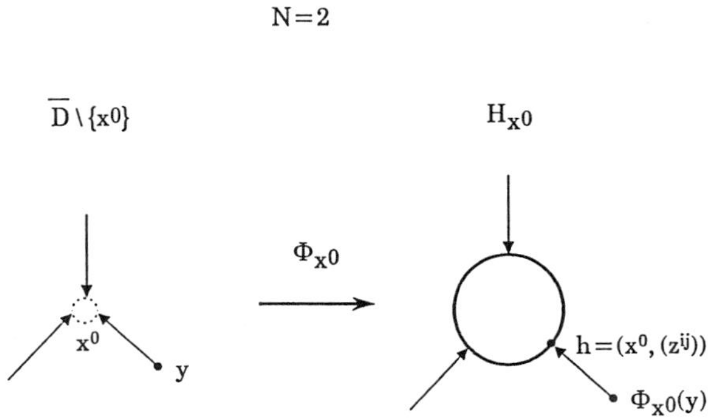

Figure 9-4

and consider an injection

$$\Phi_{x^0}\colon \bar{D} \setminus \{x^0\} \ni y \mapsto (y, (z^{ij}(x^0, y))) \in \bar{D} \times M.$$

Then the function $\tilde{u}(x^0, \Phi_{x^0}^{-1}(\cdot))$, defined on $\Phi_{x^0}(\bar{D} \setminus \{x^0\})$, can be extended to a continuous function $\hat{u}(x^0, \cdot)$ on the closure

$$H_{x^0} = \overline{\Phi_{x^0}(\bar{D} \setminus \{x^0\})}$$

of $\Phi_{x^0}(\bar{D} \setminus \{x^0\})$ in $\bar{D} \times M$. In fact, by using Taylor's formula, we have, in a neighborhood of x^0,

$$\begin{aligned} u(y) &= u(x^0) + \sum_{i=1}^{N} \frac{\partial u}{\partial x_i}(x^0)(\chi_i(y) - \chi_i(x^0)) \\ &\quad + \sum_{i,j=1}^{N} \int_0^1 \frac{\partial^2 u}{\partial x_i\, \partial x_j}(x^0 + \theta(y - x^0))(1 - \theta)\, d\theta \\ &\quad \times (\chi_i(y) - \chi_i(x^0))(\chi_j(y) - \chi_j(x^0)), \end{aligned}$$

and hence

$$\begin{aligned} \tilde{u}(x^0, y) &= \sum_{i,j=1}^{N} \int_0^1 \frac{\partial^2 u}{\partial x_i\, \partial x_j}(x^0 + \theta(y - x^0))(1 - \theta)\, d\theta\, z^{ij}(x^0, y) \\ &\to \hat{u}(x^0, h) = \frac{1}{2} \sum_{i,j=1}^{N} \frac{\partial^2 u}{\partial x_i\, \partial x_j}(x^0) z^{ij}. \end{aligned} \tag{6}$$

as $\Phi_{x^0}(y) = (y, (z^{ij}(x^0, y))) \to h = (x^0, (z^{ij}))$ (cf. Figure 9-4).

We define a non-negative measure $\hat{p}_t(x^0, \cdot)$ on H_{x^0} by

$$\hat{p}_t(x^0, \hat{E}) = \tilde{p}_t(x^0, \Phi_{x^0}^{-1}(\hat{E})), \qquad \hat{E} \in \mathscr{B}_{H_{x^0}}. \tag{7}$$

Then it follows from inequality (5) that for all sufficiently small $t > 0$ we have

$$\hat{p}_t(x^0, H_{x^0}) \le \tilde{p}_t(x^0, \bar{D}) \le \mathfrak{A}\left(\sum_{i=1}^{N} (\chi_i - \chi_i(x^0))^2\right)(x^0) + 1.$$

Hence, applying Theorem 3.3.7 to our situation, we obtain that there exists a sequence $\{t_n\}$, $t_n \downarrow 0$, such that the measures $\hat{p}_{t_n}(x^0, \cdot)$ converge weakly to a finite non-negative Borel measure $\hat{p}(x^0, \cdot)$ on H_{x^0}.

Therefore, in view of (6) and (7), we can pass to the limit in formula (4′) to obtain the following:

$$\begin{aligned}
\mathfrak{A}u(x^0) &= c(x^0)u(x^0) + \sum_{i=1}^{N} b^i(x^0)\frac{\partial u}{\partial x_i}(x^0) \\
&\quad + \lim_{t \downarrow 0} \int_{\bar{D}\setminus\{x^0\}} \tilde{p}_t(x^0, dy)\tilde{u}(x^0, y) \\
&= c(x^0)u(x^0) + \sum_{i=1}^{N} b^i(x^0)\frac{\partial u}{\partial x_i}(x^0) \\
&\quad + \lim_{n \to \infty} \int_{H_{x^0}} \hat{p}_{t_n}(x^0, dh)\hat{u}(x^0, h) \\
&= c(x^0)u(x^0) + \sum_{i=1}^{N} b^i(x^0)\frac{\partial u}{\partial x_i}(x^0) \\
&\quad + \int_{H_{x^0}} \hat{p}(x^0, dh)\hat{u}(x^0, h).
\end{aligned} \tag{4″}$$

To rewrite the last term of formula (4″), we define a non-negative Borel measure $\tilde{p}(x^0, \cdot)$ on $\bar{D} \setminus \{x^0\}$ by

$$\tilde{p}(x^0, E) = \hat{p}(x^0, \Phi_{x^0}(E)), \qquad E \in \mathscr{B}_{\bar{D}\setminus\{x^0\}},$$

and let

$$Z: \bar{D} \times M \ni h = (y, (z^{ij})) \mapsto (z^{ij}) \in M.$$

Then we have

$$\begin{aligned}
&\int_{H_{x^0}} \hat{p}(x^0, dh)\hat{u}(x^0, h) \\
&\quad = \int_{H_{x^0}\setminus\Phi_{x^0}(\bar{D}\setminus\{x^0\})} \hat{p}(x^0, dh)\hat{u}(x^0, h) \\
&\qquad + \int_{\Phi_{x^0}(\bar{D}\setminus\{x^0\})} \hat{p}(x^0, dh)\hat{u}(x^0, h) \\
&\quad = \frac{1}{2}\sum_{i,j=1}^{N} \int_{H_{x^0}\setminus\Phi_{x^0}(\bar{D}\setminus\{x^0\})} \hat{p}(x^0, dh) Z^{ij}(h) \frac{\partial^2 u}{\partial x_i\,\partial x_j}(x^0) \\
&\qquad + \int_{\bar{D}\setminus\{x^0\}} \tilde{p}(x^0, dy)\tilde{u}(x^0, y) \\
&\quad = \sum_{i,j=1}^{N} a^{ij}(x^0)\frac{\partial^2 u}{\partial x_i\,\partial x_j}(x^0) \\
&\qquad + \int_D e(x^0, dy)\left[u(y) - u(x^0) - \sum_{i=1}^{N}\frac{\partial u}{\partial x_i}(x^0)(\chi_i(y) - \chi_i(x^0))\right], \qquad (8)
\end{aligned}$$

where:

$$a^{ij}(x^0) = \frac{1}{2}\int_{H_{x^0}\setminus\Phi_{x^0}(\bar{D}\setminus\{x^0\})} \hat{p}(x^0, dh) Z^{ij}(h),$$

and

$$\begin{cases} e(x^0, \{x^0\}) = 0, \\ e(x^0, E) = \displaystyle\int_{E\setminus\{x^0\}} \tilde{p}(x^0, dy)\left(\frac{1}{d(x^0, y)}\right), \qquad E \in \mathscr{B}_{\bar{D}}. \end{cases}$$

Therefore, combining formulas (4″) and (8), we obtain expression (2) for $\mathfrak{A}u$ in the interior D of $\bar{D}$.

Property (3) follows immediately from our construction of a^{ij}, b^i, c and e.

The proof of Theorem 9.4.1 is now complete. ∎

Theorem 9.4.1 tells us that the infinitesimal generator $\mathfrak{A}$ of a Feller semigroup $\{T_t\}_{t\geq 0}$ on $\bar{D}$ is written in the interior D of $\bar{D}$ as the sum of a degenerate elliptic differential operator of second order and an integro-differential operator.

Intuitively the above result may be interpreted as follows. By Theorems 9.2.6 and 9.1.6, there correspond to a Feller semigroup $\{T_t\}_{t \geq 0}$ a unique transition function p_t and a Markov process $\mathscr{X} = (x_t, \mathscr{F}, \mathscr{F}_t, P_x)$ in the following manner:

$$T_t f(x) = \int_{\bar{D}} p_t(x, dy) f(y), \qquad f \in C(\bar{D});$$

$$p_t(x, E) = P_x\{x_t \in E\}, \qquad E \in \mathscr{B}_{\bar{D}}.$$

In view of Theorem 9.1.9 and Remarks 9.1.10, it will be true that if the paths of $\mathscr{X}$ are *continuous*, then the transition function p_t has local character such as condition (N) of Theorem 9.1.9; hence the infinitesimal generator $\mathfrak{A}$ is *local*, that is, the value $\mathfrak{A}u(x^0)$ at an interior point x^0 is determined by the values of u in an arbitrary small neighborhood of x^0. But Theorem 4.3.1 tells us that a linear operator is local if and only if it is a differential operator. Therefore we have an assurance of the following assertion:

The infinitesimal generator $\mathfrak{A}$ of a Feller semigroup $\{T_t\}_{t \geq 0}$ on $\bar{D}$ is a differential operator in the interior D of $\bar{D}$ if the paths of its corresponding Markov process $\mathscr{X}$ are continuous.

In the general case when the paths of $\mathscr{X}$ may have discontinuities such as jumps, the infinitesimal generator $\mathfrak{A}$ takes the form of the sum of a differential operator and an integro-differential (non-local) operator, as proved in Theorem 9.4.1.

9.5. Infinitesimal Generators of Feller Semigroups —(2)—

In this section, we shall derive an explicit formula on the *boundary* ∂D of $\bar{D}$ for the infinitesimal generator $\mathfrak{A}$ of a Feller semigroup $\{T_t\}_{t \geq 0}$ on $\bar{D}$.

Let D be a bounded domain in $\mathbf{R}^N$ with C^∞ boundary ∂D, and choose for each point x' of ∂D a neighborhood U of x' in $\mathbf{R}^N$ and a local coordinate system $(x_1, \ldots, x_{N-1}, x_N)$ on U such that:

(a) $x \in U \cap D \Leftrightarrow x \in U,\ x_N(x) > 0;$
$x \in U \cap \partial D \Leftrightarrow x \in U,\ x_N(x) = 0.$

(b) The functions $(x_1, \ldots, x_{N-1})$, restricted to $U \cap \partial D$, form a local coordinate system of ∂D on $U \cap \partial D$.

One may suppose that the functions $x_1, \ldots, x_{N-1}$, x_N can be extended respectively to C^∞ functions $\chi_1, \ldots, \chi_{N-1}, \chi_N$ on $\mathbf{R}^N$ so that

$$\begin{cases} d(x', y) = \chi_N(y) + \sum_{i=1}^{N-1} (\chi_i(y) - \chi_i(x'))^2 > 0 \\ \text{if } x' \in U \cap \partial D \text{ and } y \in \bar{D} \setminus \{x'\}. \end{cases} \tag{1}$$

The next theorem tells us that every C^2 function in the domain $D(\mathfrak{A})$ must obey a boundary condition at each point of ∂D.

9.5.1 Theorem. *Let D be a bounded domain in $\mathbf{R}^N$ with C^∞ boundary ∂D, and let $\{T_t\}_{t \geq 0}$ be a Feller semigroup on $\bar{D}$ and $\mathfrak{A}$ its infinitesimal generator. Then every function u in $D(\mathfrak{A}) \cap C^2(\bar{D})$ satisfies at each point x' of ∂D a boundary condition of the form*

$$\begin{aligned} &\sum_{i,j=1}^{N-1} \alpha^{ij}(x') \frac{\partial^2 u}{\partial x_i \, \partial x_j}(x') + \sum_{i=1}^{N-1} \beta^i(x') \frac{\partial u}{\partial x_i}(x') \\ &\quad + \gamma(x')u(x') + \mu(x') \frac{\partial u}{\partial x_N}(x') - \delta(x')\mathfrak{A}u(x') \\ &\quad + \int_{\bar{D}} \nu(x', dy)\left[u(y) - u(x') - \sum_{i=1}^{N-1} \frac{\partial u}{\partial x_i}(x')(\chi_i(y) - \chi_i(x'))\right] \\ &= 0 \end{aligned} \tag{2}$$

where:

1. *The matrix $(\alpha^{ij}(x'))$ is symmetric and positive semi-definite.*
2. $\gamma(x') \leq 0$.
3. $\mu(x') \geq 0$.
4. $\delta(x') \geq 0$.
5. *$\nu(x', \cdot)$ is a non-negative Borel measure on $\bar{D}$ such that* (3)

$$\begin{cases} \nu(x', \bar{D} \setminus W) < \infty, \\ \displaystyle\int_{W \cap \bar{D}} \nu(x', dy)\left[\chi_N(y) + \sum_{j=1}^{N-1} (\chi_j(y) - \chi_j(x'))^2\right] < \infty, \end{cases}$$

for any neighborhood W of x' in $\mathbf{R}^N$.

Proof. The proof is essentially the same as that of Theorem 9.4.1.

1) By Theorem 9.2.6, there corresponds to a Feller semigroup $\{T_t\}_{t\geq 0}$ on $\bar{D}$ a unique Feller and uniformly stochastically continuous transition function p_t on $\bar{D}$ in the following manner:

$$T_t f(x) = \int_{\bar{D}} p_t(x, dy) f(y), \qquad f \in C(\bar{D}).$$

Thus we have

$$\begin{aligned}
&\frac{1}{t}(T_t u(x') - u(x')) \\
&\quad = \frac{1}{t}(p_t(x', \bar{D}) - 1)u(x') \\
&\qquad + \frac{1}{t}\sum_{i=1}^{N-1} \int_{\bar{D}} p_t(x', dy)(\chi_i(y) - \chi_i(x')) \frac{\partial u}{\partial x_i}(x') \\
&\qquad + \frac{1}{t}\int_{\bar{D}} p_t(x', dy)\left[u(y) - u(x') - \sum_{i=1}^{N-1} \frac{\partial u}{\partial x_i}(x')(\chi_i(y) - \chi_i(x'))\right] \\
&\quad = \gamma_t(x')u(x') + \sum_{i=1}^{N-1} \beta_t^i(x') \frac{\partial u}{\partial x_i}(x') \\
&\qquad + \frac{1}{t}\int_{\bar{D}\setminus\{x'\}} p_t(x', dy)\tilde{u}(x', y) d(x', y),
\end{aligned} \tag{4}$$

where:

$$\gamma_t(x') = \frac{1}{t}(p_t(x', \bar{D}) - 1),$$

$$\beta_t^j(x') = \frac{1}{t}\int_{\bar{D}} p_t(x', dy)(\chi_j(y) - \chi_j(x')),$$

and

$$d(x', y) = \chi_N(y) + \sum_{i=1}^{N-1} (\chi_i(y) - \chi_i(x'))^2, \qquad y \in \bar{D},$$

$$\tilde{u}(x', y) = \frac{u(y) - u(x') - \sum_{i=1}^{N-1} \frac{\partial u}{\partial x_i}(x')(\chi_i(y) - \chi_i(x'))}{d(x', y)}, \qquad y \in \bar{D}\setminus\{x'\}.$$

We rewrite the last term of formula (4). To do so, we introduce a non-negative function

$$\ell_t(x') = \frac{1}{t} \int_{\bar{D}} p_t(x', dy) d(x', y),$$

and consider two cases.

Case A: $\ell_t(x') > 0$. In this case we can write

$$\frac{1}{t} \int_{\bar{D} \setminus \{x'\}} p_t(x', dy) \tilde{u}(x', y) d(x', y) = \ell_t(x') \int_{\bar{D} \setminus \{x'\}} \tilde{q}_t(x', dy) \tilde{u}(x', y),$$

where

$$\tilde{q}_t(x', E) = \frac{1}{t\ell_t(x')} \int_E p_t(x', dy) d(x', y), \qquad E \in \mathscr{B}_{\bar{D}}.$$

(Here and in the following $\mathscr{B}_K$ denotes the σ-algebra of all Borel sets in a metric space K.) We remark that

$$\tilde{q}_t(x', \bar{D} \setminus \{x'\}) = 1,$$

since $d(x', x') = 0$.

Case B: $\ell_t(x') = 0$. In this case, in view of (1), it follows that

$$p_t(x', \bar{D} \setminus \{x'\}) = 0.$$

Hence we can write

$$\frac{1}{t} \int_{\bar{D} \setminus \{x'\}} p_t(x', dy) \tilde{u}(x', y) d(x', y)$$

$$= \ell_t(x') \int_{\bar{D} \setminus \{x'\}} \tilde{q}_t(x', dy) \tilde{u}(x', y)$$

$$(= 0),$$

where (for example)

$$\tilde{q}_t(x', \cdot) = \text{the unit mass at a point of } D,$$

so that

$$\tilde{q}_t(x', \bar{D} \setminus \{x'\}) = 1.$$

Summing up Case A and Case B, we obtain that

$$\frac{1}{t}(T_t u(x') - u(x')) = \gamma_t(x')u(x') + \sum_{j=1}^{N-1} \beta_t^j(x') \frac{\partial u}{\partial x_j}(x')$$

$$+ \ell_t(x') \int_{\bar{D}\setminus\{x'\}} \tilde{q}_t(x', dy)\tilde{u}(x', y). \tag{4'}$$

2) Now we introduce a compactification of $\bar{D}\setminus\{x'\}$ to which the function $\tilde{u}(x', \cdot)$ may be continuously extended.

We let

$$w(x', y) = \frac{\chi_N(y)}{d(x', y)}, \qquad y \in \bar{D}\setminus\{x'\},$$

$$z^{ij}(x', y) = \frac{(\chi_i(y) - \chi_i(x'))(\chi_j(y) - \chi_j(x'))}{d(x', y)}, \qquad y \in \bar{D}\setminus\{x'\}.$$

Then the functions $w(x', \cdot)$ and $z^{ij}(x', \cdot)$ satisfy

$$\begin{cases} 0 \leq w(x', y) \leq 1, \\ |z^{ij}(x', y)| \leq 1, \\ w(x', y) + \sum_{i=1}^{N-1} z^{ii}(x', y) = 1, \end{cases}$$

and the matrix $(z^{ij}(x', \cdot))$ is symmetric and positive semi-definite. We define a compact subspace M of symmetric, positive semi-definite matrices by

$$M = \{(z^{ij})_{1 \leq i, j \leq N-1};\ z^{ij} = z^{ji},\ (z^{ij}) \geq 0,\ |z^{ij}| \leq 1\},$$

and a compact subspace H of $\bar{D} \times [0, 1] \times M$ by

$$H = \left\{(y, w, (z^{ij})) \in \bar{D} \times [0, 1] \times M;\ w + \sum_{i=1}^{N-1} z^{ii} = 1\right\}, \tag{5}$$

and consider an injection

$$\Phi_{x'}: \bar{D}\setminus\{x'\} \ni y \mapsto (y, w(x', y), (z^{ij}(x^0, y))) \in H.$$

Then the function $\tilde{u}(x', \Phi_{x'}^{-1}(\cdot))$, defined on $\Phi_{x'}(\bar{D}\setminus\{x'\})$, can be extended to a continuous function $\hat{u}(x', \cdot)$ on the closure

$$H_{x'} = \overline{\Phi_{x'}(\bar{D}\setminus\{x'\})}$$

of $\Phi_{x'}(\bar{D} \setminus \{x'\})$ in H. In fact, by using Taylor's formula, we have, in a neighborhood of x',

$$u(y) = u(x') + \sum_{i=1}^{N-1} \frac{\partial u}{\partial x_i}(x')(\chi_i(y) - \chi_i(x')) + \frac{\partial u}{\partial x_N}(x')\chi_N(y)$$

$$+ \sum_{i,j=1}^{N} \int_0^1 \frac{\partial^2 u}{\partial x_i\, \partial x_j}(x' + \theta(y - x'))(1 - \theta)\, d\theta$$

$$\times (\chi_i(y) - \chi_i(x'))(\chi_j(y) - \chi_j(x')),$$

and hence

$$\tilde{u}(x', y) = \frac{\partial u}{\partial x_N}(x')w(x', y)$$

$$+ \sum_{i,j=1}^{N} \int_0^1 \frac{\partial^2 u}{\partial x_i\, \partial x_j}(x' + \theta(y - x'))(1 - \theta)\, d\theta z^{ij}(x', y)$$

$$\to \hat{u}(x', h) = \frac{\partial u}{\partial x_N}(x')w + \frac{1}{2}\sum_{i,j=1}^{N} \frac{\partial^2 u}{\partial x_i\, \partial x_j}(x')z^{ij}, \tag{6}$$

as $\Phi_{x'}(y) = (y, w(x', y), z^{ij}(x', y)) \to h = (x', w, (z^{ij}))$ (cf. Figure 9-4).

We define a non-negative measure $\hat{q}_t(x', \cdot)$ on $H_{x'}$ by

$$\hat{q}_t(x', \hat{E}) = \tilde{q}_t(x', \Phi_{x'}^{-1}(\hat{E})), \qquad \hat{E} \in \mathscr{B}_{H_{x'}}.$$

Then formula (4′) can be written as follows:

$$\frac{1}{t}(T_t u(x') - u(x')) = \gamma_t(x')u(x') + \sum_{j=1}^{N-1} \beta_t^j(x') \frac{\partial u}{\partial x_j}(x')$$

$$+ \ell_t(x') \int_{H_{x'}} \hat{q}_t(x', dh)\hat{u}(x', h). \tag{4″}$$

We remark that the measure $\hat{q}_t(x', \cdot)$ is a probability measure on $H_{x'}$.

3) We pass to the limit in formula (4″). To do so, we introduce non-negative functions

$$\theta_m(x') = -\gamma_{1/m}(x') + \sum_{j=1}^{N-1} |\beta_{1/m}^j(x')| + \ell_{1/m}(x'), \qquad m = 1, 2, \ldots, \tag{7}$$

and consider two cases.

Case I: $\liminf_{m\to\infty} \theta_m(x') = 0$. In this case, there exists a subsequence $\{\theta_{m_k}(x')\}$ of $\{\theta_m(x')\}$ such that

$$\lim_{k\to\infty} \theta_{m_k}(x') = 0.$$

Thus, passing to the limit in formula (4″) with $t = 1/m_k$, we obtain that

$$\mathfrak{A}u(x') = 0.$$

Hence we have condition (2), taking

$$\begin{cases} \alpha^{ij}(x') = \beta^i(x') = \gamma(x') = \mu(x') = 0, \\ \delta(x') = 1, \\ \nu(x', dx) = 0. \end{cases}$$

Case II: $\liminf_{m\to\infty} \theta_m(x') > 0$. In this case, there exist a subsequence $\{\theta_{m_k}(x')\}$ of $\{\theta_m(x')\}$ and a function $\theta(x')$ such that

$$\lim_{k\to\infty} \theta_{m_k}(x') = \theta(x') > 0. \tag{8}$$

Then, dividing both sides of formula (4″) with $t = 1/m_k$ by the function $\theta_{m_k}(x')$, we obtain that

$$\bar{\delta}_k(x')\left(\frac{T_{t_k}u(x') - u(x')}{t_k}\right) = \bar{\gamma}_k(x')u(x') + \sum_{j=1}^{N-1} \bar{\beta}_k^j(x')\frac{\partial u}{\partial x_j}(x')$$
$$+ \bar{\ell}_k(x') \int_{H_{x'}} \bar{q}_k(x', dh)\hat{u}(x', h), \tag{9}$$

where:

$$t_k = \frac{1}{m_k},$$

$$\bar{\delta}_k(x') = \frac{1}{\theta_{m_k}(x')}, \qquad \bar{\gamma}_k(x') = \frac{\gamma_{t_k}(x')}{\theta_{m_k}(x')},$$

$$\bar{\beta}_k^j(x') = \frac{\beta_{t_k}^j(x')}{\theta_{m_k}(x')}, \qquad \bar{\ell}_k(x') = \frac{\ell_{t_k}(x')}{\theta_{m_k}(x')},$$

$$\bar{q}_k(x', \cdot) = \hat{q}_{t_k}(x', \cdot).$$

But we have, by (8),

$$0 \le \bar{\delta}_k(x') < \infty,$$

and further, by (7),

$$\begin{cases} 0 \le -\bar{\gamma}_k(x') \le 1, \qquad -1 \le \bar{\beta}_k^j(x') \le 1, \qquad 0 \le \bar{\ell}_k(x') \le 1, \\ -\bar{\gamma}_k(x') + \sum_{j=1}^{N-1} |\bar{\beta}_k^j(x')| + \bar{\ell}_k(x') = 1. \end{cases}$$

We remark that the measures $\bar{q}_k(x', \cdot)$ are probability measures on $H_{x'}$.

Since the metric spaces $[0, +\infty]$, $[0, 1]$, $[-1, 1]$ are compact and the space of probability measures on $H_{x'}$ is also compact (cf. Theorem 3.3.7), we can pass to the limit in formula (9) to obtain the following:

$$\begin{aligned} \delta(x')\mathfrak{A}u(x') = \gamma(x')u(x') &+ \sum_{j=1}^{N-1} \beta^j(x') \frac{\partial u}{\partial x_j}(x') \\ &+ \ell(x') \int_{H_{x'}} \hat{q}(x', dh)\hat{u}(x', h). \end{aligned} \tag{10}$$

Here the functions δ, γ, β^j and ℓ satisfy

$$\begin{cases} 0 \le \delta(x') < \infty, \\ 0 \le -\gamma(x') \le 1, \\ -1 \le \beta^j(x') \le 1, \\ 0 \le \ell(x') \le 1, \end{cases}$$

and

$$-\gamma(x') + \sum_{j=1}^{N-1} |\beta^j(x')| + \ell(x') = 1, \tag{11}$$

and the measure $\hat{q}(x', \cdot)$ is a probability measure on $H_{x'}$.

To rewrite the last term of formula (10), we define a non-negative Borel measure $\tilde{q}(x', \cdot)$ on $\bar{D} \setminus \{x'\}$ by

$$\tilde{q}(x', E) = \hat{q}(x', \Phi_{x'}(E)), \qquad E \in \mathscr{B}_{\bar{D}\setminus\{x'\}},$$

and let

$$W: \bar{D} \times [0, 1] \times M \ni h = (y, w, (z^{ij})) \mapsto w \in [0, 1],$$

$$Z: \bar{D} \times [0, 1] \times M \ni h = (y, w, (z^{ij})) \mapsto (z^{ij}) \in M.$$

Then, in view of (6), it follows that

$$\begin{aligned}
&\ell(x') \int_{H_{x'}} \hat{q}(x', dh)\hat{u}(x', h) \\
&= \ell(x') \int_{H_{x'} \setminus \Phi_{x'}(\overline{D} \setminus \{x'\})} \hat{q}(x', dh)\hat{u}(x', h) \\
&\quad + \ell(x') \int_{\Phi_{x'}(\overline{D} \setminus \{x'\})} \hat{q}(x', dh)\hat{u}(x', h) \\
&= \ell(x') \left\{ \int_{H_{x'} \setminus \Phi_{x'}(\overline{D} \setminus \{x'\})} \hat{q}(x', dh) W(h) \frac{\partial u}{\partial x_N}(x') \right. \\
&\quad \left. + \frac{1}{2} \sum_{i,j=1}^{N} \int_{H_{x'} \setminus \Phi_{x'}(\overline{D} \setminus \{x'\})} \hat{q}(x', dh) Z^{ij}(h) \frac{\partial^2 u}{\partial x_i \, \partial x_j}(x') \right\} \\
&\quad + \ell(x') \int_{\overline{D} \setminus \{x'\}} \tilde{q}(x', dy)\tilde{u}(x', y) \\
&= \mu(x') \frac{\partial u}{\partial x_N} + \sum_{i,j=1}^{N-1} \alpha^{ij}(x') \frac{\partial^2 u}{\partial x_i \, \partial x_j}(x') \\
&\quad + \int_{\overline{D}} \nu(x', dy) \left[u(y) - u(x') - \sum_{i=1}^{N-1} \frac{\partial u}{\partial x_i}(x')(\chi_i(y) - \chi_i(x')) \right],
\end{aligned} \tag{12}$$

where:

$$\mu(x') = \ell(x') \int_{H_{x'} \setminus \Phi_{x'}(\overline{D} \setminus \{x'\})} \hat{q}(x', dh) W(h), \tag{13}$$

$$\alpha^{ij}(x') = \frac{\ell(x')}{2} \int_{H_{x'} \setminus \Phi_{x'}(\overline{D} \setminus \{x'\})} \hat{q}(x', dh) Z^{ij}(h), \tag{14}$$

and

$$\begin{cases} \nu(x', \{x'\}) = 0, \\ \nu(x', E) = \ell(x') \displaystyle\int_{E \setminus \{x'\}} \tilde{q}(x', dy) \left(\frac{1}{d(x', y)} \right), \qquad E \in \mathscr{B}_{\overline{D}}. \end{cases} \tag{15}$$

Therefore, combining formulas (10) and (12), we obtain condition (2) in Case II.

Property (3) follows from our construction of α^{ij}, β^i, γ, μ, δ and ν.

5) Finally we show that condition (2) is consistent, that is, condition (2) does not take the form $0 = 0$.

In case I, we have taken

$$\delta(x') = 1.$$

In case II, we assume that

$$\begin{cases} \gamma(x') = \beta^i(x') = 0, \\ \nu(x', \cdot) = 0. \end{cases}$$

Then we have, by equation (11),

$$\ell(x') = 1,$$

and hence, by formula (15),

$$\hat{q}(x', \Phi_{x'}(\bar{D} \setminus \{x'\})) = \tilde{q}(x', \bar{D} \setminus \{x'\}) = 0.$$

This implies that

$$\hat{q}(x', H_{x'} \setminus \Phi_{x'}(\bar{D} \setminus \{x'\})) = 1,$$

since the measure $\hat{q}(x', \cdot)$ is a probability measure on $H_{x'}$. Therefore, in view of (5), it follows from formulas (13) and (14) that

$$\begin{aligned} &\mu(x') + 2 \sum_{i=1}^{N-1} \alpha^{ii}(x') \\ &= \ell(x') \int_{H_{x'} \setminus \Phi_{x'}(\bar{D} \setminus \{x'\})} \hat{q}(x', dh)\left(W(h) + \sum_{i=1}^{N-1} Z^{ii}(h) \right) \\ &= \ell(x')\hat{q}(x', H_{x'} \setminus \Phi_{x'}(\bar{D} \setminus \{x'\}) \\ &= 1. \end{aligned}$$

The proof of Theorem 9.5.1 is now complete. ∎

9.5.2 Remark. We can reconstruct the functions α^{ij}, β^i, γ, μ and δ so that they are bounded and Borel measurable on ∂D (cf. Bony–Courrège–Priouret [1], Théorème XIII).

Probabilistically, Theorems 9.4.1 and 9.5.1 may be interpreted as follows: a particle in a *Markov process* $\mathscr{X}$ on $\bar{D}$ is governed by an integro-differential operator of the form (9.4.2) in the interior D of $\bar{D}$, and it obeys a boundary condition of the form (2) on the boundary ∂D of $\bar{D}$.

Analytically, via a version of the Hille–Yosida theorem (Theorem 9.3.1), Theorems 9.4.1 and 9.5.1 may be interpreted as follows: a *Feller semigroup* $\{T_t\}_{t \geq 0}$ on $\bar{D}$ is described by an integro-differential operator of the form (9.4.2) and a boundary condition of the form (2). Hence we are reduced to the study of *boundary value problems* in the theory of partial differential equations.

9.6. Feller Semigroups and Boundary Value Problems

By virtue of Theorems 9.4.1 and 9.5.1, we can reduce the study of Feller semigroups to the study of boundary value problems. In this section, we shall prove general existence theorems for Feller semigroups in terms of boundary value problems in the case when the measures $e(x^0, \cdot)$ in formula (9.4.2) and the measures $\nu(x', \cdot)$ in formula (9.5.2) *identically vanish* in D and on ∂D, respectively (see formulas (1) and (3) below). In other words, we shall confine ourselves to a class of Feller semigroups whose infinitesimal generators have no integro-differential operator term in formulas (9.4.2) and (9.5.2).

We start by formulating our problem precisely. Let D be a bounded domain in $\mathbf{R}^N$ with smooth boundary ∂D, and choose for each point x' of ∂D a neighborhood U of x' in $\mathbf{R}^N$ and a local coordinate system $(x_1, \ldots, x_{N-1}, x_N)$ on U such that (cf. Figure 9-5):

1. $x \in U \cap D \Leftrightarrow x \in U,\ x_N(x) > 0;$
   $x \in U \cap \partial D \Leftrightarrow x \in U,\ x_N(x) = 0.$
2. The functions $(x_1, \ldots, x_{N-1})$, restricted to $U \cap \partial D$, form a local coordinate system of ∂D on $U \cap \partial D$.

We may take

$$x_N(x) = \operatorname{dist}(x, \partial D), \qquad x \in \mathbf{R}^N.$$

Then we have

$$\operatorname{grad} x_N(x') = \text{the unit interior normal } \mathbf{n} \text{ to } \partial D \text{ at } x',$$

and hence

$$\frac{\partial}{\partial x_N} = \frac{\partial}{\partial \mathbf{n}}.$$

Let A be a second-order *elliptic differential operator* with real coefficients such that

$$Au(x) = \sum_{i,j=1}^{N} a^{ij}(x) \frac{\partial^2 u}{\partial x_i\, \partial x_j}(x) + \sum_{i=1}^{N} b^i(x) \frac{\partial u}{\partial x_i}(x) + c(x)u(x) \tag{1}$$

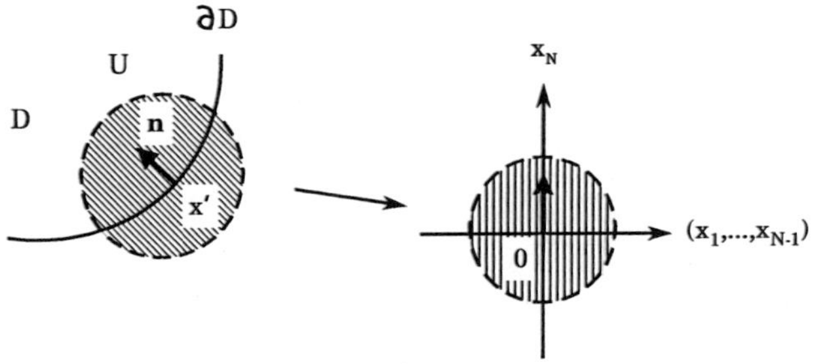

Figure 9-5

where:

1. $a^{ij} \in C^\infty(\mathbf{R}^N)$, $a^{ij} = a^{ji}$ and there exists a constant $a_0 > 0$ such that

$$\sum_{i,j=1}^{N} a^{ij}(x)\xi_i\xi_j \geq a_0|\xi|^2, \qquad x \in \mathbf{R}^N, \xi \in \mathbf{R}^N. \tag{2}$$

2. $b^i \in C^\infty(\mathbf{R}^N)$.
3. $c \in C^\infty(\mathbf{R}^N)$ and $c \leq 0$ on $\bar{D}$.

The functions a^{ij}, b^i and c are called the diffusion coefficients, the drift coefficients and the termination coefficient, respectively.

Let L be a boundary condition such that

$$\begin{aligned} Lu(x') = &\sum_{i,j=1}^{N-1} \alpha^{ij}(x') \frac{\partial^2 u}{\partial x_i\, \partial x_j}(x') + \sum_{i=1}^{N-1} \beta^i(x') \frac{\partial u}{\partial x_i}(x') \\ &+ \gamma(x')u(x') + \mu(x') \frac{\partial u}{\partial \mathbf{n}}(x') - \delta(x')Au(x') \end{aligned} \tag{3}$$

where:

1. The α^{ij} are the components of a C^∞ symmetric contravariant tensor of type $\binom{2}{0}$ on ∂D and

$$\sum_{i,j=1}^{N-1} \alpha^{ij}(x')\eta_i\eta_j \geq 0, \qquad x' \in \partial D, \eta = \sum_{j=1}^{N-1} \eta_j\, dx_j \in T^*_{x'}(\partial D),$$

where $T^*_{x'}(\partial D)$ is the cotangent space of ∂D at x'.

2. $\beta^i \in C^\infty(\partial D)$. (4)
3. $\gamma \in C^\infty(\partial D)$ and $\gamma \leq 0$ on ∂D.
4. $\mu \in C^\infty(\partial D)$ and $\mu \geq 0$ on ∂D.
5. $\delta \in C^\infty(\partial D)$ and $\delta \geq 0$ on ∂D.

The condition L will be called a *Ventcel' boundary condition.* Its terms

$$\sum_{i,j=1}^{N-1} \alpha^{ij} \frac{\partial^2 u}{\partial x_i \, \partial x_j} + \sum_{i=1}^{N-1} \beta^i \frac{\partial u}{\partial x_i}, \; \gamma u, \; \mu \frac{\partial u}{\partial \mathbf{n}}, \; \delta A u$$

are supposed to correspond to the diffusion along the boundary, absorption, reflection and viscosity phenomena, respectively (cf. Figure 0-4).

We are interested in the following:

Problem. *Given analytic data (A, L), can we construct a Feller semigroup $\{T_t\}_{t \geq 0}$ on $\bar{D}$ whose infinitesimal generator $\mathfrak{A}$ is characterized by (A, L)?*

9.6.1 Remark. In the case $N = 1$, this problem is completely solved both from probabilistic and analytic viewpoints by Feller [1], [2], [3], Dynkin [1], [2], Itô-McKean Jr. [1] and Ray [1]. So we shall consider the case $N \geq 2$.

In this section, we shall prove general existence theorems for Feller semigroups on ∂D and then on $\bar{D}$ (Theorems 9.6.15 and 9.6.22) if the boundary value problem

$$\begin{cases} (\alpha - A)u = 0 & \text{in } D, \\ (\lambda - L)u = \varphi & \text{on } \partial D, \end{cases} \tag{$*$}$$

is solvable for sufficiently many functions φ in $C(\partial D)$. Here α and λ are positive constants.

First we consider the following Dirichlet problem: for given functions f and φ defined in D and on ∂D, respectively, find a function u in D such that

$$\begin{cases} (\alpha - A)u = f & \text{in } D, \\ u|_{\partial D} = \varphi & \text{on } \partial D. \end{cases} \tag{D'}$$

Theorem 8.1.1 tells us that problem (D') has a unique solution u in $C^{2+\theta}(\bar{D})$ for any $f \in C^\theta(\bar{D})$ and $\varphi \in C^{2+\theta}(\partial D)$. Therefore we can introduce linear operators

$$\begin{aligned} G_\alpha^0 &: C^\theta(\bar{D}) \to C^{2+\theta}(\bar{D}), \\ H_\alpha &: C^{2+\theta}(\partial D) \to C^{2+\theta}(\bar{D}), \end{aligned} \qquad (\alpha > 0),$$

as follows:

1) For any $f \in C^\theta(\bar{D})$, the function $G_\alpha^0 f \in C^{2+\theta}(\bar{D})$ is the unique solution of the problem

$$\begin{cases} (\alpha - A)G_\alpha^0 f = f & \text{in } D, \\ G_\alpha^0 f|_{\partial D} = 0 & \text{on } \partial D. \end{cases} \tag{5}$$

2) For any $\varphi \in C^{2+\theta}(\partial D)$, the function $H_\alpha \varphi \in C^{2+\theta}(\bar{D})$ is the unique solution of the problem

$$\begin{cases} (\alpha - A)H_\alpha \varphi = 0 & \text{in } D, \\ H_\alpha \varphi|_{\partial D} = \varphi & \text{on } \partial D. \end{cases} \tag{6}$$

The operator G_α^0 is called the *Green operator* and the operator H_α is called the *harmonic operator.*

Then we have:

9.6.2 Lemma. *The operator G_α^0 $(\alpha > 0)$, considered from $C(\bar{D})$ into itself, is non-negative and continuous with norm*

$$\|G_\alpha^0\| = \|G_\alpha^0 1\| = \sup_{x \in \bar{D}} G_\alpha^0 1(x).$$

Proof. Let f be an arbitrary function in $C^\theta(\bar{D})$ such that $f \geq 0$ on $\bar{D}$. Then, applying Theorem 7.1.1 (the weak maximum principle) with $A = A - \alpha$ to the function $-G_\alpha^0 f$, we obtain from formula (5) that

$$G_\alpha^0 f \geq 0 \qquad \text{on } \bar{D}.$$

This proves the non-negativity of G_α^0.

Since G_α^0 is non-negative, we have, for all $f \in C^\theta(\bar{D})$,

$$-G_\alpha^0 \|f\| \leq G_\alpha^0 f \leq G_\alpha^0 \|f\| \qquad \text{on } \bar{D}.$$

This implies the continuity of G_α^0 with norm

$$\|G_\alpha^0\| = \|G_\alpha^0 1\|.$$

The proof is complete. ■

Similarly, we obtain from formula (6) the following:

9.6.3 Lemma. *The operator H_α $(\alpha > 0)$, considered from $C(\partial D)$ into $C(\bar{D})$, is non-negative and continuous with norm*

$$\|H_\alpha\| = \|H_\alpha 1\| = \sup_{x \in \bar{D}} H_\alpha 1(x).$$

More precisely, we have:

9.6.4 Theorem. (i) (a) *The operator* G_α^0 $(\alpha > 0)$ *can be uniquely extended to a non-negative, bounded linear operator on* $C(\bar{D})$ *into itself, denoted again* G_α^0, *with norm*

$$\|G_\alpha^0\| = \|G_\alpha^0 1\| \leq \frac{1}{\alpha}. \tag{7}$$

(b) *For all* $f \in C(\bar{D})$, *we have*

$$G_\alpha^0 f|_{\partial D} = 0.$$

(c) *For all* $\alpha, \beta > 0$, *the* resolvent equation *holds*:

$$G_\alpha^0 f - G_\beta^0 f + (\alpha - \beta) G_\alpha^0 G_\beta^0 f = 0, \qquad f \in C(\bar{D}). \tag{8}$$

(d) *For any* $f \in C(\bar{D})$, *we have*

$$\lim_{\alpha \to +\infty} \alpha G_\alpha^0 f(x) = f(x), \qquad x \in D. \tag{9}$$

Furthermore, if $f|_{\partial D} = 0$, *then this convergence is uniform in* $x \in \bar{D}$, *that is,*

$$\lim_{\alpha \to +\infty} \alpha G_\alpha^0 f = f \qquad \text{in } C(\bar{D}). \tag{9'}$$

(e) *The operator* G_α^0 *maps* $C^{k+\theta}(\bar{D})$ *into* $C^{k+2+\theta}(\bar{D})$ *for any non-negative integer* k.

(ii) (a′) *The operator* H_α $(\alpha > 0)$ *can be uniquely extended to a non-negative, bounded linear operator on* $C(\partial D)$ *into* $C(\bar{D})$, *denoted again* H_α, *with norm* $\|H_\alpha\| = 1$.

(b′) *For all* $\varphi \in C(\partial D)$, *we have*

$$H_\alpha \varphi|_{\partial D} = \varphi.$$

(c′) *For all* $\alpha, \beta > 0$, *we have*

$$H_\alpha \varphi - H_\beta \varphi + (\alpha - \beta) G_\alpha^0 H_\beta \varphi = 0, \qquad \varphi \in C(\partial D). \tag{10}$$

(d′) *The operator* H_α *maps* $C^{k+\theta+2}(\partial D)$ *into* $C^{k+2+\theta}(\bar{D})$ *for any non-negative integer* k.

Proof. (i) (a) Making use of mollifiers (cf. Section 4.2), we find that the space $C^\theta(\bar{D})$ is dense in $C(\bar{D})$ and further that non-negative functions can be approximated by non-negative C^∞ functions. Hence, by Lemma 9.6.2, it follows that the operator G_α^0: $C^\theta(\bar{D}) \to C^{2+\theta}(\bar{D})$ can be uniquely extended to a non-negative, bounded linear operator $G_\alpha^0\colon C(\bar{D}) \to C(\bar{D})$ with norm $\|G_\alpha^0\| = \|G_\alpha^0 1\|$.

Further, since the function $G_\alpha^0 1$ satisfies

$$\begin{cases} (A-\alpha)G_\alpha^0 1 = -1 & \text{in } D, \\ G_\alpha^0 1|_{\partial D} = 0 & \text{on } \partial D, \end{cases}$$

applying Theorem 7.1.2 with $A = A - \alpha(\alpha > 0)$, we obtain that

$$\|G_\alpha^0\| = \|G_\alpha^0 1\| \leq \frac{1}{\alpha}.$$

(b) This follows from formula (5), since the space $C^\theta(\bar{D})$ is dense in $C(\bar{D})$ and the operator $G_\alpha^0\colon C(\bar{D}) \to C(\bar{D})$ is bounded.

(c) We find from the uniqueness property of solutions of problem (D') that equation (8) holds for all $f \in C^\theta(\bar{D})$. Hence it holds for all $f \in C(\bar{D})$, since the space $C^\theta(\bar{D})$ is dense in $C(\bar{D})$ and the operators G_α^0 are bounded.

(d) First let f be an arbitrary function in $C^{2+\theta}(\bar{D})$ satisfying $f|_{\partial D} = 0$. Then it follows from the uniqueness property of solutions of problem (D') that for all $\alpha, \beta > 0$ we have

$$f - \alpha G_\alpha^0 f = G_\alpha^0((\beta - A)f) - \beta G_\alpha^0 f.$$

Thus we have, by estimate (7),

$$\|f - \alpha G_\alpha^0 f\| \leq \frac{1}{\alpha}\|(\beta - A)f\| + \frac{\beta}{\alpha}\|f\|.$$

so that

$$\lim_{\alpha \to +\infty} \|f - \alpha G_\alpha^0 f\| = 0.$$

Now let f be an arbitrary function in $C(\bar{D})$ satisfying $f|_{\partial D} = 0$. By means of mollifiers, we can find a sequence $\{f_j\}$ in $C^{2+\theta}(\bar{D})$ such that

$$\begin{cases} f_j \to f & \text{in } C(\bar{D}) \text{ as } j \to \infty, \\ f_j|_{\partial D} = 0 & \text{on } \partial D. \end{cases}$$

Then we have, by estimate (7),

$$\begin{aligned} \|f - \alpha G_\alpha^0 f\| &\leq \|f - f_j\| + \|f_j - \alpha G_\alpha^0 f_j\| + \|\alpha G_\alpha^0 f_j - \alpha G_\alpha^0 f\| \\ &\leq 2\|f - f_j\| + \|f_j - \alpha G_\alpha^0 f_j\|. \end{aligned}$$

and hence

$$\limsup_{\alpha \to +\infty} \|f - \alpha G_\alpha^0 f\| \leq 2\|f - f_j\|.$$

This proves assertion (9′), since $\|f - f_j\| \to 0$ as $j \to \infty$.

To prove assertion (9), let f be an arbitrary function in $C(\bar{D})$ and x an arbitrary point of D. Take a function $\psi \in C(\bar{D})$ such that

$$\begin{cases} 0 \leq \psi \leq 1 & \text{on } \bar{D}, \\ \psi = 0 & \text{in a neighborhood of } x, \\ \psi = 1 & \text{near the boundary } \partial D. \end{cases}$$

Then it follows from the non-negativity of G_α^0 and estimate (7) that

$$0 \leq \alpha G_\alpha^0 \psi(x) + \alpha G_\alpha^0(1 - \psi)(x) = \alpha G_\alpha^0 1(x) \leq 1. \tag{11}$$

But, applying assertion (9′) to the function $1 - \psi$, we have

$$\lim_{\alpha \to +\infty} \alpha G_\alpha^0(1 - \psi)(x) = (1 - \psi)(x) = 1.$$

In view of inequalities (11), this implies that

$$\lim_{\alpha \to +\infty} \alpha G_\alpha^0 \psi(x) = 0.$$

Thus, since $-\|f\|\psi \leq f\psi \leq \|f\|\psi$ on $\bar{D}$, it follows that

$$|\alpha G_\alpha^0(f\psi)(x)| \leq \|f\|\alpha G_\alpha^0 \psi(x) \to 0 \qquad \text{as } \alpha \to +\infty.$$

Therefore, applying assertion (9′) to the function $(1 - \psi)f$, we obtain that

$$f(x) = ((1 - \psi)f)(x) = \lim_{\alpha \to +\infty} \alpha G_\alpha^0((1 - \psi)f)(x) = \lim_{\alpha \to +\infty} \alpha G_\alpha^0 f(x).$$

(e) This is an immediate consequence of part (iii) of Theorem 8.1.1.

(ii) (a′) Since the space $C^{2+\theta}(\partial D)$ is dense in $C(\partial D)$, by Lemma 9.6.3, it follows that the operator $H_\alpha: C^{2+\theta}(\partial D) \to C^{2+\theta}(\bar{D})$ can be uniquely extended to a non-negative, bounded linear operator $H_\alpha: C(\partial D) \to C(\bar{D})$. Further, applying Theorem 7.1.2 with $A = A - \alpha$ (cf. Remark 7.1.3), we have

$$\|H_\alpha\| = \|H_\alpha 1\| = 1.$$

(b′) This follows from formula (6), since the space $C^{2+\theta}(\partial D)$ is dense in $C(\partial D)$ and the operator $H_\alpha: C(\partial D) \to C(\bar{D})$ is bounded.

(c′) We find from the uniqueness property of solutions of problem (D') that formula (10) holds for all $\varphi \in C^{2+\theta}(\partial D)$. Hence it holds for all $\varphi \in C(\partial D)$, since the space $C^{2+\theta}(\partial D)$ is dense in $C(\partial D)$ and the operators G_α^0 and H_α are bounded.

(d′) This is an immediate consequence of part (iii) of Theorem 8.1.1.

The proof of Theorem 9.6.4 is now complete. ■

Next we consider problem (∗) in the framework of the spaces of continuous functions. To do so, we introduce three operators associated with problem (∗).

(I) First we introduce a linear operator

$$A: C(\bar{D}) \to C(\bar{D})$$

as follows:

1. The domain $D(A)$ of A is the space $C^2(\bar{D})$.

2. $$Au = \sum_{i,j=1}^{N} a^{ij} \frac{\partial^2 u}{\partial x_i \, \partial x_j} + \sum_{i=1}^{N} b^i \frac{\partial u}{\partial x_i} + cu, \qquad u \in D(A).$$

Then we have:

9.6.5 Lemma. *The operator A has its minimal closed extension $\bar{A}$ in $C(\bar{D})$.*

Proof. We apply part (i) of Theorem 9.3.3 to the operator A.

Since the matrix $(a^{ij}(x))$ is positive semi-definite, it follows that if $u \in C^2(\bar{D})$ takes its positive maximum at a point x_0 of D, then we have

$$\sum_{i,j=1}^{N} a^{ij}(x_0) \frac{\partial^2 u}{\partial x_i \, \partial x_j}(x_0) \leq 0, \qquad \frac{\partial u}{\partial x_i}(x_0) = 0 \qquad (1 \leq i \leq N);$$

and hence

$$\begin{aligned} Au(x_0) &= \sum_{i,j=1}^{N} a^{ij}(x_0) \frac{\partial^2 u}{\partial x_i \, \partial x_j}(x_0) + c(x_0)u(x_0) \\ &\leq 0. \end{aligned}$$

This implies that the operator A satisfies condition (β) of Theorem 9.3.3 with $K_0 = D$ and $K = \bar{D}$. Therefore Lemma 9.6.5 follows from an application of the same theorem. ■

9.6.6 Remark. Since the injection $C(\bar{D}) \to \mathscr{D}'(D)$ is continuous, we have the formula

$$\bar{A}u = \sum_{i,j=1}^{N} a^{ij} \frac{\partial^2 u}{\partial x_i \, \partial x_j} + \sum_{i=1}^{N} b^i \frac{\partial u}{\partial x_i} + cu, \qquad u \in D(\bar{A}),$$

where the right-hand side is taken in the sense of *distributions*.

The extended operators $G_\alpha^0: C(\bar{D}) \to C(\bar{D})$ and $H_\alpha: C(\partial D) \to C(\bar{D})$ $(\alpha > 0)$ still satisfy formulas (5) and (6) respectively in the following sense:

9.6.7 Lemma. (*i*) *For any* $f \in C(\bar{D})$, *we have*

$$\begin{cases} G_\alpha^0 f \in D(\bar{A}), \\ (\alpha I - \bar{A}) G_\alpha^0 f = f \quad \text{in } D. \end{cases}$$

(ii) *For any* $\varphi \in C(\partial D)$, *we have*

$$\begin{cases} H_\alpha \varphi \in D(\bar{A}), \\ (\alpha I - \bar{A}) H_\alpha \varphi = 0 \quad \text{in } D. \end{cases}$$

Here $D(\bar{A})$ *is the domain of* $\bar{A}$.

Proof. (i) Choose a sequence $\{f_j\}$ in $C^\theta(\bar{D})$ such that $f_j \to f$ in $C(\bar{D})$ as $j \to \infty$. Then it follows from the boundedness of G_α^0 that

$$G_\alpha^0 f_j \to G_\alpha^0 f \qquad \text{in } C(\bar{D}),$$

and also

$$(\alpha - A) G_\alpha^0 f_j = f_j \to f \qquad \text{in } C(\bar{D}).$$

Hence we have

$$\begin{cases} G_\alpha^0 f \in D(\bar{A}), \\ (\alpha I - \bar{A}) G_\alpha^0 f = f \quad \text{in } D, \end{cases}$$

since the operator $\bar{A}: C(\bar{D}) \to C(\bar{D})$ is closed.

(ii) Similarly, part (ii) is proved, sincee the space $C^{2+\theta}(\partial D)$ is dense in $C(\partial D)$ and the operator $H_\alpha: C(\partial D) \to C(\bar{D})$ is bounded. ∎

9.6.8 Corollary. *Every function u in $D(\bar{A})$ can be written in the following form:*

$$u = G_\alpha^0((\alpha I - \bar{A})u) + H_\alpha(u|_{\partial D}), \qquad \alpha > 0. \tag{12}$$

Proof. We let

$$w = u - G_\alpha^0((\alpha I - \bar{A})u) - H_\alpha(u|_{\partial D}).$$

Then it follows from Lemma 9.6.7 that the function w is in $D(\bar{A})$ and satisfies

$$\begin{cases} (\alpha I - \bar{A})w = 0 & \text{in } D, \\ w|_{\partial D} = 0 & \text{on } \partial D. \end{cases}$$

Thus, in view of Remark 9.6.6, we can apply Theorem 8.2.3 with $A = A - \alpha$ and $s = 0$ to obtain that

$$w = 0.$$

This proves formula (12). ■

(II) Secondly we introduce a linear operator

$$LG_\alpha^0: C(\bar{D}) \to C(\partial D)$$

as follows:

1. The domain $D(LG_\alpha^0)$ of LG_α^0 is the space
$$D(LG_\alpha^0) = \{f \in C(\bar{D});\ G_\alpha^0 f \in C^2(\bar{D})\}.$$
2. $LG_\alpha^0 f = L(G_\alpha^0 f), \qquad f \in D(LG_\alpha^0).$

We remark that the domain $D(LG_\alpha^0)$ contains $C^\theta(\bar{D})$.

Then we have:

9.6.9 Lemma. *The operator LG_α^0 ($\alpha > 0$) can be uniquely extended to a non-negative, bounded linear operator $\overline{LG_\alpha^0}: C(\bar{D}) \to C(\partial D)$.*

Proof. Let f be an arbitrary function in $D(LG_\alpha^0)$ such that $f \geq 0$ on $\bar{D}$. Then we have

$$\begin{cases} G_\alpha^0 f \in C^2(\bar{D}), \\ G_\alpha^0 f \geq 0 & \text{on } \bar{D}, \\ G_\alpha^0 f|_{\partial D} = 0 & \text{on } \partial D, \end{cases}$$

and hence

$$LG_\alpha^0 f = \mu \frac{\partial}{\partial \mathbf{n}} (G_\alpha^0 f) - \delta A G_\alpha^0 f$$

$$= \mu \frac{\partial}{\partial \mathbf{n}} (G_\alpha^0 f) + \delta f \geq 0 \qquad \text{on } \partial D.$$

This proves that the operator LG_α^0 is non-negative.

By the non-negativity of LG_α^0, we have, for all $f \in D(LG_\alpha^0)$,

$$-LG_\alpha^0 \|f\| \leq LG_\alpha^0 f \leq LG_\alpha^0 \|f\| \qquad \text{on } \partial D.$$

This implies the boundedness of LG_α^0 with norm

$$\|LG_\alpha^0\| = \|LG_\alpha^0 1\|.$$

Recall that the space $C^\theta(\bar{D})$ is dense in $C(\bar{D})$ and that non-negative functions can be approximated by non-negative C^∞ functions. Hence we find that the operator LG_α^0 can be uniquely extended to a non-negative, bounded linear operator $\overline{LG_\alpha^0}: C(\bar{D}) \to C(\partial D)$. ■

The next lemma states a fundamental relationship between the operators $\overline{LG_\alpha^0}$ and $\overline{LG_\beta^0}$ for $\alpha, \beta > 0$.

9.6.10 Lemma. *For any $f \in C(\bar{D})$, we have*

$$\overline{LG_\alpha^0} f - \overline{LG_\beta^0} f + (\alpha - \beta)\overline{LG_\alpha^0} G_\beta^0 f = 0, \qquad \alpha, \beta > 0. \tag{13}$$

Proof. Choose a sequence $\{f_j\}$ in $C^\theta(\bar{D})$ such that $f_j \to f$ in $C(\bar{D})$ as $j \to \infty$. Then, using the resolvent equation (8) with $f = f_j$, we have

$$LG_\alpha^0 f_j - LG_\beta^0 f_j + (\alpha - \beta) LG_\alpha^0 G_\beta^0 f_j = 0.$$

Hence formula (13) follows by letting $j \to \infty$, since the operators $\overline{LG_\alpha^0}$, $\overline{LG_\beta^0}$ and G_β^0 are all bounded. ■

(III) Finally we introduce a linear operator

$$LH_\alpha: C(\partial D) \to C(\partial D)$$

as follows:

1. The domain $D(LH_\alpha)$ of LH_α is the space $C^{2+\theta}(\partial D)$.
2. $LH_\alpha \psi = L(H_\alpha \psi), \qquad \psi \in D(LH_\alpha).$

Then we have:

9.6.11 Lemma. *The operator LH_α ($\alpha > 0$) has its minimal closed extension $\overline{LH_\alpha}$ in $C(\partial D)$.*

Proof. We apply part (i) of Theorem 9.3.3 to the operator LH_α. To do so, it suffices to show that the operator LH_α satisfies condition (β') with $K = \partial D$ (or condition (β) with $K = K_0 = \partial D$) of the same theorem.

Suppose that a function ψ in $D(LH_\alpha) = C^{2+\theta}(\partial D)$ takes its positive maximum at some point x' of ∂D. Since the function $H_\alpha \psi$ is in $C^{2+\theta}(\bar{D})$ and satisfies

$$\begin{cases} (A - \alpha)H_\alpha \psi = 0 & \text{in } D, \\ H_\alpha \psi|_{\partial D} = \psi & \text{on } \partial D, \end{cases}$$

applying Theorem 7.1.1 (the weak maximum principle) with $A = A - \alpha$ to the function $H_\alpha \psi$, we find that the function $H_\alpha \psi$ takes its positive maximum at $x' \in \partial D$. Thus we can apply Lemma 7.1.7 with $\Sigma_3 = \partial D$ to obtain that

$$\frac{\partial}{\partial \mathbf{n}}(H_\alpha \psi)(x') < 0. \tag{14}$$

Hence we have, by hypotheses (4),

$$\begin{aligned} LH_\alpha \psi(x') &= \sum_{i,j=1}^{N-1} \alpha^{ij}(x') \frac{\partial^2 \psi}{\partial x_i \, \partial x_j}(x') + \mu(x') \frac{\partial}{\partial \mathbf{n}}(H_\alpha \psi)(x') \\ &\quad + \gamma(x')\psi(x') - \alpha\delta(x')\psi(x') \\ &\leq 0. \end{aligned}$$

This verifies condition (β') of Theorem 9.3.3. Therefore Lemma 9.6.11 follows from an application of the same theorem. ■

9.6.12 Remark. In view of assertion (9.3.8), it follows that:

If a function $\psi \in D(\overline{LH_\alpha})$ takes its positive maximum at some point x' of ∂D, then we have (15)

$$\overline{LH_\alpha}\psi(x') \leq 0.$$

The next lemma states a fundamental relationship between the operators $\overline{LH_\alpha}$ and $\overline{LH_\beta}$ for $\alpha, \beta > 0$.

9.6.13 Lemma. *The domain $D(\overline{LH_\alpha})$ of $\overline{LH_\alpha}$ does not depend on $\alpha > 0$; so we denote by $\mathscr{D}$ the common domain. Then we have*

$$\overline{LH_\alpha}\psi - \overline{LH_\beta}\psi + (\alpha - \beta)\overline{LG_\alpha^0}H_\beta\psi = 0, \qquad \alpha, \beta > 0, \psi \in \mathscr{D}. \tag{16}$$

Proof. Let ψ be an arbitrary function in $D(\overline{LH_\beta})$, and choose a sequence $\{\psi_j\}$ in $D(LH_\beta) = C^{2+\theta}(\partial D)$ such that

$$\begin{cases} \psi_j \to \psi & \text{in } C(\partial D), \\ LH_\beta\psi_j \to \overline{LH_\beta}\psi & \text{in } C(\partial D). \end{cases}$$

Then it follows from the boundedness of H_β and $\overline{LG_\alpha^0}$ that

$$LG_\alpha^0(H_\beta\psi_j) = \overline{LG_\alpha^0}(H_\beta\psi_j) \to \overline{LG_\alpha^0}(H_\beta\psi) \qquad \text{in } C(\partial D).$$

Therefore, using formula (10) with $\varphi = \psi_j$, we obtain that

$$\begin{aligned} LH_\alpha\psi_j = LH_\beta\psi_j - (\alpha - \beta)LG_\alpha^0(H_\beta\psi_j) \\ \to \overline{LH_\beta}\psi - (\alpha - \beta)\overline{LG_\alpha^0}(H_\beta\psi) \qquad \text{in } C(\partial D). \end{aligned}$$

This implies that

$$\begin{cases} \psi \in D(\overline{LH_\alpha}), \\ \overline{LH_\alpha}\psi = \overline{LH_\beta}\psi - (\alpha - \beta)\overline{LG_\alpha^0}(H_\beta\psi). \end{cases}$$

Conversely, interchanging α and β, we have

$$D(\overline{LH_\alpha}) \subset D(\overline{LH_\beta}),$$

and so

$$D(\overline{LH_\alpha}) = D(\overline{LH_\beta}).$$

This proves the lemma. ■

In view of Remark 9.6.6, it follows that every function $f \in C(\bar{D}) \subset L^2(D)$ satisfies the equation

$$(\alpha - A)G_\alpha^0 f = f \qquad \text{in } D$$

in the sense of distributions. Hence, applying Theorem 5.6.5 with $A = A - \alpha$ to the function $G_\alpha^0 f$, we find that the boundary condition $L(G_\alpha^0 f)$ can be defined as a distribution on ∂D.

Similarly, we find that the boundary condition $L(H_\alpha \psi)$, for any $\psi \in C(\partial D)$, can be defined as a distribution on ∂D, since

$$(\alpha - A)H_\alpha \psi = 0 \qquad \text{in } D.$$

More precisely, we can prove:

9.6.14 Lemma. (i) *If we define a linear operator*

$$\widetilde{LG_\alpha^0}\colon C(\bar{D}) \to \mathscr{D}'(\partial D)$$

by

$$\widetilde{LG_\alpha^0} f = L(G_\alpha^0 f), \qquad f \in C(\bar{D}),$$

then we have

$$\overline{LG_\alpha^0} \subset \widetilde{LG_\alpha^0}.$$

(ii) *If we define a linear operator*

$$\widetilde{LH_\alpha}\colon C(\partial D) \to \mathscr{D}'(\partial D)$$

by

$$\widetilde{LH_\alpha}\psi = L(H_\alpha \psi), \qquad \psi \in C(\partial D),$$

then we have

$$\overline{LH_\alpha} \subset \widetilde{LH_\alpha}.$$

Proof. (i) Let f be an arbitrary function in $D(\overline{LG_\alpha^0}) = C(\bar{D})$, and choose a sequence $\{f_j\}$ in $C^\theta(\bar{D}) \subset D(LG_\alpha^0)$ such that

$$f_j \to f \qquad \text{in } C(\bar{D}).$$

Then we have

$$\begin{cases} G_\alpha^0 f_j \to G_\alpha^0 f & \text{in } C(\bar{D}), \\ (\alpha - A)G_\alpha^0 f_j = f_j \to f = (\alpha - A)G_\alpha^0 f & \text{in } C(\bar{D}). \end{cases}$$

Thus, applying Theorem 5.6.5 with $A = A - \alpha$ and $s = \sigma = 0$, we obtain that

$$LG_\alpha^0 f_j \to \widetilde{LG_\alpha^0} f \qquad \text{in } \mathscr{D}'(\partial D).$$

On the other hand, by the boundedness of $\overline{LG_\alpha^0}$, it follows that

$$LG_\alpha^0 f_j \to \overline{LG_\alpha^0} f \qquad \text{in } C(\partial D).$$

Hence we have

$$\overline{LG_\alpha^0 f} = LG_\alpha^0 f.$$

This proves part (i).

Similarly, part (ii) follows from the closedness of $\overline{LH_\alpha}$. ■

Now we can prove a general existence theorem for Feller semigroups on ∂D in terms of boundary value problem (∗). The next theorem tells us that the operator $\overline{LH_\alpha}$ is the infinitesimal generator of some Feller semigroup on ∂D if and only if problem (∗) is solvable for sufficiently many functions φ in $C(\partial D)$.

9.6.15 Theorem. (i) *If the operator $\overline{LH_\alpha}$ ($\alpha > 0$) is the infinitesimal generator of a Feller semigroup on ∂D, then for each constant $\lambda > 0$ the boundary value problem*

$$\begin{cases} (\alpha - A)u = 0 & \text{in } D, \\ (\lambda - L)u = \varphi & \text{on } \partial D, \end{cases} \tag{$*$}$$

has a solution $u \in C^{2+\theta}(\bar{D})$ for any φ in some dense subset of $C(\partial D)$.

(ii) *Conversely if, for some constant $\lambda \geq 0$, boundary value problem (∗) has a solution $u \in C^{2+\theta}(\bar{D})$ for any φ in some dense subset of $C(\partial D)$, then the operator $\overline{LH_\alpha}$ is the infinitesimal generator of some Feller semigroup on ∂D.*

Proof. (i) If the operator $\overline{LH_\alpha}$ generates a Feller semigroup on ∂D, applying part (i) of Theorem 9.3.1 with $K = \partial D$ to the operator $\mathfrak{A} = \overline{LH_\alpha}$, we obtain that

$$R(\lambda I - \overline{LH_\alpha}) = C(\partial D) \qquad \text{for each } \lambda > 0.$$

This implies that the range $R(\lambda I - LH_\alpha)$ is a dense subset of $C(\partial D)$ for each $\lambda > 0$. But, if $\varphi \in C(\partial D)$ is in the range $R(\lambda I - LH_\alpha)$, and if $\varphi = (\lambda I - LH_\alpha)\psi$ with $\psi \in C^{2+\theta}(\partial D)$, then the function $u = H_\alpha \psi \in C^{2+\theta}(\bar{D})$ is a solution of problem (∗). This proves part (i).

(ii) We apply part (ii) of Theorem 9.3.3 with $K = \partial D$ to the operator LH_α. To do so, it suffices to show that the operator LH_α satisfies condition (γ) of the same theorem, since it satisfies condition (β'), as is shown in the proof of Lemma 9.6.11.

By the uniqueness property of solutions of problem (D'), it follows that any function $u \in C^{2+\theta}(\bar{D})$ which satisfies the equation

$$(\alpha - A)u = 0 \qquad \text{in } D$$

can be written in the form

$$u = H_\alpha(u|_{\partial D}), \qquad u|_{\partial D} \in C^{2+\theta}(\partial D) = D(LH_\alpha).$$

Thus we find that if there exists a solution $u \in C^{2+\theta}(\bar{D})$ of problem $(*)$ for $\varphi \in C(\partial D)$, then we have

$$(\lambda I - LH_\alpha)(u|_{\partial D}) = \varphi,$$

and so

$$\varphi \in R(\lambda I - LH_\alpha).$$

Therefore if, for some constant $\lambda \geq 0$, problem $(*)$ has a solution $u \in C^{2+\theta}(\bar{D})$ for any φ in some dense subset of $C(\partial D)$, then the range $R(\lambda I - LH_\alpha)$ is dense in $C(\partial D)$. This verifies condition (γ) (with $\alpha_0 = \lambda$) of Theorem 9.3.3. Hence part (ii) follows from an application of the same theorem.

Theorem 9.6.15 is proved. ■

Further, we give a general existence theorem for Feller semigroups on $\bar{D}$ in terms of Feller semigroups on ∂D. In other words, we construct a Feller semigroup on $\bar{D}$ by making use of Feller semigroups on ∂D.

First we give a precise meaning to the boundary conditions Lu for functions u in $D(\bar{A})$.

We let

$$D(L) = \{u \in D(\bar{A});\ u|_{\partial D} \in \mathscr{D}\},$$

where $\mathscr{D}$ is the common domain of the operators $\overline{LH_\alpha}$, $\alpha > 0$. We remark that the space $D(L)$ contains $C^{2+\theta}(\bar{D})$, since $C^{2+\theta}(\partial D) = D(LH_\alpha) \subset \mathscr{D}$. Corollary 9.6.8 tells us that every function u in $D(L) \subset D(\bar{A})$ can be written in the form

$$u = G_\alpha^0((\alpha I - \bar{A})u) + H_\alpha(u|_{\partial D}), \qquad \alpha > 0. \tag{12}$$

Then we define

$$Lu = \overline{LG_\alpha^0}((\alpha I - \bar{A})u) + \overline{LH_\alpha}(u|_{\partial D}). \tag{17}$$

The next lemma justifies definition (17) of Lu for $u \in D(L)$.

9.6.16 Lemma. *The right-hand side of formula* (17) *depends only on u, not on the choice of expression* (12).

Proof. Suppose that

$$u = G_\alpha^0((\alpha I - \bar{A})u) + H_\alpha(u|_{\partial D})$$
$$= G_\beta^0((\beta I - \bar{A})u) + H_\beta(u|_{\partial D}),$$

where $\alpha, \beta > 0$. Then it follows from formula (13) with $f = (\alpha I - \bar{A})u$ and formula (16) with $\psi = u|_{\partial D}$ that

$$\begin{aligned}
&\overline{LG_\alpha^0}((\alpha I - \bar{A})u) + \overline{LH_\alpha}(u|_{\partial D})\\
&\quad = \overline{LG_\beta^0}((\alpha I - \bar{A})u) - (\alpha - \beta)\overline{LG_\alpha^0}\,G_\beta^0((\alpha I - \bar{A})u)\\
&\qquad + \overline{LH_\beta}(u|_{\partial D}) - (\alpha - \beta)\overline{LG_\alpha^0}\,H_\beta(u|_{\partial D})\\
&\quad = \overline{LG_\beta^0}((\beta I - \bar{A})u) + \overline{LH_\beta}(u|_{\partial D})\\
&\qquad + (\alpha - \beta)\{\overline{LG_\beta^0}u - \overline{LG_\alpha^0}\,G_\beta^0(\alpha I - \bar{A})u - \overline{LG_\alpha^0}\,H_\beta(u|_{\partial D})\}. \qquad (18)
\end{aligned}$$

But we obtain from formula (13) with $f = u$ that

$$\begin{aligned}
&\overline{LG_\beta^0}u - \overline{LG_\alpha^0}(G_\beta^0(\alpha I - \bar{A})u) - \overline{LG_\alpha^0}H_\beta(u|_{\partial D})\\
&\quad = \overline{LG_\beta^0}u - \overline{LG_\alpha^0}(G_\beta^0(\beta I - \bar{A})u + H_\beta(u|_{\partial D}) + (\alpha - \beta)G_\beta^0 u)\\
&\quad = \overline{LG_\beta^0}u - \overline{LG_\alpha^0}u - (\alpha - \beta)\overline{LG_\alpha^0}\,G_\beta^0 u\\
&\quad = 0. \qquad (19)
\end{aligned}$$

Therefore, combining formulas (18) and (19), we have

$$\overline{LG_\alpha^0}((\alpha I - \bar{A})u) + \overline{LH_\alpha}(u|_{\partial D}) = \overline{LG_\beta^0}((\beta I - \bar{A})u) + \overline{LH_\beta}(u|_{\partial D}).$$

This proves the lemma. ■

We introduce a definition on the boundary condition L.

9.6.17 Definition. A Ventcel' boundary condition L is said to be *transversal* on ∂D if it satisfies

$$\mu(x') + \delta(x') > 0 \qquad \text{on } \partial D. \qquad (20)$$

Intuitively, the transversality condition implies that either reflection or viscosity phenomenon occurs at each point of ∂D. Probabilistically, this means that every Markov process on ∂D is the "trace" on ∂D of trajectories of some Markov process on $\bar{D}$ (cf. Ueno [1]).

The next theorem tells us that the transversality condition for L permits us to "piece together" a Markov process (Feller semigroup) on ∂D with A-diffusion in D to construct a Markov process (Feller semigroup) on $\bar{D} = D \cup \partial D$.

9.6.18 Theorem. *Define a linear operator*

$$\mathfrak{A}: C(\bar{D}) \to C(\bar{D})$$

as follows:

1. *The domain* $D(\mathfrak{A})$ *of* $\mathfrak{A}$ *is the space*

$$D(\mathfrak{A}) = \{u \in D(\bar{A});\, u|_{\partial D} \in \mathscr{D},\, Lu = 0\}. \tag{21}$$

2. $\mathfrak{A}u = \bar{A}u, \qquad u \in D(\mathfrak{A}).$

Suppose that the boundary condition L *is transversal on* ∂D *and that the operator* $\overline{LH}_\alpha$ $(\alpha > 0)$ *is the infinitesimal generator of some Feller semigroup on* ∂D. *Then the operator* $\mathfrak{A}$ *is the infinitesimal generator of some Feller semigroup on* $\bar{D}$, *and the Green operator* $G_\alpha = (\alpha I - \mathfrak{A})^{-1}$ *is given by the following*:

$$G_\alpha f = G_\alpha^0 f - H_\alpha(\overline{LH}_\alpha^{-1}\overline{LG_\alpha^0} f), \qquad f \in C(\bar{D}). \tag{22}$$

Proof. We apply part (ii) of Theorem 9.3.1 to the operator $\mathfrak{A}$. The proof is divided into several steps.

(1) First we prove:

If for some $\alpha > 0$ *the operator* $\overline{LH}_\alpha$ *generates a Feller semigroup on* ∂D, *then for* any $\beta > 0$ *the operator* $\overline{LH}_\beta$ *generates a Feller semigroup on* ∂D.

We apply Corollary 9.3.4 with $K = \partial D$ to the operator $\overline{LH}_\beta$. By formula (16), it follows that the operator $\overline{LH}_\beta$ can be written as

$$\overline{LH}_\beta = \overline{LH}_\alpha + M_{\alpha\beta},$$

where $M_{\alpha\beta} = (\alpha - \beta)\overline{LG_\alpha^0}H_\beta$ is a bounded linear operator on $C(\partial D)$ into itself. Further assertion (15) implies that the operator $\overline{LH}_\beta$ satisfies condition (β') of Theorem 9.3.3. Therefore it follows from an application of Corollary 9.3.4 that the operator $\overline{LH}_\beta$ generates a Feller semigroup on ∂D.

(2) Next we prove:

If the operator $\overline{LH_\alpha}$ $(\alpha > 0)$ *is the infinitesimal generator of some Feller semigroup on* ∂D *and if the boundary condition* L *is* transversal *on* ∂D, *then the equation*

$$\overline{LH_\alpha}\psi = \varphi \tag{23}$$

has a unique solution ψ *in* $D(\overline{LH_\alpha})$ *for any* $\varphi \in C(\partial D)$; *hence the inverse* $\overline{LH_\alpha}^{-1}$ *of* $\overline{LH_\alpha}$ *can be defined on the whole space* $C(\partial D)$. *Further the operator* $-\overline{LH_\alpha}^{-1}$ *is non-negative and bounded on* $C(\partial D)$.

Applying Theorem 7.1.1 with $A = A - \alpha$ to the function $H_\alpha 1$, we obtain that the function $H_\alpha 1$ takes its positive maximum 1 only on the boundary ∂D. Thus we can apply Lemma 7.1.7 with $\Sigma_3 = \partial D$ to obtain that

$$\frac{\partial}{\partial \mathbf{n}}(H_\alpha 1) < 0 \qquad \text{on } \partial D. \tag{24}$$

Hence the transversality condition (20) gives that

$$LH_\alpha 1 = \mu \frac{\partial}{\partial \mathbf{n}}(H_\alpha 1) + \gamma - \alpha\delta < 0 \qquad \text{on } \partial D,$$

and so

$$k_\alpha = -\sup_{x' \in \partial D} LH_\alpha 1(x') > 0.$$

Further, using Corollary 9.3.2 with $K = \partial D$, $\mathfrak{A} = \overline{LH_\alpha}$ and $c = k_\alpha$, we obtain that the operator $\overline{LH_\alpha} + k_\alpha I$ is the infinitesimal generator of some Feller semigroup on ∂D. Therefore, since $k_\alpha > 0$, it follows from an application of part (i) of Theorem 9.3.1 with $\mathfrak{A} = \overline{LH_\alpha} + k_\alpha I$ that the equation

$$-\overline{LH_\alpha}\psi = (k_\alpha I - (\overline{LH_\alpha} + k_\alpha I))\psi = \varphi$$

has a unique solution $\psi \in D(\overline{LH_\alpha})$ for any $\varphi \in C(\partial D)$, and further the operator $-\overline{LH_\alpha}^{-1} = (k_\alpha I - (\overline{LH_\alpha} + k_\alpha I))^{-1}$ is non-negative and bounded on $C(\partial D)$ with norm

$$\| -\overline{LH_\alpha}^{-1} \| = \| (k_\alpha I - (\overline{LH_\alpha} + k_\alpha I))^{-1} \| \leq \frac{1}{k_\alpha}.$$

(3) By assertion (23), we can define the right-hand side of formula (22) for all $\alpha > 0$. Now we prove

$$G_\alpha = (\alpha I - \mathfrak{A})^{-1}, \qquad \alpha > 0. \tag{25}$$

In view of Lemmas 9.6.7, 9.6.13 and 9.6.14, it follows that for any $f \in C(\bar{D})$ we have

$$\begin{cases} G_\alpha f = G_\alpha^0 f - H_\alpha(\overline{LH_\alpha}^{-1}\overline{LG_\alpha^0}f) \in D(\bar{A}), \\ G_\alpha f|_{\partial D} = -\overline{LH_\alpha}^{-1}(\overline{LG_\alpha^0}f) \in D(\overline{LH_\alpha}) = \mathscr{D}, \\ LG_\alpha f = \overline{LG_\alpha^0}f - \overline{LH_\alpha}(\overline{LH_\alpha}^{-1}\overline{LG_\alpha^0}f) = 0, \end{cases}$$

and

$$(\alpha I - \bar{A})G_\alpha f = f.$$

This proves that

$$\begin{cases} G_\alpha f \in D(\mathfrak{A}), \\ (\alpha I - \mathfrak{A})G_\alpha f = f, \end{cases}$$

that is,

$$(\alpha I - \mathfrak{A})G_\alpha = I \qquad \text{on } C(\bar{D}).$$

Therefore, in order to prove formula (25), it suffices to show the injectivity of the operator $\alpha I - \mathfrak{A}$ for $\alpha > 0$.

Suppose that

$$u \in D(\mathfrak{A}), \qquad (\alpha I - \mathfrak{A})u = 0.$$

Then, by Corollary 9.6.8, the function u can be written as

$$u = H_\alpha(u|_{\partial D}), \qquad u|_{\partial D} \in \mathscr{D} = D(\overline{LH_\alpha}).$$

Thus we have

$$\overline{LH_\alpha}(u|_{\partial D}) = Lu = 0.$$

In view of assertion (23), this implies that

$$u|_{\partial D} = 0,$$

so that

$$u = 0 \qquad \text{in } D.$$

(4) The non-negativity of G_α ($\alpha > 0$) follows immediately from formula (22), since the operators G_α^0, H_α, $-\overline{LH_\alpha}$ and $\overline{LG_\alpha^0}$ are all non-negative.

(5) We prove that the operator G_α is bounded on $C(\bar{D})$ with norm

$$\|G_\alpha\| \leq \frac{1}{\alpha}, \qquad \alpha > 0. \tag{26}$$

To do so, it suffices to show that

$$G_\alpha 1 \leq \frac{1}{\alpha} \qquad \text{on } \bar{D},$$

since G_α is non-negative on $C(\bar{D})$.

First it follows from the uniqueness property of solutions of problem (D') that

$$\alpha G_\alpha^0 1 + H_\alpha 1 = 1 + G_\alpha^0 c \qquad \text{on } \bar{D}. \tag{27}$$

Applying the operator L to the both sides of equality (27), we obtain that

$$\begin{aligned} -LH_\alpha 1 &= -L1 - LG_\alpha^0 c + \alpha LG_\alpha^0 1 \\ &= -\gamma - \mu \frac{\partial}{\partial \mathbf{n}}(G_\alpha^0 c) + \alpha LG_\alpha^0 1 \\ &\geq \alpha LG_\alpha^0 1 \qquad \text{on } \partial D, \end{aligned}$$

since $G_\alpha^0 c|_{\partial D} = 0$ and $G_\alpha^0 c \leq 0$ on $\bar{D}$. Hence we have, by the non-negativity of $\overline{LH_\alpha}^{-1}$,

$$-\overline{LH_\alpha}^{-1} LG_\alpha^0 1 \leq \frac{1}{\alpha} \qquad \text{on } \partial D. \tag{28}$$

Using formula (22) with $f = 1$, inequality (28) and equality (27), we obtain that

$$\begin{aligned} G_\alpha 1 &= G_\alpha^0 1 - H_\alpha(\overline{LH_\alpha}^{-1} LG_\alpha^0 1) \\ &\leq G_\alpha^0 1 + \frac{1}{\alpha} H_\alpha 1 \\ &= \frac{1}{\alpha} + \frac{1}{\alpha} G_\alpha^0 c \\ &\leq \frac{1}{\alpha} \qquad \text{on } \bar{D}, \end{aligned}$$

since the operators H_α and G_α^0 are non-negative.

(6) Finally we prove:

The domain $D(\mathfrak{A})$ is everywhere dense in $C(\bar{D})$. (29)

6-1) Before the proof, we need some lemmas on the behavior of the operators G_α^0, H_α and $\overline{LH_\alpha}^{-1}$ as $\alpha \to +\infty$.

9.6.19 Lemma. *For all $f \in C(\bar{D})$, we have*

$$\lim_{\alpha \to +\infty} [\alpha G_\alpha^0 f + H_\alpha(f|_{\partial D})] = f \qquad \text{in } C(\bar{D}). \tag{30}$$

Proof. Choose a constant $\beta > 0$ and let

$$g = f - H_\beta(f|_{\partial D}).$$

Then, using formula (10) with $\varphi = f|_{\partial D}$, we obtain that

$$\alpha G_\alpha^0 g - g = [\alpha G_\alpha^0 f + H_\alpha(f|_{\partial D}) - f] - \beta G_\alpha^0 H_\beta(f|_{\partial D}). \tag{31}$$

But we have, by estimate (7),

$$\lim_{\alpha \to +\infty} G_\alpha^0 H_\beta(f|_{\partial D}) = 0 \qquad \text{in } C(\bar{D}),$$

and, by assertion (9′),

$$\lim_{\alpha \to +\infty} \alpha G_\alpha^0 g = g \qquad \text{in } C(\bar{D}),$$

since $g|_{\partial D} = 0$. Therefore formula (30) follows by letting $\alpha \to +\infty$ in formula (31). ▼

9.6.20 Lemma. *The function $(\partial/\partial \mathbf{n})(H_\alpha 1)|_{\partial D}$ diverges to $-\infty$ uniformly and monotonically as $\alpha \to +\infty$.*

Proof. First, formula (10) with $\varphi = 1$ gives that

$$H_\alpha 1 = H_\beta 1 - (\alpha - \beta) G_\alpha^0 H_\beta 1, \qquad \alpha, \beta > 0.$$

Thus, in view of the non-negativity of G_α^0 and H_α, it follows that

$$\alpha \geq \beta > 0 \qquad \Rightarrow \qquad H_\alpha 1 \leq H_\beta 1 \qquad \text{on } \bar{D}.$$

Since $H_\alpha 1|_{\partial D} = H_\beta 1|_{\partial D} = 1$, this implies that the functions $(\partial/\partial \mathbf{n})(H_\alpha 1)|_{\partial D}$ are monotonically non-increasing in $\alpha > 0$. Further, using formula (9) with $f = H_\beta 1$, we find that the function

$$H_\alpha 1(x) = H_\beta 1(x) - \left(1 - \frac{\beta}{\alpha}\right) \alpha G_\alpha^0 H_\beta 1(x)$$

converges to zero monotonically as $\alpha \to +\infty$, for each $x \in D$.

Now, for any given constant $K > 0$, we can construct a function $u \in C^2(\bar{D})$ such that

$$\begin{cases} u|_{\partial D} = 1 & \text{on } \partial D, & (32.a) \\ \left.\dfrac{\partial u}{\partial \mathbf{n}}\right|_{\partial D} \leq -K & \text{on } \partial D. & (32.b) \end{cases}$$

In fact, it follows from part (d′) of Theorem 9.6.4 that for any integer $m > 0$ the function

$$u = (H_{\alpha_0} 1)^m, \qquad \alpha_0 > 0,$$

belongs to $C^\infty(\bar{D})$ and satisfies condition (32.*a*). Further we have

$$\begin{aligned} \left.\frac{\partial u}{\partial \mathbf{n}}\right|_{\partial D} &= m \frac{\partial}{\partial \mathbf{n}} (H_{\alpha_0} 1)|_{\partial D} \\ &\leq m \sup_{x' \in \partial D} \frac{\partial}{\partial \mathbf{n}} (H_{\alpha_0} 1)(x'). \end{aligned}$$

In view of inequality (24), this implies that the function $u = (H_{\alpha_0} u)^m$ satisfies condition (32.*b*) for sufficiently large m.

Take a function $u \in C^2(\bar{D})$ which satisfies conditions (32.*a*) and (32.*b*), and choose a neighborhood U of ∂D, relative to $\bar{D}$, with C^∞ boundary ∂U such that (cf. Figure 9-6)

$$u \geq \frac{1}{2} \qquad \text{on } U. \tag{33}$$

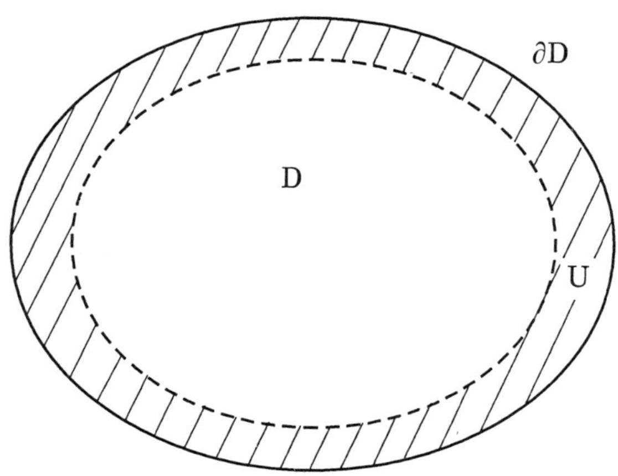

Figure 9-6

Recall that the function $H_\alpha 1$ converges to zero in D monotonically as $\alpha \to +\infty$. Since $u|_{\partial D} = H_\alpha 1|_{\partial D} = 1$, using Dini's theorem, we can find a constant $\alpha > 0$ (depending on u and hence on K) such that

$$\begin{cases} H_\alpha 1 \leq u & \text{on } \partial U \setminus \partial D, \quad (34.a) \\ \alpha > 2\|Au\|. & \quad (34.b) \end{cases}$$

It follows from inequalities (33) and (34.*b*) that

$$\begin{aligned} (A - \alpha)(H_\alpha 1 - u) &= \alpha u - Au \\ &\geq \frac{\alpha}{2} - \|Au\| \\ &> 0 \quad \text{in } U. \end{aligned}$$

Thus, applying Theorem 7.1.1 with $A = A - \alpha$ to the function $H_\alpha 1 - u$, we obtain that the function $H_\alpha 1 - u$ may take its positive maximum only on the boundary ∂U. But, conditions (32.*a*) and (34.*a*) imply that

$$H_\alpha 1 - u \leq 0 \qquad \text{on } \partial U = (\partial U \setminus \partial D) \cup \partial D.$$

Therefore we have

$$H_\alpha 1 \leq u \qquad \text{on } \bar{U} = U \cup \partial U,$$

and hence

$$\frac{\partial}{\partial \mathbf{n}}(H_\alpha 1)|_{\partial D} \leq \left.\frac{\partial u}{\partial \mathbf{n}}\right|_{\partial D} \leq -K \qquad \text{on } \partial D,$$

since $u|_{\partial D} = H_\alpha 1|_{\partial D} = 1$. This completes the proof of Lemma 9.6.20. ▼

9.6.21 Lemma. $\lim_{\alpha \to +\infty} \|-\overline{LH_\alpha}^{-1}\| = 0.$

Proof. In view of Lemma 9.6.20, the transversality condition (20) implies that the function

$$LH_\alpha 1(x') = \mu(x') \frac{\partial}{\partial \mathbf{n}}(H_\alpha 1)(x') - \alpha\delta(x') + \gamma(x'), \qquad x' \in \partial D,$$

diverges to $-\infty$ monotonically as $\alpha \to +\infty$. By Dini's theorem, this convergence is uniform in $x' \in \partial D$. Hence the function $1/LH_\alpha 1(x')$ converges to zero uniformly in $x' \in \partial D$ as $\alpha \to +\infty$. This gives that

$$\begin{aligned}\| -\overline{LH_\alpha}^{-1} \| &= \| -\overline{LH_\alpha}^{-1} 1 \| \\ &\leq \left\| \frac{1}{LH_\alpha 1} \right\| \to 0 \qquad \text{as } \alpha \to +\infty,\end{aligned}$$

since we have

$$1 = \frac{-LH_\alpha 1(x')}{|LH_\alpha 1(x')|} \leq \left\| \frac{1}{LH_\alpha 1} \right\| \cdot (-LH_\alpha 1(x')), \qquad x' \in \partial D. \qquad \blacktriangledown$$

6-2) *Proof of assertion* (29). Since the space $C^{2+\theta}(\bar{D})$ is dense in $C(\bar{D})$, it suffices to prove that

$$\lim_{\alpha \to +\infty} \| \alpha G_\alpha f - f \| = 0, \qquad f \in C^{2+\theta}(\bar{D}).$$

First we remark that

$$\begin{aligned}\| \alpha G_\alpha f - f \| &= \| \alpha G_\alpha^0 f - \alpha H_\alpha (\overline{LH_\alpha}^{-1} L G_\alpha^0 f) - f \| \\ &\leq \| \alpha G_\alpha^0 f + H_\alpha (f|_{\partial D}) - f \| \\ &\quad + \| -\alpha H_\alpha (\overline{LH_\alpha}^{-1} L G_\alpha^0 f) - H_\alpha (f|_{\partial D}) \| \\ &\leq \| \alpha G_\alpha^0 f + H_\alpha (f|_{\partial D}) - f \| + \| -\alpha \overline{LH_\alpha}^{-1} L G_\alpha^0 f - f|_{\partial D} \|.\end{aligned}$$

Thus, in view of formula (30), it suffices to show that

$$\lim_{\alpha \to +\infty} [\alpha \overline{LH_\alpha}^{-1} L G_\alpha^0 f + f|_{\partial D}] = 0 \qquad \text{in } C(\partial D). \tag{35}$$

Take a constant β such that $0 < \beta < \alpha$, and write

$$f = G_\beta^0 g + H_\beta \varphi$$

where (cf. formula (12))

$$\begin{cases} g = (\beta - A) f \in C^\theta(\bar{D}), \\ \varphi = f|_{\partial D} \in C^{2+\theta}(\partial D). \end{cases}$$

Then, using equations (8) (with $f = g$) and (10), we obtain that

$$\begin{aligned} G_\alpha^0 f &= G_\alpha^0 G_\beta^0 g + G_\alpha^0 H_\beta \varphi \\ &= \frac{1}{\alpha - \beta}(G_\beta^0 g - G_\alpha^0 g + H_\beta \varphi - H_\alpha \varphi). \end{aligned}$$

Hence we have

$$\begin{aligned} &\| -\alpha \overline{LH_\alpha}^{-1} LG_\alpha^0 f - f|_{\partial D} \| \\ &= \left\| \frac{\alpha}{\alpha - \beta}(-\overline{LH_\alpha}^{-1})(LG_\beta^0 g - LG_\alpha^0 g + LH_\beta \varphi) + \frac{\alpha}{\alpha - \beta}\varphi - \varphi \right\| \\ &\le \frac{\alpha}{\alpha - \beta} \| -\overline{LH_\alpha}^{-1} \| \cdot \| LG_\beta^0 g + LH_\beta \varphi \| \\ &\quad + \frac{\alpha}{\alpha - \beta} \| -\overline{LH_\alpha}^{-1} \| \cdot \| LG_\alpha^0 \| \cdot \| g \| + \frac{\beta}{\alpha - \beta} \| \varphi \|. \end{aligned} \tag{36}$$

By Lemma 9.6.21, it follows that the first term on the right-hand side of (36) converges to zero as $\alpha \to +\infty$. For the second term, using formula (13) with $f = 1$, and the non-negativity of G_β^0 and LG_α^0, we find that

$$\begin{aligned} \| LG_\alpha^0 \| &= \| LG_\alpha^0 1 \| \\ &= \| LG_\beta^0 1 - (\alpha - \beta) LG_\alpha^0 G_\beta^0 1 \| \\ &\le \| LG_\beta^0 1 \|. \end{aligned}$$

Hence the second term also converges to zero as $\alpha \to +\infty$. It is clear that the third term converges to zero as $\alpha \to +\infty$. This completes the proof of assertion (35) and hence of assertion (29).

(7) Summing up, we have proved that the operator $\mathfrak{A}$, defined by formula (21), satisfies conditions (a) through (d) in Theorem 9.3.1. Hence it follows from an application of the same theorem that the operator $\mathfrak{A}$ is the infinitesimal generator of some Feller semigroup on $\bar{D}$.

The proof of Theorem 9.6.18 is now complete. ■

Combining Theorem 9.6.15 and Theorem 9.6.18, we can prove general existence theorems for Feller semigroups in terms of boundary value problem (∗):

9.6.22 Theorem. *Let the differential operator A satisfy condition* (2), *and let the boundary condition L satisfy condition* (4) *and be transversal on ∂D.*

Suppose that the following two conditions are satisfied:

[I] (Existence) *For some constants* $\alpha \geq 0$ *and* $\lambda \geq 0$, *the boundary value problem*

$$\begin{cases} (\alpha - A)u = 0 & \text{in } D, \\ (\lambda - L)u = \varphi & \text{on } \partial D, \end{cases} \tag{$*$}$$

has a solution u *in* $C(\bar{D})$ *for any* φ *in some dense subset of* $C(\partial D)$.

[II] (Uniqueness) *For some constant* $\alpha > 0$, *we have*:

$$u \in C(\bar{D}),\ (\alpha - A)u = 0 \text{ in } D,\ Lu = 0 \text{ on } \partial D \quad \Rightarrow \quad u = 0 \text{ in } D.$$

Then there exists a Feller semigroup $\{T_t\}_{t \geq 0}$ *on* $\bar{D}$ *whose infinitesimal generator* $\mathfrak{A}$ *is characterized as follows:*

1. *The domain* $D(\mathfrak{A})$ *of* $\mathfrak{A}$ *is the space*

$$D(\mathfrak{A}) = \{u \in C(\bar{D});\ Au \in C(\bar{D}),\ Lu = 0\}. \tag{37}$$

2. $\mathfrak{A}u = Au, \qquad u \in D(\mathfrak{A})$.

Here Au and Lu are taken in the sense of distributions.

Proof. Part (ii) of Theorem 9.6.15 tells us that if condition [I] is satisfied, then the operator $\overline{LH_\alpha}$ is the infinitesimal generator of some Feller semigroup on ∂D; hence Theorem 9.6.18 applies.

It remains to show that if condition [II] is satisfied, then the two definitions (21) and (37) of $D(\mathfrak{A})$ coincide:

$$\begin{aligned} D(\mathfrak{A}) &\equiv \{u \in D(\bar{A});\ u|_{\partial D} \in \mathscr{D},\ Lu = 0\} \\ &= \{u \in C(\bar{D});\ Au \in C(\bar{D}),\ Lu = 0\}. \end{aligned} \tag{38}$$

In view of Remark 9.6.6, Lemmas 9.6.14 and 9.6.16, it follows that

$$D(\mathfrak{A}) \subset \{u \in C(\bar{D});\ Au \in C(\bar{D}),\ Lu = 0\}.$$

Conversely, let u be an arbitrary function in $C(\bar{D})$ such that $Au \in C(\bar{D})$ and $Lu = 0$. We let

$$w = u - G_\alpha((\alpha - A)u).$$

Then we have, by formula (25),

$$\begin{cases} (\alpha - A)w = 0 & \text{in } D, \\ Lw = 0 & \text{on } \partial D. \end{cases}$$

Thus condition [II] gives us that $w = 0$, that is,

$$u = G_\alpha((\alpha - A)u) \in D(\mathfrak{A}).$$

This proves assertion (38), and the theorem. ■

In general, there is a close relationship between the uniqueness and regularity properties of solutions of boundary value problems. Indeed, we obtain the following:

9.6.23 Corollary. *Let A and L be as in Theorem* 9.6.22, *and suppose that condition* [I] *and the following condition (replacing condition* [II]) *are satisfied*:

[III] (Regularity) *For some constant* $\alpha > 0$, *we have*:

$$u \in C(\bar{D}),\ (\alpha - A)u = 0 \text{ in } D,\ Lu \in C^\infty(\partial D) \quad \Rightarrow \quad u \in C^\infty(\bar{D}).$$

Then there exists a Feller semigroup $\{T_t\}_{t \geq 0}$ on $\bar{D}$ whose infinitesimal generator $\mathfrak{A}$ enjoys property (37), *and coincides with the minimal closed extension in $C(\bar{D})$ of the restriction of A to the space $\{u \in C^2(\bar{D}); Lu = 0\}$.*

Proof. 1) First we show that conditions [I] and [III] imply condition [II]; hence Theorem 9.6.22 applies.

Suppose that

$$u \in C(\bar{D}), \qquad (\alpha - A)u = 0 \text{ in } D, \qquad Lu = 0 \text{ on } \partial D.$$

Then we obtain from condition [III] that $u \in C^\infty(\bar{D})$. Thus, by the uniqueness property of solutions of Dirichlet problem (D'), it follows that the function u can be written as

$$u = H_\alpha(u|_{\partial D}), \qquad u|_{\partial D} \in C^\infty(\partial D) \subset D(LH_\alpha).$$

Hence we have

$$LH_\alpha(u|_{\partial D}) = Lu = 0 \qquad \text{on } \partial D. \tag{39}$$

But, combining part (ii) of Theorem 9.6.15 and assertion (23), we find that if condition [I] is satisfied and if the boundary condition L is transversal on ∂D, then the minimal closed extension $\overline{LH_\alpha}$ of LH_α is bijective for each $\alpha > 0$. Thus we have, by (39),

$$u|_{\partial D} = 0,$$

and so

$$u = 0 \qquad \text{in } D.$$

This proves that condition [II] is satisfied.

2) Next we show that if condition [III] is satisfied, then we have:

$$f \in C^\infty(\bar{D}) \quad \Rightarrow \quad G_\alpha f \in C^\infty(\bar{D}). \tag{40}$$

Part (e) of Theorem 9.6.4 tells us that $G_\alpha^0 f \in C^\infty(\bar{D})$ whenever $f \in C^\infty(\bar{D})$. We let

$$w = H_\alpha(\overline{LH_\alpha}^{-1}(LG_\alpha^0 f)).$$

Then it follows from Lemmas 9.6.7 and 9.6.14 that

$$\begin{cases} (\alpha - A)w = 0 & \text{in } D, \\ \quad Lw = \overline{LH_\alpha}\,\overline{LH_\alpha}^{-1}((LG_\alpha^0 f)) = LG_\alpha^0 f \in C^\infty(\partial D). \end{cases}$$

Thus condition [III] gives us that $w \in C^\infty(\bar{D})$. In view of formula (22), this implies that

$$G_\alpha^0 f = G_\alpha^0 f - w \in C^\infty(\bar{D}).$$

3) Finally we show that the operator $\mathfrak{A}$, defined by formula (37), coincides with the minimal closed extension in $C(\bar{D})$ of the restriction of A to the space $\{u \in C^2(\bar{D});\ Lu = 0\}$.

Let u be an arbitrary element of $D(\mathfrak{A})$. We choose a sequence $\{f_n\}$ in $C^\infty(\bar{D})$ such that

$$f_n \to (\alpha - \mathfrak{A})u \qquad \text{in } C(\bar{D}),$$

and let

$$u_n = G_\alpha f_n.$$

Then we have, by (25) and (40),

$$u_n \in D(\mathfrak{A}) \cap C^\infty(\bar{D}).$$

Further, since the operator $G_\alpha: C(\bar{D}) \to C(\bar{D})$ is bounded, it follows that

$$u_n = G_\alpha f_n \to G_\alpha((\alpha I - \mathfrak{A})u) = u \qquad \text{in } C(\bar{D}),$$

and also

$$Au_n = \alpha u_n - f_n \to \alpha u - (\alpha I - \mathfrak{A})u = \mathfrak{A}u \qquad \text{in } C(\bar{D}).$$

This proves that:

$$\begin{aligned} \text{The graph of } \mathfrak{A} &\equiv \{(u, \mathfrak{A}u);\ u \in D(\mathfrak{A})\} \\ &= \text{the closure in } C(\bar{D}) \times C(\bar{D}) \text{ of the graph} \\ &\quad\ \{(u, Au);\ u \in C^2(\bar{D}),\ Lu = 0\}. \end{aligned}$$

Corollary 9.6.23 is proved. ■

Notes

Section 9.1: Our treatment of Markov processes follows the expositions of Dynkin [1], [2] and Blumenthal-Getoor [1].

Section 9.2: The semigroup approach to Markov processes can be traced back to the work of Kolmogorov [1]. It was substantially developed in the early 1950s, with Feller [2], [3] doing the pioneering work. Our presentation here follows the book of Dynkin [2] and also part of Lamperti's [2].

Section 9.3: Our treatment of Feller semigroups is taken from Sato-Ueno [1]. Theorem 9.3.3 was proved independently by Itô [1] and Wentzell (Ventcel') [1].

Sections 9.4-9.5: Theorems 9.4.1 and 9.5.1 are essentially due to Wentzell (Ventcel') [1]. Our proof of these theorems follows Bony-Courrège-Priouret [1], where the infinitesimal generators of Feller semigroups are studied in great detail in terms of the maximum principle. For the probabilistic meaning of Ventcel' boundary conditions, the reader might refer to Dynkin-Yushkevich [1].

Section 9.6: The results discussed here are adapted from Sato-Ueno [1] and Bony-Courrège-Priouret [1], while Theorem 9.6.22 and Corollary 9.6.23 are due to Taira [6]. We remark that in Taira [6] Theorem 9.6.22 and Corollary 9.6.23 are proved for some *degenerate* elliptic differential operators of second order. However, in this section, we confined ourselves to the elliptic case. This makes it possible to develop the basic machinery of Taira [6] with a minimum of bother and the principal ideas can be presented more concretely and explicitly.

It seems that our method of construction of Feller semigroups is, in spirit, not far removed from the probabilistic method used by Watanabe [1].

10 Construction of Feller Semigroups

In the last chapter we reduced the problem of construction of Feller semigroups to the problem of unique solvability for the boundary value problem

$$\begin{cases} (\alpha - A)u = 0 & \text{in } D, \\ (\lambda - L)u = \varphi & \text{on } \partial D, \end{cases} \tag{$*$}$$

and gave existence theorems for Feller semigroups.

In this chapter we prove existence and uniqueness theorems for problem (∗), and construct Feller semigroups. Our proof of the existence and uniqueness theorems for problem (∗) is based on the maximum principles discussed in Section 7.1 and the *a priori* estimates stated in Sections 6.9–6.10. We use these estimates on one hand to prove regularity theorems for problem (∗), and on the other hand to show that the index of problem (∗) is equal to *zero*, by using a variant of the Agmon-Nirenberg method developed in Section 8.4. Combining the regularity theorems and the maximum principles, we can obtain the uniqueness theorems and hence the existence theorems for problem (∗), since the index of problem (∗) is equal to zero. Intuitively, our results may be stated as follows: if a Markovian particle goes through the set where no reflection phenomenon occurs in *finite* time, then there exists a Feller semigroup corresponding to such a diffusion phenomenon.

10.1. Statements of Results

Let D be a bounded domain in $\mathbf{R}^N$ with C^∞ boundary ∂D, and let A be a second-order *elliptic* differential operator with real coefficients as in Section 9.6:

$$Au(x) = \sum_{i,j=1}^{N} a^{ij}(x) \frac{\partial^2 u}{\partial x_i \, \partial x_j}(x) + \sum_{i=1}^{N} b^i(x) \frac{\partial u}{\partial x_i}(x) + c(x)u(x),$$

where

1. $a^{ij} \in C^\infty(\mathbf{R}^N)$, $a^{ij} = a^{ji}$ and there exists a constant $a_0 > 0$ such that

$$\sum_{i,j=1}^{N} a^{ij}(x)\xi_i \xi_j \geq a_0 |\xi|^2, \qquad x \in \mathbf{R}^N, \ \xi \in \mathbf{R}^N. \tag{1}$$

2. $b^i \in C^\infty(\mathbf{R}^N)$.
3. $c \in C^\infty(\mathbf{R}^N)$ and $c \leq 0$ on $\bar{D}$.

Let L be a Ventcel' boundary condition as in Section 9.6:

$$Lu(x') = \sum_{i,j=1}^{N-1} \alpha^{ij}(x') \frac{\partial^2 u}{\partial x_i \, \partial x_j}(x') + \sum_{i=1}^{N-1} \beta^i(x') \frac{\partial u}{\partial x_i}(x')$$

$$+ \gamma(x')u(x') + \mu(x') \frac{\partial u}{\partial \mathbf{n}}(x') - \delta(x')Au(x'),$$

where

1. The α^{ij} are the components of a C^∞ symmetric contravariant tensor of type $\binom{2}{0}$ on ∂D and

$$\sum_{i,j=1}^{N-1} \alpha^{ij}(x')\eta_i \eta_j \geq 0, \qquad x' \in \partial D, \ \eta = \sum_{j=1}^{N-1} \eta_j \, dx_j \in T^*_{x'}(\partial D),$$

where $T^*_{x'}(\partial D)$ is the cotangent space of ∂D at x'.
2. $\beta^i \in C^\infty(\partial D)$. (2)
3. $\gamma \in C^\infty(\partial D)$ and $\gamma \leq 0$ on ∂D.
4. $\mu \in C^\infty(\partial D)$ and $\mu \geq 0$ on ∂D.
5. $\delta \in C^\infty(\partial D)$ and $\delta \geq 0$ on ∂D.
6. $\mathbf{n}$ is the unit interior normal to ∂D at x'.

To state hypotheses for L, we introduce some notation and definitions.

As in Section 6.9, we say that a tangent vector $v = \sum_{j=1}^{N-1} v^j(\partial/\partial x_j) \in T_{x'}(\partial D)$ is *subunit* for the operator $L^0 = \sum_{i,j=1}^{N-1} \alpha^{ij}(\partial^2/\partial x_i\, \partial x_j)$ if it satisfies

$$\left(\sum_{j=1}^{N-1} v^j\eta_j\right)^2 \leq \sum_{i,j=1}^{N-1} \alpha^{ij}(x')\eta_i\eta_j, \qquad \eta = \sum_{j=1}^{N-1} \eta_j\, dx_j \in T^*_{x'}(\partial D).$$

If $\rho > 0$, we define a "non-Euclidean" ball $B_{L^0}(x', \rho)$ of radius ρ about x' as follows (cf. Figure 0-8):

$B_{L^0}(x', \rho) =$ the set of all points $y \in \partial D$ which can be joined to x' by a Lipschitz path $v\colon [0, \rho] \to \partial D$ for which the tangent vector $\dot{v}(t)$ of ∂D at $v(t)$ is subunit for L^0 for almost every t.

Also we let

$B_E(x', \rho) =$ the ordinary Euclidean ball of radius ρ about x'.

Recall that the boundary condition L is said to be *transversal* on ∂D if it satisfies

$$\mu(x') + \delta(x') > 0 \qquad \text{on } \partial D. \tag{3}$$

Now we can state our main result:

10.1.1 Theorem. *Let the differential operator A satisfy condition* (1) *and let the boundary condition L satisfy condition* (2) *and be transversal on ∂D. Suppose that*:

(*A*.1) *There exist constants $0 < \varepsilon \leq 1$ and $C > 0$ such that for all sufficiently small $\rho > 0$ we have*

$$B_E(x', \rho) \subset B_{L^0}(x', C\rho^\varepsilon), \qquad x' \in M = \{x' \in \partial D;\ \mu(x') = 0\}.$$

Then there exists a Feller semigroup $\{T_t\}_{t \geq 0}$ on $\bar{D}$ whose infinitesimal generator $\mathfrak{A}$ is characterized as follows:

1. *The domain $D(\mathfrak{A})$ of $\mathfrak{A}$ is the space*

$$D(\mathfrak{A}) = \{u \in C(\bar{D});\ Au \in C(\bar{D}),\ Lu = 0\}.$$

2. $$\mathfrak{A}u = Au, \qquad u \in D(\mathfrak{A}).$$

Furthermore, the generator $\mathfrak{A}$ coincides with the minimal closed extension in $C(\bar{D})$ of the restriction of A to the space $\{u \in C^2(\bar{D});\ Lu = 0\}$.

10.1.2 Remark. Theorem 7.2.1 tells us that the non-Euclidean ball $B_{L^0}(x', \rho)$ may be interpreted as the set of all points where a Markovian particle with generator L^0, starting at x', diffuses during the time interval $[0, \rho]$. Hence the intuitive meaning of hypothesis $(A.1)$ is that a Markovian particle with generator L^0 goes through the set M where no reflection phenomenon occurs in *finite* time (cf. Figure 0-9).

Further, we consider the case when $\alpha^{ij} \equiv 0$ on ∂D:

$$Lu(x') = \sum_{i=1}^{N-1} \beta^i(x') \frac{\partial u}{\partial x_i}(x') + \gamma(x')u(x') + \mu(x') \frac{\partial u}{\partial \mathbf{n}}(x') - \delta(x')Au(x'). \tag{4}$$

Here $\beta = \sum_{i=1}^{N-1} \beta^i(\partial/\partial x_i)$ is a C^∞ vector field on ∂D.
Then we can prove the following:

10.1.3 Theorem. *Let A and L be as in Theorem 10.1.1, L being of the form* (4). *Suppose that*:

$(A.2)$ The vector field β is non-zero on the set $M = \{x' \in \partial D; \mu(x') = 0\}$ and any maximal integral curve of β is not entirely *contained in M.*

Then we have the same conclusion as in Theorem 10.1.1.

10.1.4 Remark. The vector field β is the drift vector field. Hence Theorem 7.2.1 tells us that hypothesis $(A.2)$ has an intuitive meaning similar to hypothesis $(A.1)$.

10.2. Proof of Theorem 10.1.1

We apply Corollary 9.6.23. The next theorem allows us to verify conditions [I] and [III] of the same corollary.

10.2.1 Theorem. *Let A and L be as in Theorem 10.1.1. Suppose that hypothesis $(A.1)$ is satisfied. Then there exists a constant $0 < \kappa \leq 1$ such that, for any $\alpha > 0$, the boundary value problem*

$$\begin{cases} (A - \alpha)u = f & \text{in } D, \\ Lu = \varphi & \text{on } \partial D, \end{cases} \tag{$*$}$$

has a unique solution $u \in H^{s-2+\kappa}(\bar{D})$ *for any* $f \in H^{s-2}(\bar{D})$ *and* $\varphi \in H^{s-5/2}(\partial D)$ $(s \geq 3)$.

Furthermore, we have, for any $t < s - 2 + \kappa$,

$$u \in H^t(\bar{D}),\ (A - \alpha)u \in H^{s-2}(\bar{D}),\ Lu \in H^{s-5/2}(\partial D) \Rightarrow u \in H^{s-2+\kappa}(\bar{D}). \qquad (1)$$

Here $\alpha \geq 0$.

Granting Theorem 10.2.1 for the moment, we shall prove Theorem 10.1.1.

In view of the Sobolev imbedding theorem, Theorem 10.2.1 implies the following results:

1) For any $\alpha > 0$, the boundary value problem

$$\begin{cases} (\alpha - A)u = 0 & \text{in } D, \\ -Lu = \varphi & \text{on } \partial D, \end{cases}$$

has a unique solution $u \in C^\infty(\bar{D})$ for any $\varphi \in C^\infty(\partial D)$.

2) For any $\alpha \geq 0$, we have

$$u \in C(\bar{D}),\ (\alpha - A)u = 0,\ Lu \in C^\infty(\partial D) \quad \Rightarrow \quad u \in C^\infty(\bar{D}).$$

These results verify conditions [I] (with $\lambda = 0$) and [III] of Corollary 9.6.23. Hence Theorem 10.1.1 follows from an application of the same corollary.

Proof of Theorem 10.2.1. We divide the proof into five steps.

(1) First we reduce the study of problem $(*)$ to that of a pseudo-differential operator on the boundary, just as in Section 8.3.

Applying Theorem 8.2.4 to the operator $A - \alpha$ $(\alpha \geq 0)$, we obtain the following:

(a) The Dirichlet problem

$$\begin{cases} (A - \alpha)w = 0 & \text{in } D, \\ \gamma_0 w = \varphi & \text{on } \partial D, \end{cases}$$

has a unique solution w in $H^t(\bar{D})$ for any $\varphi \in H^{t-1/2}(\partial D)$ $(t \in \mathbf{R})$.

(b) The mapping

$$P(\alpha)\colon H^{t-1/2}(\partial D) \to H^t(\bar{D}),$$

defined by $w = P(\alpha)\varphi$, is an isomorphism of $H^{t-1/2}(\partial D)$ onto the space $N(A - \alpha, t) = \{u \in H^t(\bar{D}); (A - \alpha)u = 0 \text{ in } D\}$ for all $t \in \mathbf{R}$; and its inverse is the trace operator γ_0 on ∂D.

We let

$$T(\alpha)\colon C^\infty(\partial D) \to C^\infty(\partial D)$$
$$\varphi \mapsto LP(\alpha)\varphi.$$

Since we have

$$LP(\alpha)\varphi = \sum_{i,j=1}^{N-1} \alpha^{ij} \frac{\partial^2 \varphi}{\partial x_i \, \partial x_j} + \sum_{i=1}^{N-1} \beta^i \frac{\partial \varphi}{\partial x_i} + \gamma\varphi$$
$$+ \mu \frac{\partial}{\partial \mathbf{n}} (P(\alpha)\varphi)|_{\partial D} - \alpha\delta\varphi,$$

it follows that the operator $T(\alpha)$ can be written in the form

$$T(\alpha) = Q(\alpha) + \mu\Pi(\alpha),$$

where

$$\begin{cases} Q(\alpha)\varphi = \displaystyle\sum_{i,j=1}^{N-1} \alpha^{ij} \frac{\partial^2 \varphi}{\partial x_i \, \partial x_j} + \sum_{i=1}^{N-1} \beta^i \frac{\partial \varphi}{\partial x_i} + (\gamma - \alpha\delta)\varphi, \\ \Pi(\alpha)\varphi = \dfrac{\partial}{\partial \mathbf{n}} (P(\alpha)\varphi)|_{\partial D}. \end{cases}$$

But the operator $Q(\alpha)$ is a second-order differential operator on ∂D, and its symbol is given by

$$-\sum_{i,j=1}^{N-1} \alpha^{ij}(x')\xi_i\xi_j + \sqrt{-1} \sum_{i=1}^{N-1} \beta^i(x')\xi_i + (\gamma(x') - \alpha\delta(x')).$$

Note that

$$\sum_{i,j=1}^{N-1} \alpha^{ij}(x')\xi_i\xi_j \geq 0 \qquad \text{on the cotangent bundle } T^*(\partial D).$$

Furthermore, since the operator $P(\alpha)$ is of the form (8.2.15), arguing as in the proof of Theorem 8.2.2, we find that the operator $\Pi(\alpha)$ is a classical pseudo-differential operator of first order on ∂D, and its symbol is given by

$$[p_1(x', \xi') + \sqrt{-1}\, q_1(x', \xi')] + \text{terms of order} \leq 0 \text{ depending on } \alpha$$

where:

$$p_1(x', \xi') = \frac{(4A_2(x')a_0(x', \xi') - a_1(x', \xi')^2)^{1/2}}{2A_2(x')}, \tag{2}$$

$$q_1(x', \xi') = -\frac{a_1(x', \xi')}{2A_2(x')} \tag{3}$$

(cf. formulas (8.2.2) and (8.2.3)). We remark that

$p_1(x', \xi') < 0$ on the bundle $T^*(\partial D) \setminus \{0\}$ of non-zero cotangent vectors.

Therefore, we obtain that the operator $T(\alpha) = Q(\alpha) + \mu\Pi(\alpha)$ is a classical pseudo-differential operator of second order on ∂D and its symbol is given by

$$\begin{aligned}&\left[-\sum_{i,j=1}^{N-1} \alpha^{ij}(x')\xi_i\xi_j\right]\\ &\quad + \left[\mu(x')p_1(x', \xi') + \sqrt{-1}\left(\mu(x')q_1(x', \xi') + \sum_{i=1}^{N-1} \beta^i(x')\xi_i\right)\right] \\ &\quad + \text{terms of order} \leq 0 \text{ depending on } \alpha.\end{aligned} \tag{4}$$

Since the operator $T(\alpha)\colon C^\infty(\partial D) \to C^\infty(\partial D)$ extends to a continuous linear operator $T(\alpha)\colon H^s(\partial D) \to H^{s-2}(\partial D)$ for all $s \in \mathbf{R}$, we can introduce a densely defined, closed linear operator

$$\mathcal{T}(\alpha)\colon H^{s-5/2+\kappa}(\partial D) \to H^{s-5/2}(\partial D)$$

as follows:

(α) The domain $D(\mathcal{T}(\alpha))$ of $\mathcal{T}(\alpha)$ is the space

$$D(\mathcal{T}(\alpha)) = \{\varphi \in H^{s-5/2+\kappa}(\partial D);\ T(\alpha)\varphi \in H^{s-5/2}(\partial D)\}.$$

(β) $$\mathcal{T}(\alpha)\varphi = T(\alpha)\varphi, \qquad \varphi \in D(\mathcal{T}(\alpha)).$$

Here κ is a positive constant and will be fixed later on (see Lemma 10.2.2 below).

Then, arguing as in Section 8.3, we can prove that the problems of existence, uniqueness and regularity of solutions of problem (∗) are reduced to the same problems for the operator $\mathcal{T}(\alpha)$, respectively (cf. Theorems 8.3.4–8.3.9).

(2) The next lemma is the essential step in the proof of Theorem 10.2.1.

10.2.2 Lemma. *Let A and L be as in Theorem* 10.1.1, *and suppose that hypothesis* (*A*.1) *is satisfied. Then there exists a constant* $0 < \kappa \leq 1$ *such that we have, for all* $s \in \mathbf{R}$,

$$\varphi \in \mathscr{D}'(\partial D),\ T(\alpha)\varphi \in H^s(\partial D) \quad \Rightarrow \quad \varphi \in H^{s+\kappa}(\partial D). \tag{5}$$

Furthermore, for any $t < s + \kappa$, *there exists a constant* $C_{s,t} > 0$ *such that*

$$|\varphi|_{H^{s+\kappa}(\partial D)} \leq C_{s,t}(|T(\alpha)\varphi|_{H^s(\partial D)} + |\varphi|_{H^t(\partial D)}). \tag{6}$$

Thus, the operator $T(\alpha)$ is *hypoelliptic*, with loss of $2 - \kappa$ derivatives.

Proof. Our proof mimics that of Theorem 2.4.2 of Oleĭnik-Radkevič [1], so we only give a sketch of the proof.

2-1) First we prove the following *energy estimate*.

10.2.3 Proposition. *Let A and L be as in Theorem 10.1.1, and let (U, ψ) be a chart on ∂D with $\psi(x') = (x_1, \ldots, x_{N-1})$. Then, for every compact $K \subset U$ and $s \geq 0$, there exists a constant $C_{K,s} > 0$ such that*

$$\sum_{j=1}^{N-1}\left(\left|\sum_{i=1}^{N-1}\alpha^{ij}D_i\varphi\right|^2_{H^s(\partial D)} + \left|\sum_{\ell,m=1}^{N-1}\frac{\partial\alpha^{\ell m}}{\partial x_j}D_\ell D_m\varphi\right|^2_{H^{s-1}(\partial D)}\right)$$

$$\leq C_{K,s}(|T(\alpha)\varphi|^2_{L^2(\partial D)} + |\varphi|^2_{H^{2s}(\partial D)}), \qquad \varphi \in C_K^\infty(U). \tag{7}$$

Proof. We shall denote by the letter C a generic positive constant depending only on K and s.

Recall that

$$T(\alpha) = Q(\alpha) + \mu\Pi(\alpha).$$

We rewrite the differential operator $Q(\alpha)$ as follows:

$$Q(\alpha)\varphi = -\sum_{i,j=1}^{N-1} D_j(\alpha^{ij}D_i\varphi) + \sum_{i=1}^{N-1}\left(\sum_{j=1}^{N-1} D_j\alpha^{ij} + \sqrt{-1}\,\beta^i\right)D_i\varphi + (\gamma - \alpha\delta)\varphi.$$

Then we have, by integration by parts,

$$\mathrm{Re}(Q(\alpha)\varphi, \varphi) = -\sum_{i,j=1}^{N-1}(\alpha^{ij}D_i\varphi, D_j\varphi) + (h\varphi, \varphi)$$

with a function $h \in C^\infty(\partial D)$. Here (,) is the inner product of $L^2(\partial D)$. Thus, in view of the Schwarz inequality, it follows that

$$\sum_{i,j=1}^{N-1} (\alpha^{ij} D_i \varphi, D_j \varphi) \le -\operatorname{Re}(Q(\alpha)\varphi, \varphi) + C|\varphi|^2_{L^2(\partial D)}, \qquad \varphi \in C^\infty(\partial D). \tag{8}$$

On the other hand, the operator $\mu\Pi(\alpha)$ is a first order pseudo-differential operator with principal symbol

$$\mu(x')(p_1(x', \xi') + \sqrt{-1}\, q_1(x', \pi')),$$

and

$$\mu(x') p_1(x', \xi') \le 0 \qquad \text{on } T^*(\partial D).$$

Hence, applying the sharp Gårding inequality (Theorem 6.9.2) to the operator $-\mu\Pi(\alpha)$, we obtain that

$$-\operatorname{Re}(\mu\Pi(\alpha)\varphi, \varphi) \ge -C|\varphi|^2_{L^2(\partial D)}, \qquad \varphi \in C_K^\infty(U). \tag{9}$$

Therefore, combining estimates (8) and (9), we have

$$\sum_{i,j=1}^{N-1} (\alpha^{ij} D_i \varphi, D_j \varphi) \le -\operatorname{Re}(T(\alpha)\varphi, \varphi) + C|\varphi|^2_{L^2(\partial D)}, \qquad \varphi \in C_K^\infty(U). \tag{10}$$

The desired estimate (7) follows from estimate (10), just as in the proof of Theorem 2.6.1 of Oleĭnik-Radkevič [1].

The proof of Proposition 10.2.3 is complete.

2-2) Next we prove a local version of estimate (6).

If P is a pseudo-differential operator with complete symbol $p(x, \xi)$, we denote by $P^{(j)}$ (resp. $P_{(j)}$) a pseudo-differential operator with complete symbol $(\partial p/\partial \xi_j)(x, \xi)$ (resp. $1/\sqrt{-1}(\partial p/\partial x_j)(x, \xi)$).

10.2.4 Proposition. *Let A and L be as in Theorem 10.1.1, and suppose that hypothesis (A.1) is satisfied. Then, for any point x_0' of ∂D, one can find a neighborhood $U(x_0')$ of x_0' such that:*

For every compact $K \subset U(x_0')$, there exists a constant $0 < \kappa(K) \le 1$ such that for any $s \in \mathbf{R}$ and $t < s + \kappa(K)$ we have

$$\sum_{j=1}^{N-1} (|T(\alpha)^{(j)}\varphi|^2_{H^{s+\kappa/2}(\partial D)} + |T(\alpha)_{(j)}\varphi|^2_{H^{s-1+\kappa/2}(\partial D)}) + |\varphi|^2_{H^{s+\kappa}(\partial D)}$$

$$\le C_{K,s,t}(|T(\alpha)\varphi|^2_{H^s(\partial D)} + |\varphi|^2_{H^t(\partial D)}), \qquad \varphi \in C_K^\infty(U(x_0')), \tag{11}$$

with a constant $C_{K,s,t} > 0$.

10.2.5 Remark. The constant $\kappa(K)$ in the proposition can be chosen as $\kappa(K) = 1$ if $\mu(x_0') > 0$ and as $\kappa(K) = \varepsilon$ if $\mu(x_0') = 0$ where ε is the constant in hypothesis $(A.1)$ (see the proof of Proposition 10.2.4 below).

Proof of Proposition 10.2.4. i) First we prove estimate (11) in the case $\mu(x_0') = 0$. In doing so, we make essential use of Theorem 6.9.4 due to Fefferman-Phong [1].

Let x_0' be an arbitrary point of ∂D such that $\mu(x_0') = 0$. Since hypothesis $(A.1)$ is satisfied, one can find a neighborhood $U(x_0')$ of x_0' such that for all sufficiently small $\rho > 0$ we have

$$B_E(x', \rho) \subset B_{L^0}(x', 2C\rho^{\varepsilon}), \qquad x' \in U(x_0').$$

Thus, applying Theorem 6.9.4 to the operator $Q(\alpha)$, we obtain that:

For every compact $K \subset U(x_0')$, there exist constants $c_K > 0$ and $C_K > 0$ such that

$$-\operatorname{Re}(Q(\alpha)\varphi, \varphi) \geq c_K |\varphi|^2_{H^{\varepsilon}(\partial D)} - C_K |\varphi|^2_{L^2(\partial D)}, \qquad \varphi \in C_K^{\infty}(U(x_0')). \tag{12}$$

On the other hand, applying the sharp Gårding inequality (Theorem 6.9.2) to the operator $-\mu\Pi(\alpha)$, we have

$$-\operatorname{Re}(\mu\Pi(\alpha)\varphi, \varphi) \geq -C_K' |\varphi|^2_{L^2(\partial D)}, \qquad \varphi \in C_K^{\infty}(U(x_0')), \tag{13}$$

with a constant $C_K' > 0$.

Hence, combining estimates (12) and (13), we obtain that

$$-\operatorname{Re}(T(\alpha)\varphi, \varphi) \geq c_K |\varphi|^2_{H^{\varepsilon}(\partial D)} - (C_K + C_K') |\varphi|^2_{L^2(\partial D)}.$$

By virtue of the Schwarz inequality, this gives that

$$|\varphi|^2_{H^{\varepsilon}(\partial D)} \leq C_K''(|T(\alpha)\varphi|^2_{H^s(\partial D)} + |\varphi|^2_{L^2(\partial D)}), \qquad \varphi \in C_K^{\infty}(U(x_0')), \tag{14}$$

with a constant $C_K'' > 0$.

Therefore, using estimates (7) and (14), we can obtain estimate (11) with $\kappa(K) = \varepsilon$, just as in the proof of Theorem 2.6.2 of Oleĭnik-Radkevič [1].

ii) Now we prove estimate (11) in the case $\mu(x_0') > 0$. In doing so, we make essential use of Theorem 6.10.1 due to Hörmander [6].

In view of formula (4), we have the following:

(a) The principal symbol of $T(\alpha)$ is equal to

$$-\sum_{i,\, j=1}^{N-1} \alpha^{ij}(x')\xi_i\xi_j.$$

(b) The subprincipal symbol of $T(\alpha)$ on the characteristic set $\Sigma = \{(x', \xi') \in T^*(\partial D) \setminus \{0\}; \sum_{i,j=1}^{N-1} \alpha^{ij}(x')\xi_i\xi_j = 0\}$ is equal to

$$\mu(x')p_1(x', \xi') \\ + \sqrt{-1}\left(\mu(x')q_1(x', \xi') + \sum_{i=1}^{N-1} \beta^i(x')\xi_i - \sum_{i,j=1}^{N-1} \frac{\partial \alpha^{ij}}{\partial x_j}(x')\xi_i\right). \tag{15}$$

Since $\sum_{i,j} \alpha^{ij}(x')\xi_i\xi_j \geq 0$ and $\mu(x')p_1(x', \xi') \leq 0$ on $T^*(\partial D)$, we find that all the hypotheses of Theorem 6.10.1 are satisfied for the operator $-T(\alpha)$.

Let x'_0 be an arbitrary point of ∂D such that $\mu(x'_0) > 0$. Then we can find a neighborhood $U(x'_0)$ of x'_0 such that

$$\mu(x'_0)p_1(x', \xi') < 0, \qquad x' \in U(x'_0), \qquad \xi' \in T^*_{x'}(\partial D) \setminus \{0\},$$

since $p_1(x', \xi') < 0$ on $T^*(\partial D) \setminus \{0\}$. In view of formula (15), this implies that condition (ii) of Theorem 6.10.1 is satisfied. Hence, applying the same theorem to the operator $-T(\alpha)$, we obtain that:

For every compact $K \subset U(x'_0)$, there exists a constant $C_K > 0$ such that

$$|\varphi|^2_{H^1(\partial D)} \leq C_K(|T(\alpha)\varphi|^2_{L^2(\partial D)} + |\varphi|^2_{L^2(\partial D)}), \qquad \varphi \in C^\infty_K(U(x'_0)). \tag{16}$$

Therefore, using estimates (7) and (16), we can obtain estimate (11) with $\kappa(K) = 1$, just as in the proof of Theorem 2.6.2 of Oleĭnik-Radkevič [1].

The proof of Proposition 10.2.4 is complete.

2-3) *End of the proof of Lemma 10.2.2* By Proposition 10.2.4, we can cover the boundary ∂D by a finite number of local charts $\{U_j, \psi_j)\}_{j=1}^d$ in each of which estimate (11) holds for all $\varphi \in C^\infty_0(U_j)$. Let $\{\varphi_j\}_{j=1}^d$ be a partition of unity subordinate to the covering $\{U_j\}_{j=1}^d$, and choose a function $\theta_j \in C^\infty_0(U_j)$ such that $\theta_j = 1$ on supp φ_j. We let

$$\kappa = \min_{1 \leq j \leq d} \kappa(\text{supp } \theta_j),$$

where $\kappa(\text{supp } \theta_j)$ is the constant with $K = \text{supp } \theta_j$ in Proposition 10.2.4. Note that $0 < \kappa \leq 1$.

Now let φ be an arbitrary element of $\mathscr{D}'(\partial D)$ such that $T(\alpha)\varphi \in H^s(\partial D)$. Then one may suppose that $\varphi \in H^t(\partial D)$ for some $t < s + \kappa$. Thus, to prove the lemma, it suffices to show the following:

$$\varphi_j\varphi \in H^t(\partial D),\ T(\alpha)\varphi \in H^s(\partial D) \qquad \Rightarrow \qquad \varphi_j\varphi \in H^{s+\kappa}(\partial D). \tag{5'}$$

$$|\varphi_j\varphi|^2_{H^{s+\kappa}(\partial D)} \leq C(|T(\alpha)\varphi|^2_{H^s(\partial D)} + |\varphi_j\varphi|^2_{H^t(\partial D)}). \tag{6'}$$

Here and in the following the letter C denotes a generic positive constant depending only on s and t.

We choose constants m and k such that $0 < m < s + \kappa - t$ and $k = [s] + 1$ where $[s]$ stands for the integral part of s. Then, applying the first inequality of Theorem 5.7.5 with $s_1 = \kappa$ to $\varphi_j\varphi$, and further estimate (11) with $s = 0$ and $t = t - s(< \kappa)$ to $(\varphi_j\varphi) * \chi_\varepsilon$, we obtain that

$$
\begin{aligned}
&|\varphi_j\varphi|^2_{H^{(s+\kappa, m, \rho)}(\partial D)} \\
&\leq C\left(\int_0^1 |(\varphi_j\varphi) * \chi_\varepsilon|^2_{H^\kappa(\partial D)}\left(1 + \frac{\rho^2}{\varepsilon^2}\right)^{-m} \varepsilon^{-2s}\frac{d\varepsilon}{\varepsilon} + |\varphi_j\varphi|^2_{H^t(\partial D)}\right) \\
&\leq C\left(\int_0^1 |T(\alpha)((\varphi_j\varphi) * \chi_\varepsilon)|^2_{L^2(\partial D)}\left(1 + \frac{\rho^2}{\varepsilon^2}\right)^{-m} \varepsilon^{-2s}\frac{d\varepsilon}{\varepsilon}\right. \\
&\quad \left. + \int_0^1 |(\varphi_j\varphi) * \chi_\varepsilon|^2_{H^{t-s}(\partial D)}\left(1 + \frac{\rho^2}{\varepsilon^2}\right)^{-m} \varepsilon^{-2s}\frac{d\varepsilon}{\varepsilon} + |\varphi_j\varphi|^2_{H^t(\partial D)}\right).
\end{aligned} \tag{17}
$$

But, using the second inequality of Theorem 5.7.5 with $s_1 = t - s$, we can estimate the second term on the last inequality of (17) as follows.

$$
\begin{aligned}
\int_0^1 |(\varphi_j\varphi) * \chi_\varepsilon|^2_{H^{t-s}(\partial D)}\left(1 + \frac{\rho^2}{\varepsilon^2}\right)^{-m} \varepsilon^{-2s}\frac{d\varepsilon}{\varepsilon} &\leq C|\varphi_j\varphi|^2_{H^{(t, m, \rho)}(\partial D)} \\
&\leq C|\varphi_j\varphi|^2_{H^t(\partial D)}.
\end{aligned} \tag{18}
$$

Further, in light of the *pseudo-local property* for pseudo-differential operators, we can estimate the first term on the last inequality of (17) as follows (cf. Oleĭnik-Radkevič [1], inequality (2.4.46)):

$$
\begin{aligned}
&\int_0^1 |T(\alpha)((\varphi_j\varphi) * \chi_\varepsilon)|^2_{L^2(\partial D)}\left(1 + \frac{\rho^2}{\varepsilon^2}\right)^{-m} \varepsilon^{-2s}\frac{d\varepsilon}{\varepsilon} \\
&\qquad \leq C(|T(\alpha)\varphi|^2_{H^s(\partial D)} + |\varphi_j\varphi|^2_{H^t(\partial D)}).
\end{aligned} \tag{19}
$$

Therefore, carrying estimates (18) and (19) into estimate (17), we have

$$
|\varphi_j\varphi|^2_{H^{(s+\kappa, m, \rho)}(\partial D)} \leq C(|T(\alpha)\varphi|^2_{H^s(\partial D)} + |\varphi_j\varphi|^2_{H^t(\partial D)}).
$$

By virtue of Lemma 5.7.4, this proves assertions (5′) and (6′).

The proof of Lemma 10.2.2 is complete. ▼

(3) *Regularity* result (1) for problem (∗) is an immediate consequence of regularity result (5) for the operator $T(\alpha)$.

(4) We prove the following *uniqueness* result:

For any $\alpha > 0$, *we have*

$$u \in H^{s-2+\kappa}(\bar{D}),\ (A - \alpha)u = 0 \text{ in } D,\ Lu = 0 \text{ on } \partial D \quad \Rightarrow \quad u = 0 \text{ in } D. \tag{20}$$

Regularity result (1) tells us that

$$u \in H^{s-2+\kappa}(\bar{D}),\ (A - \alpha)u = 0 \text{ in } D,\ Lu = 0 \text{ on } \partial D \quad \Rightarrow \quad u \in C^{\infty}(\bar{D}).$$

Therefore, uniqueness result (20) is an immediate consequence of the following maximum principle:

10.2.6 Proposition. *Let A and L be as in Theorem 10.1.1. Then we have, for any* $\alpha > 0$,

$$u \in C^2(\bar{D}),\ (A - \alpha)u \geq 0 \text{ in } D,\ Lu \geq 0 \text{ on } \partial D \quad \Rightarrow \quad u \leq 0 \text{ on } \bar{D}.$$

Proof. If u is a constant, then we have

$$0 \leq (A - \alpha)u = (c - \alpha)u \qquad \text{in } D.$$

This implies that u is non-positive, since $c \leq 0$ in D and $\alpha > 0$.

Now we consider the case when u is not a constant. Assume to the contrary that

$$\max_{\bar{D}} u > 0.$$

Then, applying Theorem 7.1.1 (the weak maximum principle) to the operator $A - \alpha$, we obtain that there exists a point x_0' of ∂D such that

$$\begin{cases} u(x_0') = \max_{\bar{D}} u, \\ u(x) < u(x_0'), \qquad x \in D. \end{cases}$$

Thus it follows from an application of Lemma 7.1.7 with $\Sigma_3 = \partial D$ that

$$\frac{\partial u}{\partial \mathbf{n}}(x_0') < 0.$$

Further we have

$$\begin{cases} \dfrac{\partial u}{\partial x_i}(x'_0) = 0 \qquad (1 \le i \le N-1), \\ Au(x'_0) \ge \alpha u(x'_0) > 0, \end{cases}$$

and also

$$\sum_{i,j=1}^{N-1} \alpha^{ij}(x'_0) \frac{\partial^2 u}{\partial x_i \, \partial x_j}(x'_0) \le 0,$$

since the matrices $(\alpha^{ij}(x'_0))$ and $((-\partial^2 u/\partial x_i \, \partial x_j)(x'_0))$ are positive semi-definite. Hence, in view of the *transversality* condition (10.1.3) for L, it follows that

$$Lu(x'_0) = \sum_{i,j=1}^{N-1} \alpha^{ij}(x'_0) \frac{\partial^2 u}{\partial x_i \, \partial x_j}(x'_0) + \gamma(x'_0)u(x'_0) + \mu(x'_0)\frac{\partial u}{\partial \mathbf{n}}(x'_0) - \delta(x'_0)Au(x'_0) < 0.$$

This contradicts the hypothesis $Lu \ge 0$ on ∂D. ▼

(5) Finally we prove the following *existence* result:

For any $\alpha > 0$, *problem* $(*)$ *has a solution* $u \in H^{s-2+\kappa}(\bar{D})$ *for any* $f \in H^{s-2}(\bar{D})$ *and* $\varphi \in H^{s-5/2}(\partial D)$ $(s \ge 3)$. (21)

We make use of Theorem 8.4.1 (and its proof).

5-1) Replacing the parameter α in problem $(*)$ by the differential operator $-\partial^2/\partial y^2$ on the unit circle $S = \mathbf{R}/2\pi\mathbf{Z}$, we consider the following boundary value problem:

$$\begin{cases} \left(A + \dfrac{\partial^2}{\partial y^2}\right)\tilde{u} = \tilde{f} & \text{in} \quad D \times S, \\ L\tilde{u} = \tilde{\varphi} & \text{on } \partial D \times S. \end{cases} \tag{$\tilde{*}$}$$

Then Theorem 8.4.1 tells us that:

If the index of problem $(\tilde{*})$ is finite, *then the index of problem* $(*)$ *is equal to* zero *for all* $\alpha \ge 0$. (22)

5-2) We reduce the study of problem $(\tilde{*})$ to that of a pseudo-differential operator on the boundary, just as in problem $(*)$.

Applying Theorem 8.2.4 to the operator $\Lambda = A + \partial^2/\partial y^2$, we obtain the following results:

(ã) The Dirichlet problem

$$\begin{cases} \Lambda\tilde{w} = 0 & \text{in} \quad D \times S, \\ \gamma_0 \tilde{w} = \tilde{\varphi} & \text{on } \partial D \times S, \end{cases}$$

has a unique solution $\tilde{w}$ in $H^t(\bar{D} \times S)$ for any $\tilde{\varphi} \in H^{t-1/2}(\partial D \times S)$ $(t \in \mathbf{R})$.

($\tilde{\text{b}}$) The mapping

$$\tilde{P}: H^{t-1/2}(\partial D \times S) \to H^t(\bar{D} \times S),$$

defined by $\tilde{w} = \tilde{P}\tilde{\varphi}$, is an isomorphism of $H^{t-1/2}(\partial D \times S)$ onto the space $N(\Lambda, t) = \{\tilde{u} \in H^t(\bar{D} \times S); \Lambda\tilde{u} = 0 \text{ in } D \times S\}$ for all $t \in \mathbf{R}$; and its inverse is the trace operator γ_0 on $\partial D \times S$.

We let

$$\tilde{T}: C^\infty(\partial D \times S) \to C^\infty(\partial D \times S),$$
$$\tilde{\varphi} \mapsto L\tilde{P}\tilde{\varphi}.$$

Then the operator $\tilde{T}$ can be decomposed as follows:

$$\tilde{T} = \tilde{Q} + \mu\tilde{\Pi}$$

where

$$\begin{cases} \tilde{Q}\tilde{\varphi} = \displaystyle\sum_{i,j=1}^{N-1} \alpha^{ij} \frac{\partial^2 \tilde{\varphi}}{\partial x_i\, \partial x_j} + \delta \frac{\partial^2 \tilde{\varphi}}{\partial y^2} + \sum_{i=1}^{N-1} \beta^i \frac{\partial \tilde{\varphi}}{\partial x_i} + \gamma\tilde{\varphi}, \\ \tilde{\Pi}\tilde{\varphi} = \dfrac{\partial}{\partial \mathbf{n}} (\tilde{P}\tilde{\varphi})|_{\partial D \times S}. \end{cases}$$

The operator $\tilde{Q}$ is a second-order differential operator on $\partial D \times S$ and its symbol is given by

$$-\sum_{i,j=1}^{N-1} \alpha^{ij}(x')\xi_i\xi_j - \delta(x')\eta^2 + \sqrt{-1} \sum_{i=1}^{N-1} \beta^i(x')\xi_i + \gamma(x'),$$

where η is the dual variable of y in the cotangent bundle $T^*(S)$. We remark that

$$\sum_{i,j=1}^{N-1} \alpha^{ij}(x')\xi_i\xi_j + \delta(x')\eta^2 \geq 0 \qquad \text{on the cotangent bundle } T^*(\partial D \times S).$$

Furthermore, arguing as in the proof of Theorem 8.2.2, we find that the operator $\tilde{\Pi}$ is a classical pseudo-differential operator of first order on $\partial D \times S$, and its principal symbol is given by (cf. formulas (2), (3))

$$\tilde{p}_1(x', \xi', y, \eta) + \sqrt{-1}\,\tilde{q}_1(x', \xi', y, \eta)$$

where:

$$\tilde{p}_1(x', \xi', y, \eta) = \frac{(4A_2(x')(a_0(x', \xi') - \eta^2) - a_1(x', \xi')^2)^{1/2}}{2A_2(x')}, \tag{23}$$

$$\tilde{q}_1(x', \xi', y, \eta) = q_1(x', \xi') = -\frac{a_1(x', \xi')}{2A_2(x')}. \tag{24}$$

Note that

$\tilde{p}_1(x', \xi', y, \eta) < 0$ on the bundle $T^*(\partial D \times S) \setminus \{0\}$ of non-zero cotangent vectors.

Therefore, the operator $\tilde{T} = \tilde{Q} + \mu\tilde{\Pi}$ is a classical pseudo-differential operator of second order on $\partial D \times S$ and its symbol is given by

$$\left[-\sum_{i,j=1}^{N-1} \alpha^{ij}(x')\xi_i\xi_j - \delta(x')\eta^2\right] + \Bigg[\mu(x')\tilde{p}_1(x', \xi', y, \eta)$$

$$+ \sqrt{-1}\,(\mu(x')\tilde{q}_1(x', \xi', y, \eta) + \sum_{i=1}^{N-1} \beta^i(x')\xi_i)\Bigg] \tag{25}$$

$$+ \text{ terms of order } \leq 0.$$

Since the operator $\tilde{T}\colon C^\infty(\partial D \times S) \to C^\infty(\partial D \times S)$ extends to a continuous linear operator $\tilde{T}\colon H^s(\partial D \times S) \to H^{s-2}(\partial D \times S)$ for all $s \in \mathbf{R}$, we can associate with problem ($\tilde{*}$) a densely defined, closed linear operator

$$\tilde{\mathscr{T}}\colon H^{s-5/2+\kappa}(\partial D \times S) \to H^{s-5/2}(\partial D \times S)$$

as follows:

($\tilde{\alpha}$) The domain $D(\tilde{\mathscr{T}})$ of $\tilde{\mathscr{T}}$ is the space

$$D(\tilde{\mathscr{T}}) = \{\tilde{\varphi} \in H^{s-5/2+\kappa}(\partial D \times S);\ \tilde{T}\tilde{\varphi} \in H^{s-5/2}(\partial D \times S)\}.$$

($\tilde{\beta}$) $$\tilde{\mathscr{T}}\tilde{\varphi} = \tilde{T}\tilde{\varphi}, \qquad \tilde{\varphi} \in D(\tilde{\mathscr{T}}).$$

Then, just as in problem ($*$), it is easy to see that the study of problem ($\tilde{*}$) is reduced to that of the operator $\tilde{\mathscr{T}}$.

5-3) We show that:

If hypothesis $(A.1)$ is satisfied, then the operator $\tilde{\mathcal{T}}$ is a *Fredholm operator*, that is, the index of problem $(\tilde{*})$ is *finite*. (26)

First we remark that the *transversality* condition (10.1.3) for L implies that

$$\delta(x') > 0 \qquad \text{on } M = \{x' \in \partial D;\ \mu(x') = 0\}.$$

Hence we find that if hypothesis $(A.1)$ is satisfied, then the following hypothesis is satisfied:

$(\widetilde{A.1})$ There exists a constant $\tilde{C} > 0$ such that for all sufficiently small $\rho > 0$ we have

$$B_E((x', y), \rho) \subset B_{\widetilde{L^0}}((x', y), \tilde{C}\rho^\varepsilon), \qquad (x', y) \in \tilde{M},$$

where $\widetilde{L^0} = \sum_{i,j=1}^{N-1} \alpha^{ij}(\partial^2/\partial x_i\, \partial x_j) + \delta(\partial^2/\partial y^2)$ and $\tilde{M} = \{(x', y) \in \partial D \times S;\ \mu(x') = 0\} = M \times S$.

Therefore, arguing as in the proof of Lemma 10.2.2, we can obtain the following:

10.2.7 Lemma. *Let A and L be as in Theorem 10.1.1, and suppose that hypothesis $(A.1)$ is satisfied. Then we have, for all $s \in \mathbf{R}$,*

$$\tilde{\varphi} \in \mathscr{D}'(\partial D \times S),\ \tilde{T}\tilde{\varphi} \in H^{s-5/2}(\partial D \times S) \quad \Rightarrow \quad \tilde{\varphi} \in H^{s-5/2+\kappa}(\partial D \times S).$$

Furthermore, for any $t < s - 5/2 + \kappa$, there exists a constant $\tilde{C}_{s,t} > 0$ such that

$$|\tilde{\varphi}|_{H^{s-5/2+\kappa}(\partial D \times S)} \leq \tilde{C}_{s,t}(|\tilde{T}\tilde{\varphi}|_{H^{s-5/2}(\partial D \times S)} + |\tilde{\varphi}|_{H^t(\partial D \times S)}). \tag{27}$$

Here κ is the same constant as in Lemma 10.2.2.

Since the injection $H^{s-5/2+\kappa}(\partial D \times S) \to H^t(\partial D \times S)$ is compact, applying Theorem 3.7.6 with

$$X = H^{s-5/2+\kappa}(\partial D \times S),$$

$$Y = H^{s-5/2}(\partial D \times S),$$

$$Z = H^t(\partial D \times S),$$

$$T = \tilde{\mathcal{T}},$$

we obtain from estimate (27) that:

1. $\dim N(\tilde{\mathcal{T}}) < \infty$.
2. The range $R(\tilde{\mathcal{T}})$ is closed in $H^{s-5/2}(\partial D \times S)$.

On the other hand, it follows from an application of the closed range theorem (Theorem 3.4.6) that

$$\operatorname{codim} R(\tilde{\mathscr{T}}) = \dim N(\tilde{\mathscr{T}}^*). \tag{28}$$

Further, applying Lemma 8.4.8 to the operator $\tilde{T}$, we find that

$$N(\tilde{\mathscr{T}}^*) = \{\tilde{\psi} \in H^{-s+5/2}(\partial D \times S);\ \tilde{T}^*\tilde{\psi} = 0\},$$

where $\tilde{T}^*$, the adjoint of $\tilde{T}$, is a classical pseudo-differential operator of second order on $\partial D \times S$.

But, it follows from formula (25) that the symbol of $\tilde{T}^*$ is given by

$$\left[-\sum_{i,j=1}^{N-1} \alpha^{ij}(x')\xi_i\xi_j - \delta(x')\eta^2\right] + \left[\mu(x')\tilde{p}_1(x', \xi', y, \eta)\right.$$
$$\left. + \sqrt{-1}\left(-\mu(x')\tilde{q}_1(x', \xi', y, \eta) - \sum_{i=1}^{N-1} \beta^i(x')\xi_i + 2\sum_{i,j=1}^{N-1} \frac{\partial \alpha^{ij}}{\partial x_j}(x')\xi_i\right)\right]$$
$$+ \text{ terms of order} \leq 0.$$

Hence we can obtain the following result, analogous to Lemma 10.2.7.

10.2.8 Lemma. *Let A and L be as in Theorem 10.1.1, and suppose that hypothesis $(A.1)$ is satisfied. Then we have, for all $s \in \mathbf{R}$,*

$$\tilde{\psi} \in \mathscr{D}'(\partial D \times S),\ \tilde{T}^*\tilde{\psi} \in H^{-s+5/2-\kappa}(\partial D \times S) \quad \Rightarrow \quad \tilde{\psi} \in H^{-s+5/2}(\partial D \times S).$$

Furthermore, for any $t < -s + 5/2$, there exists a constant $\tilde{C}^*_{s,t} > 0$ such that

$$|\tilde{\psi}|_{H^{-s+5/2}(\partial D \times S)} \leq \tilde{C}^*_{s,t}(|\tilde{T}^*\tilde{\psi}|_{H^{-s+5/2-\kappa}(\partial D \times S)} + |\tilde{\psi}|_{H^t(\partial D \times S)}). \tag{29}$$

Therefore, applying Theorem 3.7.6 to the operator $\tilde{\mathscr{T}}^*$, we obtain from estimate (29) that

$$\dim N(\tilde{\mathscr{T}}^*) < \infty.$$

In view of formula (28), this proves that:

3. $\operatorname{codim} R(\tilde{\mathscr{T}}) < \infty$.

Summing up, we have proved assertion (26).

5-4) Finally we show that:

If hypothesis $(A.1)$ is satisfied, then we have

$$\operatorname{codim} R(\mathscr{T}(\alpha)) = 0 \qquad \text{for all } \alpha > 0. \tag{30}$$

This assertion implies existence result (21).

Combining assertions (22) and (26), we find that if hypothesis (A.1) is satisfied, then we have

$$\operatorname{ind} \mathscr{T}(\alpha) = \dim N(\mathscr{T}(\alpha)) - \operatorname{codim} R(\mathscr{T}(\alpha)) = 0 \qquad \text{for all } \alpha \geq 0. \quad (31)$$

But it follows from uniqueness result (20) that

$$\dim N(\mathscr{T}(\alpha)) = 0 \qquad \text{for all } \alpha \geq 0.$$

In view of assertion (31), this proves assertion (30), and hence existence result (21). ▼

The proof of Theorem 10.2.1 and hence that of Theorem 10.1.1 is now complete. ■

10.3. Proof of Theorem 10.1.3

We verify conditions [I] and [III] of Corollary 9.6.23; then Theorem 10.1.3 follows from an application of the same corollary. In doing so, we make essential use of Theorem 6.10.3 due to Melin-Sjöstrand [1].

(1) First we verify condition [III]. To do so, it suffices to prove the following *regularity* result, analogous to regularity result (10.2.1).

10.3.1 Theorem. *Let A and L be as in Theorem 10.1.3. Suppose that hypothesis (A.2) is satisfied. Then we have, for any $s \geq 2$ and $t < s - 1$,*

$$u \in H^t(\bar{D}),\ (A - \alpha)u \in H^{s-2}(\bar{D}),\ Lu \in H^{s-3/2}(\partial D) \quad \Rightarrow \quad u \in H^{s-1}(\bar{D}). \quad (1)$$

Here $\alpha \geq 0$.

Proof. As in the proof of Theorem 10.2.1, we are reduced to the study of the following pseudo-differential operator

$$T(\alpha): C^\infty(\partial D) \to C^\infty(\partial D),$$
$$\varphi \mapsto LP(\alpha)\varphi.$$

Since we have

$$LP(\alpha)\varphi = \sum_{i=1}^{N-1} \beta^i \frac{\partial \varphi}{\partial x_i} + \gamma\varphi + \mu \frac{\partial}{\partial \mathbf{n}} (P(\alpha)\varphi)|_{\partial D} - \alpha\delta\varphi,$$

it follows that the operator $T(\alpha)$ can be written as

$$T(\alpha) = \sum_{i=1}^{N-1} \beta^i \frac{\partial}{\partial x_i} + (\gamma - \alpha\delta) + \mu\Pi(\alpha).$$

Hence the symbol of $T(\alpha)$ is given by

$$\left[\mu(x')p_1(x', \xi') + \sqrt{-1}\left(\mu(x')q_1(x', \xi') + \sum_{i=1}^{N-1} \beta^i(x')\xi_i\right)\right]$$

$$+ \text{ terms of order} \leq 0 \text{ depending on } \alpha.$$

But we find from formulas (10.2.2) and (10.2.3) that:

(a) $p_1(x', \xi') < 0$ on $T^*(\partial D) \setminus \{0\}$.
(b) $q_1(x', \xi')$ is a *polynomial* of degree one in the variable ξ'.

Thus it is easy to verify that hypothesis $(A.2)$ implies hypotheses (B) and (C) of Theorem 6.10.3 for the operator $T(\alpha)$.

Therefore, applying Theorem 6.10.3 to the operator $T(\alpha)$, we can obtain the following result, analogous to Lemma 10.2.2.

10.3.2 Lemma. *Let A and L be as in Theorem 10.1.3, and suppose that hypothesis $(A.2)$ is satisfied. Then we have, for all $s \in \mathbf{R}$,*

$$\varphi \in \mathscr{D}'(\partial D),\ T(\alpha)\varphi \in H^{s-3/2}(\partial D) \quad \Rightarrow \quad \varphi \in H^{s-3/2}(\partial D). \tag{2}$$

Furthermore, for any $t < s - 3/2$, there exists a constant $C_{s,t} > 0$ such that

$$|\varphi|_{H^{s-3/2}(\partial D)} \leq C_{s,t}(|T(\alpha)\varphi|_{H^{s-3/2}(\partial D)} + |\varphi|_{H^t(\partial D)}).$$

Thus, the operator $T(\alpha)$ is *globally hypoelliptic*, with loss of one derivative.

It follows from an application of Theorem 8.3.9 (with $m = 1$, $\sigma = s - 1$, $\tau = s - 2$) that regularity result (2) implies regularity result (1).

The proof of Theorem 10.3.1 is complete. ■

(2) Next we verify condition [I]. We use the same notation as in Section 9.6. Note that the Poisson operator $P(\alpha)$ is essentially the same as the harmonic operator H_α in Section 9.6.

We let

$$L_0 u(x') = \sum_{i=1}^{N-1} \beta^i(x') \frac{\partial u}{\partial x_i}(x') + \gamma(x')u(x') + \mu(x')\frac{\partial u}{\partial \mathbf{n}}(x'),$$

and consider the term $-\delta(x')Au(x')$ in $Lu(x')$ as a term of "perturbation" of $L_0 u(x')$:

$$Lu(x') = L_0 u(x') - \delta(x')Au(x').$$

Then it is easy to see that the operator $\overline{LH_\alpha}$ can be decomposed as follows:

$$\overline{LH_\alpha} = \overline{L_0 H_\alpha} - \alpha\delta I.$$

The next theorem, analogous to Theorem 10.2.1, is the essential step in the proof.

10.3.3 Theorem. *Let A and L be as in Theorem 10.1.3, and suppose that hypothesis (A.2) is satisfied. Then, for any $\alpha \geq 0$ and $\lambda > 0$, the boundary value problem*

$$\begin{cases} (A - \alpha)u = f & \text{in } D, \\ (L_0 - \lambda)u = \varphi & \text{on } \partial D, \end{cases} \qquad (*)_0$$

has a unique solution $u \in H^{s-1}(\bar{D})$ for any $f \in H^{s-2}(\bar{D})$ and $\varphi \in H^{s-3/2}(\partial D)$ $(s \geq 2)$.

Furthermore, we have, for any $t < s - 1$,

$$u \in H^t(\bar{D}),\ (A - \alpha)u \in H^{s-2}(\bar{D}),\ Lu \in H^{s-3/2}(\partial D) \quad \Rightarrow \quad u \in H^{s-1}(\bar{D}). \quad (3)$$

Granting Theorem 10.3.3 for the moment, we shall verify condition [I].

Theorem 10.3.3 tells us that the operator $\overline{L_0 H_\alpha}$ is the infinitesimal generator of some Feller semigroup on ∂D. In fact, by virtue of the Sobolev imbedding theorem, Theorem 10.3.3 implies the following:

For any $\alpha \geq 0$ and $\lambda > 0$, the boundary value problem

$$\begin{cases} (\alpha - A)u = 0 & \text{in } D, \\ (\lambda - L_0)u = \varphi & \text{on } \partial D, \end{cases}$$

has a unique solution $u \in C^\infty(\bar{D})$ for any $\varphi \in C^\infty(\partial D)$.

Hence, applying part (ii) of Theorem 9.6.15 to the boundary condition L_0, we find that the operator $\overline{L_0 H_\alpha}$ is the infinitesimal generator of some Feller semigroup on ∂D.

Further, it is clear that the operator $-\alpha\delta I$ is a bounded linear operator on $C(\partial D)$ into itself, and satisfies condition (β') of Theorem 9.3.3, since $\alpha \geq 0$ and $\delta \geq 0$ on ∂D.

Therefore, applying Corollary 9.3.4 with

$$\mathfrak{A}' = \overline{L_0 H_\alpha},$$

$$M = -\alpha\delta I,$$

we obtain that the operator $\overline{LH_\alpha} = \overline{L_0 H_\alpha} - \alpha\delta I$ is the infinitesimal generator of some Feller semigroup on ∂D. Hence it follows from part (i) of Theorem 9.6.15 that:

For each $\lambda > 0$, the boundary value problem

$$\begin{cases} (\alpha - A)u = 0 & \text{in } D, \\ (\lambda - L)u = \varphi & \text{on } \partial D, \end{cases}$$

has a solution $u \in C^{2+\theta}(\bar{D})$ for any φ in some dense subset of $C(\partial D)$.

This verifies condition [I].

Theorem 10.1.3 is proved, apart from the proof of Theorem 10.3.3.

Proof of Theorem 10.3.3. The proof is essentially the same as that of Theorem 10.2.1 except that we use Lemmas 10.3.4 and 10.3.6 below instead of Lemmas 10.2.2, 10.2.7 and 10.2.8. So we only give a sketch of the proof.

1) As in the proof of Theorem 10.3.1, we are reduced to the study of the pseudo-differential operator

$$T_0(\alpha) \equiv (L_0 - \lambda)P(\alpha) = \sum_{i=1}^{N-1} \beta^i \frac{\partial}{\partial x_i} + (\gamma - \lambda) + \mu\Pi(\alpha).$$

The symbol of $T_0(\alpha)$ is given by

$$\left[\mu(x')p_1(x', \xi') + \sqrt{-1}\left(\mu(x')q_1(x', \xi') + \sum_{i=1}^{N-1} \beta^i(x')\xi_i\right)\right]$$

$$+ \text{ terms of order } \leq 0 \text{ depending on } \alpha.$$

Hence, Lemma 10.3.2 remains valid for the operator $T_0(\alpha)$:

10.3.4 Lemma. *Let A and L be as in Theorem 10.1.3, and suppose that hypothesis (A.2) is satisfied. Then we have, for all $s \in \mathbf{R}$,*

$$\varphi \in \mathscr{D}'(\partial D),\ T_0(\alpha)\varphi \in H^{s-3/2}(\partial D) \quad \Rightarrow \quad \varphi \in H^{s-3/2}(\partial D). \tag{4}$$

Furthermore, for any $t < s - 3/2$, there exists a constant $C_{s,t} > 0$ such that

$$|\varphi|_{H^{s-3/2}(\partial D)} \leq C_{s,t}(|T_0(\alpha)\varphi|_{H^{s-3/2}(\partial D)} + |\varphi|_{H^t(\partial D)}).$$

Thus, the operator $T_0(\alpha)$ is *globally hypoelliptic*, with loss of one derivative.

Regularity result (3) for problem $(*)_0$ follows from regularity result (4) for the operator $T_0(\alpha)$.

2) Now we prove the following *uniqueness* result:

$$u \in H^{s-1}(\bar{D}),\ (A - \alpha)u = 0 \textit{ in } D,\ (L_0 - \lambda)u = 0 \textit{ on } \partial D \Rightarrow u = 0 \text{ in } D. \quad (5)$$

Regularity result (3) tells us that

$$u \in H^{s-1}(\bar{D}),\ (A - \alpha)u = 0 \text{ in } D,\ (L_0 - \lambda)u = 0 \text{ on } \partial D \quad \Rightarrow \quad u \in C^\infty(\bar{D}).$$

Therefore, uniqueness result (5) is an immediate consequence of the following maximum principle, analogous to Proposition 10.2.6.

10.3.5 Proposition. *Let A and L be as in Theorem 10.1.3. Then we have, for any $\alpha \geq 0$ and $\lambda > 0$,*

$$u \in C^2(\bar{D}),\ (A - \alpha)u \geq 0 \textit{ in } D,\ (L_0 - \lambda)u \geq 0 \textit{ on } \partial D \quad \Rightarrow \quad u \leq 0 \textit{ on } \bar{D}.$$

Proof. We have only to consider the case when u is not a constant.

Assume to the contrary that

$$\max_{\bar{D}} u > 0.$$

Then, arguing as in the proof of Proposition 10.2.6, we find that there exists a point x_0' of ∂D such that

$$\begin{cases} u(x_0') = \max_{\bar{D}} u > 0, \\ \dfrac{\partial u}{\partial \mathbf{n}}(x_0') < 0, \\ \dfrac{\partial u}{\partial x_i}(x_0') = 0 \qquad (1 \leq i \leq N-1). \end{cases}$$

Thus we have

$$\begin{aligned} (L_0 - \lambda)u(x_0') &= \mu(x_0')\frac{\partial u}{\partial \mathbf{n}}(x_0') + \gamma(x_0')u(x_0') - \lambda u(x_0') \\ &\leq -\lambda u(x_0') \\ &< 0. \end{aligned}$$

This contradicts the hypothesis $(L_0 - \lambda)u \geq 0$ on ∂D. ▼

3) Finally we prove the following *existence* result:

For any $\alpha \geq 0$ *and* $\lambda > 0$, *problem* $(*)_0$ *has a solution* $u \in H^{s-1}(\bar{D})$ *for any* $f \in H^{s-2}(\bar{D})$ *and* $\varphi \in H^{s-3/2}(\partial D)$ $(s \geq 2)$. (6)

We make use of Theorem 8.4.1 just as in the proof of Theorem 10.2.1. Instead of problem $(*)_0$, we consider the following boundary value problem:

$$\begin{cases} \left(A + \dfrac{\partial^2}{\partial y^2}\right)\tilde{u} = \tilde{f} & \text{in} \quad D \times S, \\ \\ (L_0 - \lambda)\tilde{u} = \tilde{\varphi} & \text{on } \partial D \times S. \end{cases} \qquad (\tilde{*})_0$$

The study of problem $(\tilde{*})_0$ is reduced to that of the pseudo-differential operator

$$\tilde{T}_0 \equiv (L_0 - \lambda)\tilde{P} = \sum_{i=1}^{N-1} \beta^i \frac{\partial}{\partial x_i} + (\gamma - \lambda) + \mu\tilde{\Pi}.$$

The symbol of $\tilde{T}_0$ is given by

$$\left[\mu(x')\tilde{p}_1(x', \xi', y, \eta) + \sqrt{-1}\left(\mu(x')\tilde{q}_1(x', \xi', y, \eta) + \sum_{i=1}^{N-1} \beta^i(x')\xi_i\right)\right]$$
$$+ \text{ terms of order} \leq 0.$$

But we find from formulas (10.2.23) and (10.2.24) that:

($\tilde{a}$) $\tilde{p}_1(x', \xi', y, \eta) < 0$ on $T^*(\partial D \times S) \setminus \{0\}$.
($\tilde{b}$) $\tilde{q}_1(x', \xi', y, \eta) = q_1(x', \xi')$ is a *polynomial* of degree one in the variable ξ'.

Therefore, applying Theorem 6.10.3 to the operators $\tilde{T}_0$ and $\tilde{T}_0^*$, we can obtain the following result, analogous to Lemma 10.3.4.

10.3.6 Lemma. *Let A and L be as in Theorem 10.1.3, and suppose that hypothesis (A.2) is satisfied. Then we have the following*:

(i) *For all* $s \in \mathbf{R}$, *we have*

$$\tilde{\varphi} \in \mathscr{D}'(\partial D \times S),\ \tilde{T}_0\tilde{\varphi} \in H^{s-3/2}(\partial D \times S) \quad \Rightarrow \quad \tilde{\varphi} \in H^{s-3/2}(\partial D \times S).$$

Furthermore, for any $t < s - 3/2$, *there exists a constant* $\tilde{C}_{s,t} > 0$ *such that*

$$|\tilde{\varphi}|_{H^{s-3/2}(\partial D \times S)} \leq \tilde{C}_{s,t}(|\tilde{T}_0\tilde{\varphi}|_{H^{s-3/2}(\partial D \times S)} + |\tilde{\varphi}|_{H^t(\partial D \times S)}).$$

(ii) *For all* $s \in \mathbf{R}$, *we have*

$$\tilde{\psi} \in \mathscr{D}'(\partial D \times S),\ \tilde{T}_0^* \tilde{\psi} \in H^{-s+3/2}(\partial D \times S) \quad \Rightarrow \quad \tilde{\psi} \in H^{-s+3/2}(\partial D \times S).$$

Furthermore, for any $t < -s + 3/2$, *there exists a constant* $\tilde{C}_{s,t}^* > 0$ *such that*

$$|\tilde{\psi}|_{H^{-s+3/2}(\partial D \times S)} \le \tilde{C}_{s,t}^*(|\tilde{T}_0^* \tilde{\psi}|_{H^{-s+3/2}(\partial D \times S)} + |\tilde{\psi}|_{H^t(\partial D \times S)}).$$

By virtue of Lemma 10.3.6, the proof of existence result (6) goes through as in the proof of Theorem 10.2.1. ▼

The proof of Theorem 10.3.3 and hence that of Theorem 10.1.3 is now complete. ■

10.3.7 Remark. Problem $(*)_0$ is the *oblique derivative* problem. For detailed studies of this problem, the reader might refer to Egorov-Kondrat'ev [1], Melin-Sjöstrand [1] and Taira [2], [3].

Notes

Theorems 10.1.1 and 10.1.3 are adapted from Taira [4], [6] and [7]. These results are a generalization of Théorème XIX of Bony-Courrège-Priouret [1], where Ventcel' boundary conditions are assumed to be elliptic. We confined ourselves to the case when the differential operator A is elliptic on $\bar{D}$. The reason is that when A is not elliptic on $\bar{D}$ we do not know whether the operator $T(\alpha) = LP(\alpha)$, which played a fundamental role in the proof, is a pseudo-differential operator or not.

It is thus an open problem to extend Theorems 10.1.1 and 10.1.3 to the *non-elliptic* case. We remark that Taira [7] has some result along these lines, which we now state.

Let D be a bounded domain in $\mathbf{R}^N$ with C^∞ boundary ∂D, and let A be a second-order, degenerate elliptic differential operator with real coefficients such that

$$Au(x) = \sum_{i,j=1}^N a^{ij}(x) \frac{\partial^2 u}{\partial x_i \, \partial x_j}(x) + \sum_{i=1}^N b^i(x) \frac{\partial u}{\partial x_i}(x) + c(x)u(x)$$

where:

1. $a^{ij} \in C^\infty(\mathbf{R}^N)$, $a^{ij} = a^{ij}$ and

$$\sum_{i,j=1}^N a^{ij}(x)\xi_i \xi_j \ge 0, \qquad x \in \mathbf{R}^N,\ \xi \in \mathbf{R}^N.$$

2. $b^i \in C^\infty(\mathbf{R}^N)$.
3. $c \in C^\infty(\mathbf{R}^N)$ and $c \le 0$ on $\bar{D}$.

We let

$$b(x') = \sum_{i=1}^{N} \left(b^i(x') - \sum_{j=1}^{N} \frac{\partial a^{ij}}{\partial x_j}(x') \right) n_i, \qquad x' \in \partial D,$$

where $\mathbf{n} = (n_1, \ldots, n_N)$ is the unit interior normal to ∂D at x'. We divide the boundary ∂D into the following four disjoint subsets (cf. Fichera [1], Oleĭnik-Radkevič [1], Stroock-Varadhan [2]):

$$\Sigma_3 = \left\{ x' \in \partial D;\ \sum_{i,j=1}^{N} a^{ij}(x') n_i n_j > 0 \right\}.$$

$$\Sigma_2 = \left\{ x' \in \partial D;\ \sum_{i,j=1}^{N} a^{ij}(x') n_i n_j = 0,\ b(x') < 0 \right\}.$$

$$\Sigma_1 = \left\{ x' \in \partial D;\ \sum_{i,j=1}^{N} a^{ij}(x') n_i n_j = 0,\ b(x') > 0 \right\}.$$

$$\Sigma_0 = \left\{ x' \in \partial D;\ \sum_{i,j=1}^{N} a^{ij}(x') n_i n_j = 0,\ b(x') = 0 \right\}.$$

It is easy to see that the sets Σ_0, Σ_1, Σ_2 and Σ_3 are all invariantly defined (cf. the proof of Lemma 7.1.6).

The fundamental hypothesis for A is the following (cf. Figure 10-1):

(H) *Each* Σ_i ($i = 0, 1, 2, 3$) *consists of a finite number of connected hypersurfaces.*

It is worth pointing out that we may impose a boundary condition only on the set $\Sigma_2 \cup \Sigma_3$.

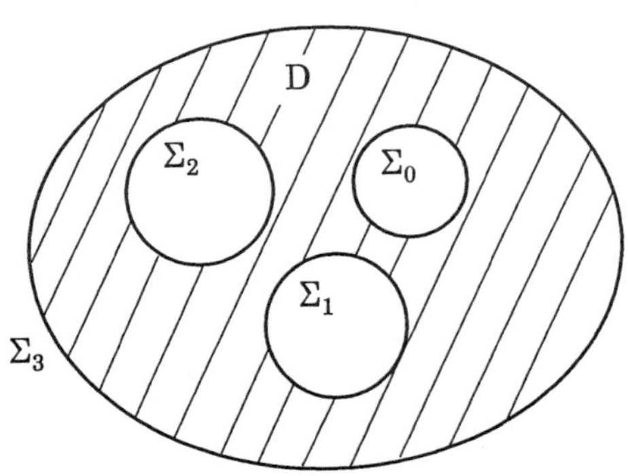

Figure 10-1

Let L be a Ventcel' boundary condition such that

$$Lu(x') = \sum_{i,j=1}^{N-1} \alpha^{ij}(x') \frac{\partial^2 u}{\partial x_i \, \partial x_j}(x') + \sum_{i=1}^{N-1} \beta^i(x') \frac{\partial u}{\partial x_i}(x')$$

$$+ \gamma(x')u(x') + \mu(x') \frac{\partial u}{\partial \mathbf{n}}(x') - \delta(x')Au(x')$$

where:

1. The α^{ij} are the components of a C^∞ symmetric contravariant tensor of type $\binom{2}{0}$ on $\Sigma_2 \cup \Sigma_3$ and

$$\sum_{i,j=1}^{N-1} \alpha^{ij}(x')\eta_i\eta_j \geq 0, \qquad x' \in \Sigma_2 \cup \Sigma_3, \, \eta \in T_{x'}(\Sigma_2 \cup \Sigma_3).$$

2. $\beta^i \in C^\infty(\Sigma_2 \cup \Sigma_3)$.
3. $\gamma \in C^\infty(\Sigma_2 \cup \Sigma_3)$ and $\gamma \leq 0$ on $\Sigma_2 \cup \Sigma_3$.
4. $\mu \in C^\infty(\Sigma_2 \cup \Sigma_3)$ and $\mu \geq 0$ on $\Sigma_2 \cup \Sigma_3$.
5. $\delta \in C^\infty(\Sigma_2 \cup \Sigma_3)$ and $\delta \geq 0$ on $\Sigma_2 \cup \Sigma_3$.

To state a hypothesis for L on Σ_3, we let

$B_{L^0}(x', \rho) =$ the set of all points $y \in \Sigma_3$ which can be joined to $x' \in \Sigma_3$ by a Lipschitz path v: $[0, \rho] \to \Sigma_3$ for which the tangent vector $\dot{v}(t)$ of Σ_3 at $v(t)$ is subunit for $L^0 = \sum_{i,j=1}^{N-1} \alpha^{ij} \frac{\partial^2}{\partial x_i \, \partial x_j}$ for almost every t.

The hypothesis for L on Σ_3 is the following:

($A.1'$) *The operator A is elliptic near Σ_3 and there exist constants $0 \leq \varepsilon_1 < 1$ and $C_1 > 0$ such that for all sufficiently small $\rho > 0$ we have*

$$B_E(x', \rho) \subset B_{L^0}(x', C_1\rho^{\varepsilon_1}), \qquad x' \in M = \{x' \in \Sigma_3; \, \mu(x') = 0\}.$$

Hypothesis ($A.1'$) has an intuitive meaning similar to hypothesis ($A.1$) in Section 10.1 (cf. Remark 10.1.2).

To state a hypothesis for L on Σ_2, we write the operator A in a neighborhood of Σ_2 in the form

$$A = A_2 \frac{\partial^2}{\partial \mathbf{n}^2} + A_1 \frac{\partial}{\partial \mathbf{n}} + A_0,$$

where A_j $(j = 0, 1, 2)$ is a differential operator of order $2 - j$ acting along the surfaces parallel to Σ_2. We remark that:

(a) $A_2 = 0$ on Σ_2.
(b) The restriction $A_0|_{\Sigma_2}$ of A_0 to Σ_2 is a second-order differential operator with non-positive principal symbol.

Since $\mu \geq 0$ and $b < 0$ on Σ_2, we can define a "non-Euclidean" ball $B_{L^0 - (\mu/b)(A_0|_{\Sigma_2})}(x', \rho)$ in the same way as $B_{L^0}(x', \rho)$, replacing Σ_3 and L^0 by Σ_2 and $L^0 - (\mu/b)(A_0|_{\Sigma_2})$, respectively.

The hypothesis for L on Σ_2 is the following:

(A.1″) There exist constants $0 \leq \varepsilon_2 < 1$ and $C_2 > 0$ such that for all sufficiently small $\rho > 0$ we have

$$B_E(x', \rho) \subset B_{L^0 - (\mu/b)(A_0|\Sigma_2)}(x', C_2 \rho^{\varepsilon_2}), \qquad x' \in \Sigma_2.$$

The intuitive meaning of hypothesis (*A*.1″) is that a Markovian particle with generator $L^0 - (\mu/b)(A_0|_{\Sigma_2})$ diffuses everywhere in Σ_2 in *finite* time.

Now we can state a generalization of Theorem 10.1.1:

Theorem. *Let the differential operator A satisfy hypothesis (H) and let the boundary condition L be transversal on $\Sigma_2 \cup \Sigma_3$. Suppose that hypotheses (A.1′) and (A.1″) are satisfied. Then there exists a Feller semigroup $\{T_t\}_{t \geq 0}$ on $\bar{D}$ whose infinitesimal generator $\mathfrak{A}$ coincides with the minimal closed extension in $C(\bar{D})$ of the restriction of A to the space $\{u \in C^\infty(\bar{D}); Lu = 0$ on $\Sigma_2 \cup \Sigma_3\}$.*

The proof of this theorem is based on the maximum principles discussed in Chapter 7 and the work of Oleĭnik-Radkevič [1] on the Dirichlet problem for degenerate elliptic differential operators of second order (cf. Notes at the end of Chapter 8).

For detailed studies of diffusion processes using stochastic differential equations, the reader is referred especially to Ikeda-Watanabe [1], and many other papers mentioned in its bibliography.

Bibliography

Abraham, R., J. E. Marsden and T. Ratiu: [1] *Manifolds, Tensor Analysis, and Applications*, Addison-Wesley, Reading, Mass., 1983.

Adams, R.: [1] *Sobolev Spaces*, Academic Press, New York, 1975.

Agmon, S.: [1] *Lectures On Elliptic Boundary Value Problems*, Van Nostrand, Princeton, 1965.

Agranovich, M. S. and M. I. Vishik: [1] "Elliptic problems with a parameter and parabolic problems of general type," *Uspehi Mat. Nauk* **19** (3) (117) (1964) 53–161 (Russian); English translation in *Russian Math. Surv.* **19** (3) (1964), 53–157.

Amano, K.: [1] "Maximum principles for degenerate elliptic-parabolic operators," *Indiana Univ. Math. J.* **29** (1979), 545–557.

Barros-Neto, J.: [1] *An Introduction to the Theory of Distributions*, Marcel Dekker, New York, 1973.

Bernstein, S. N.: [1] "Equations différentielles stochastiques," *Actualités Sci. et Ind.* **738**, pp. 5–31, *Conf. Intern. Sci. Math. Univ. Genève*, Hermann, Paris, 1938.

Blumenthal, R. M. and R. K. Getoor: [1] *Markov Processes and Potential Theory*, Academic Press, New York, 1968.

Bony, J.-M.: [1] "Principe du maximum, inégalité de Harnack et unicité du problème de Cauchy pour les opérateurs elliptiques dégénérés," *Ann. Inst. Fourier* (Grenoble) **19** (1969), 277–304.

Bony, J.-M., P. Courrège et P. Priouret: [1] "Semi-groupes de Feller sur une variété à bord compacte et problèmes aux limites intégro-différentiels du second ordre donnant lieu au principe du maximum," *Ann. Inst. Fourier* (Grenoble) **18** (1968), 369–521.

Bourbaki, N.: [1] *Topologie Générale*, Hermann, Paris, 1953–1961.
[2] *Espaces Vectoriels Topologiques*, Hermann, Paris, 1953, 1955, 1966.
[3] *Intégration*, Hermann, Paris, 1956–1965.

Boutet de Monvel, L.: [1] "Comportement d'un opérateur pseudo-différentiel sur une variété à bord," *J. Anal. Math.* **17** (1966), 241–304.

Chazarain, J. et A. Piriou: [1] *Introduction à la Théorie des Équations aux Dérivées Partielles Linéaires*, Gauthier-Villars, 1981.

de Rham, G.: [1] *Variétés Différentiables*, Hermann, Paris, 1955.

Duistermaat, J. J.: [1] *Fourier Integral Operators*, Courant Institute Lecture Notes, New York, 1973.

Duistermaat, J. J. and L. Hörmander: [1] "Fourier integral operators II," *Acta Math.* **128** (1972), 183–269.

Dynkin, E. B.: [1] *Foundations of the Theory of Markov Processes*, Fitzmatgiz, Moscow, 1959 (Russian); German translation: Springer-Verlag, Berlin, 1961; English translation: Pergamon Press, Oxford, 1960.
[2] *Markov Processes*, Vols. I, II, Springer-Verlag, Berlin, 1965.

Dynkin, E. B. and A. A. Yushkevich: [1] *Markov Process, Theorems and Problems*, Nauka, Moscow, 1967 (Russian); English translation: Plenum Press, New York, 1969.

Egorov, Ju. V.: [1] "Subelliptic operators," *Uspekhi Mat. Nauk* **30:2** (1975), 57–114; **30:3** (1975), 57–104 (Russian); English translation in *Russian Math. Surv.* **30:2** (1975), 59–118; **30:3** (1975), 55–105.

Egorov, Ju. V. and V. A. Kondrat'ev: [1] "The oblique derivative problem," *Mat. Sbornik* **78** (120) (1969), 148–176 (Russian); English translation in *Math. USSR Sbornik* **7** (1969), 139–169.

Einstein, A.: [1] *Investigations on the Theory of the Brownian Movement*, Dover, New York, 1956.

Fediĭ, V. S.: [1] "On a criterion for hypoellipticity," *Mat. Sbornik* **85** (127) (1971), 18–48 (Russian); English translation in *Math. USSR Sbornik* **14** (85) (1971), 15–45.

Fefferman, C. and D. H. Phong: [1] "Subelliptic eigenvalue problems." In *Conference on Harmonic Analysis* (1981: Chicago, Ill.), pp. 590–606. Wadsworth, Belmont, 1983.

Feller, W.: [1] "Zur Theorie der stochastischen Prozesse (Existenz- und Eindeutigkeitssätze)," *Math. Ann.* **113** (1936), 113–160.
[2] "The parabolic differential equations and the associated semigroups of transformations," *Ann. of Math.* **55** (1952), 468–519.
[3] "On second order differential equations," *Ann. of Math.* **61** (1955), 90–105.

Fichera, G.: [1] "Sulla equazioni differenziali lineari ellittico-paraboliche del secondo ordine," *Atti. Accad. Naz. Lincei Mem.* **5** (1956), 1–30.

Fortet, R.: [1] "Les fonctions aléatoires du type de Markoff associées à certaines équations linéaires aux dérivées partielles du type parabolique," *J. Math. Pures Appl.* **22** (1943), 177–243.

Fujiwara, D.: [1] "On some homogeneous boundary value problems bounded below," *J. Fac. Sci. Univ. Tokyo Sec.* IA, 17 (1970), 123–152.

Gårding, L.: [1] "Dirichlet's problem for linear elliptic partial differential equations," *Math. Scand.* **1** (1953), 55–72.

Gelfand, I. M. and G. E. Shilov: [1] *Generalized Functions*, Vols. I–III, Moscow, 1958 (Russian); English translation: Academic Press, New York, 1964, 1967, 1968.

Gilbarg, D. and N. S. Trudinger: [1] *Elliptic Partial Differential Equations of Second Order*, Springer-Verlag, Berlin, 1977.

Gohberg, I. C. and M. G. Kreĭn: [1] "The basic propositions on defect numbers, roots numbers and indices of linear operators," *Uspehi Mat. Nauk* **12:2** (74) (1957), 43–118 (Russian); English translation in *Amer. Math. Soc. Transl. Ser.* 2, **13** (1960), 185–264.

Halmos, P. R.: [1] *Measure Theory*, Van Nostrand, Princeton, 1950.

Hill, C. D.: [1] "A sharp maximum principle for degenerate elliptic-parabolic equations," *Indiana Univ. Math. J.* **20** (1970), 213–229.

Hopf, E.: [1] " Elementare Bemerkungen über die Lösungen partieller Differentialgleichungen zweiter ordnung vom elliptischen Typus," *Sitz. Ber. Preuss. Akad. Wissensch. Berlin, Math.-Phys. Kl.* **19** (1927), 147–152.

[2] "A remark on linear elliptic differential equations of second order," *Proc. Amer. Math. Soc.* **3** (1952), 791–793.

Hörmander, L.: [1] *Linear Partial Differential Operators*, Springer-Verlag, Berlin, 1963.

[2] "Pseudodifferential operators and non-elliptic boundary problems," *Ann. of Math.* **83** (1966), 129–209.

[3] "Hypoelliptic second order differential equations," *Acta Math.* **119** (1967), 147–171.

[4] "Pseudo-differential operators and hypoelliptic equations." In *Proc. Sym. Pure Math. Vol. X (Singular Integrals)*, pp. 138–183. Amer. Math. Soc., Providence, Rhode Island, 1967.

[5] "Fourier integral operators I," *Acta Math.* **127** (1971), 79–183.

[6] "A class of hypoelliptic pseudo-differential operators with double characteristics," *Math. Ann.* **217** (1975), 165–188.

[7] "Subelliptic operators." In *Seminar on Singularities of Solutions of Linear Partial Differential Equations*, pp. 127–208. *Ann. of Math. Studies*, No. 91, Princeton Univ. Press, Princeton, 1979.

Ikeda, N. and S. Watanabe: [1] *Stochastic Differential Equations and Diffusion Processes*, Kodansha, Tokyo and North-Holland, Amsterdam, 1981.

Itô, K.: [1] *Stochastic Processes*, Iwanami-Shoten, Tokyo, 1957 (Japanese).

Itô, K. and H. P. McKean, Jr.: [1] *Diffusion Processes and Their Sample Paths*, Springer-Verlag, Berlin, 1965.

Jameson, G. J. O.: [1] *Topology and Normed spaces*, Chapman and Hall, London, 1974.

Kelley, J. L.: [1] *General Topology*, Van Nostrand, New York, 1955.

Kinney, J. R.: [1] "Continuity properties of Markov processes," *Trans. Amer. Math. Soc.* **74** (1953) 289–302.

Kolmogorov, A. N.: [1] "Uber die analytischen Methoden in der Wahrscheinlichkeitsrechnung," *Math. Ann.* **104** (1931), 415–458.

Köthe, G.: [1] *Topologische Lineare Räume I*, Springer-Verlag, Berlin, 1960; English translation: Springer-Verlag, Berlin, 1969.

Kumano-go, H.: [1] *Pseudo-differential Operators*, MIT Press, Cambridge, Mass., 1981.

Lamperti, J.: [1] *Probability*, Benjamin, New York, 1966.
[2] *Stochastic Processes*, Springer-Verlag, Berlin, 1977.

Lang, S.: [1] *Differential Manifolds*, Addison-Wesley, Reading, Mass., 1972.
[2] *Real Analysis*, Addison-Wesley, Reading, Mass., 1983.

Lévy, P.: [1] *Processus Stochastiques et Mouvement Brownien*, Gauthier-Villars, Paris, 1948.

Lions, J.-L. et E. Magenes: [1] *Problèmes aux Limites Non-Homogènes et Applications*, Vols. 1, 2, Dunod, Paris, 1968; English translation: Springer-Verlag, Berlin, 1972.

Matsushima, Y.: [1] *Differentiable Manifolds*, Marcel Dekker, New York, 1972.

Melin, A.: [1] "Lower bounds for pseudo-differential operators," *Ark. för Mat.* **9** (1971), 117–140.

Melin, A. and J. Sjöstrand: [1] "Fourier integral operators with complex phase functions and parametrix for an interior boundary value problem," *Communications in Partial Differential Equations* **1** (1976), 313–400.

Munkres, J. R.: [1] *Elementary Differential Topology*, *Ann. of Math. Studies*, No. 54, Princeton Univ. Press, Princeton, 1966.

Oleĭnik, O. A.: [1] "On properties of solutions of certain boundary problems for equations of elliptic type," *Mat. Sbornik* **30** (1952), 595–702 (Russian).

Oleĭnik, O. A. and E. V. Radkevič: [1] *Second Order Equations With Nonnegative Characteristic Form*, Itogi Nauki, Moscow, 1971 (Russian); English translation: Amer. Math. Soc., Providence, Rhode Island and Plenum Press, New York, 1973.

Palais, R. ed.: [1] *Seminar on the Atiyah-Singer Index Theorem*, *Ann. of Math., Studies*, No. 57, Princeton Univ. Press, Princeton, 1963.

Peetre, J.: [1] "Rectification à l'article 'Une caractérisation des opérateurs différentiels'," *Math. Scand.* **8** (1960), 116–120.
[2] "Another approach to elliptic boundary problems," *Comm. Pure Appl. Math.* **14** (1961), 711–731.

Perrin, J.: [1] *Les Atomes*, Gallimard, Paris, 1970.

Protter, M. H. and H. F. Weinberger: [1] *Maximum Principles in Differential Equations*, Prentice-Hall, Englewood Cliffs, 1967.

Ray, D.: [1] "Stationary Markov processes with continuous paths," *Trans. Amer. Math. Soc.* **82** (1956), 452–493.

Redheffer, R. M.: [1] "The sharp maximum principle for nonlinear inequalities," *Indiana Univ. Math. J.* **21** (1971), 227–248.

Rempel, S. and B.-W. Schulze: [1] *Index Theory of Elliptic Boundary Problems*, Akademie-Verlag, Berlin, 1982.

Rudin, W.: [1] *Real and Complex Analysis*, McGraw-Hill, New York, 1966.

Sato, K. and T. Ueno: [1] "Multi-dimensional diffusion and the Markov process on the boundary," *J. Math. Kyoto Univ.* **4** (1965), 529–605.

Schechter, M.: [1] *Principles of Functional Analysis*, Academic Press, New York, 1971.

Schwartz, L.: [1] *Théorie des Destributions*, Hermann, Paris, 1966.

Seeley, R. T.: [1] "Extension of C^∞ functions defined in a half-space," *Proc. Amer. Math. Soc.* **15** (1964), 625–626.

[2] "Singular integrals and boundary value problems," *Amer. J. Math.* **88** (1966), 781–809.

[3] "Topics in pseudo-differential operators." In *Pseudo-differential Operators*, pp. 167–305. C.I.M.E. (Stresa 1968), Edizione Cremonese, Rome, 1969.

Seregin, L. V.: [1] "Continuity conditions for stochastic processes," *Teoriya Veroyat. i ee Primen.* **6** (1961), 3–30 (Russian); English translation in *Theory Prob. and its Appl.* **6** (1961), 1–26.

Stroock, D. W. and S. R. S. Varadhan: [1] "On the support of diffusion processes with applications to the strong maximum principle," *Proc. of 6th Berkeley Symp. of Prob. and Math. Stat.* Vol. III (1972), 333–359.

[2] "On degenerate elliptic-parabolic operators of second order and their associated diffusions," *Comm. Pure Appl. Math.* **25** (1972), 651–713.

Taira, K.: [1] "On non-homogeneous boundary value problems for elliptic differential operators," *Kōdai Math. Sem. Rep.* **25** (1973), 337–356.

[2] "Sur le problème de la dérivée oblique I," *J. Math. Pures Appl.* **57** (1978), 379–395.

[3] "Sur le problème de la dérivée oblique II," *Ark. för Mat.* **17** (1979), 177–191.

[4] "Sur l'existence de processus de diffusion," *Ann. Inst. Fourier* (Grenoble) **29** (1979), 99–126.

[5] "Un théorème d'existence et d'unicité des solutions pour des problèmes aux limites non-elliptiques," *Journal of Functional Analysis* **43** (1981), 166–192.

[6] "Semigroups and boundary value problems," *Duke Math. J.* **49** (1982), 287–320.

[7] "Semigroups and boundary value problems II," *Proc. Japan Acad.* **58** (1982), 277–280.

[8] "Le principe du maximum et l'hypoellipticité globale," *Séminaire Bony-Sjöstrand-Meyer* 1984–1985, Exposé No. I. Ecole Polytechnique, Palaiseau, 1985.

Taylor, M.: [1] *Pseudodifferential Operators*, Princeton Univ. Press, Princeton, 1981.

Treves, F.: [1] *Topological Vector Spaces, Distributions and Kernels*, Academic Press, New York, 1967.

Ueno, T.: [1] "The diffusion satisfying Wentzell's boundary condition and the Markov process on the boundary II," *Proc. Japan Acad.* **36** (1960), 625–629.

Vaĭnberg, B. R. and V. V. Grušin: [1] "Uniformly nonelliptic problems I, II," *Mat. Sbornik* **72** (114) (1967), 602–636; **73** (115) (1967), 126–154 (Russian); English translation in *Math. USSR Sbornik* **1** (72) (1967), 543–568; **2** (73) (1967), 111–133.

Watanabe, S.: [1] "Construction of diffusion processes with Wentzell's boundary conditions by means of Poisson point processes of Brownian excursions." In

Probability Theory, pp. 255–271. Banach Center Publications, Vol. 5, PWN-Polish Scientific Publishers, Warsaw, 1979.

Wentzell (Ventcel'), A. D.: [1] "On boundary conditions for multidimensional diffusion processes," *Teoriya Veroyat. i ee Primen.* **4** (1959), 172–185 (Russian); English translation in *Theory Prob. and its Appl.* **4** (1959), 164–177.

Wiener, N.: [1] "Differential space," *J. Math. Phys.* **2** (1923), 131–174.

Yosida, K.: [1] *Functional Analysis*, Springer-Verlag, Berlin, 1965.

List of Symbols

The number opposite each symbol is the page on which it is defined or explained.

Index

T